VOLUME ONE HUNDRED AND FOURTY TWO

PROGRESS IN MOLECULAR BIOLOGY AND TRANSLATIONAL SCIENCE

Host-Microbe Interactions

VOLUME ONE HUNDRED AND FOURTY TWO

PROGRESS IN MOLECULAR BIOLOGY AND TRANSLATIONAL SCIENCE

Host-Microbe Interactions

Edited by

Michael San Francisco

Department of Biological Sciences and Honors College Texas Tech University, Lubbock, TX, United States

Brian San Francisco

Carl R. Woese Institute for Genomic Biology University of Illinois, Urbana-Champaign, IL, United States

ELSEVIER

AMSTERDAM • BOSTON • HEIDELBERG • LONDON
NEW YORK • OXFORD • PARIS • SAN DIEGO
SAN FRANCISCO • SINGAPORE • SYDNEY • TOKYO
Academic Press is an imprint of Elsevier

Academic Press is an imprint of Elsevier
125 London Wall, London EC2Y 5AS, United Kingdom
525 B Street, Suite 1800, San Diego, CA 92101-4495, United States
50 Hampshire Street, 5th Floor, Cambridge, MA 02139, United States
The Boulevard, Langford Lane, Kidlington, Oxford OX5 1GB, United Kingdom

First edition 2016

ISBN: 978-0-12-809385-6
ISSN: 1877-1173

For information on all Academic Press publications visit our website at https://www.elsevier.com/

Publisher: Zoe Kruze
Acquisition Editor: Alex White
Editorial Project Manager: Helene Kabes
Production Project Manager: Magesh Kumar Mahalingam
Designer: Vicky Pearson Esser

Typeset by Thomson Digital

CONTENTS

CONTRIBUTORS

D. Bishop
Wound Infections Department, Naval Medical Research Center, Silver Spring, MD, United States

A.Q. Byrne
Department of Environmental Science Policy and Management, University of California, Berkeley, CA, United States

D. Carter
PAI Life Sciences, Seattle, WA, United States; Infectious Disease Research Institute, Seattle, WA, United States; Department of Global Health, University of Washington, Seattle, WA, United States

R.N. Coler
Infectious Disease Research Institute, Seattle, WA, United States; Department of Global Health, University of Washington, Seattle, WA, United States

J.A. Colmer-Hamood
Department of Immunology and Molecular Microbiology, Texas Tech University Health Sciences Center, Lubbock, TX, United States; Department of Medical Education, Texas Tech University Health Sciences Center, Lubbock, TX, United States

N. Dzvova
Department of Immunology and Molecular Microbiology, Texas Tech University Health Sciences Center, Lubbock, TX, United States

D. Fleming
Department of Surgery, Texas Tech University Health Sciences Center, Lubbock, TX, United States; Department of Immunology and Molecular Microbiology, Texas Tech University Health Sciences Center, Lubbock, TX, United States

N. German
Department of Pharmaceutical Sciences, Texas Tech University Health Sciences Center, Amarillo, TX, United States

S.A. Gray
PAI Life Sciences, Seattle, WA, United States

A.N. Hamood
Department of Immunology and Molecular Microbiology, Texas Tech University Health Sciences Center, Lubbock, TX, United States; Department of Surgery, Texas Tech University Health Sciences Center, Lubbock, TX, United States

X. Hao
Institute of Urban Environment, Chinese Academy of Sciences, Xiamen, China

N. Hugouvieux-Cotte-Pattat
Microbiology Adaptation and Pathogenesis, CNRS, University of Lyon, University Claude Bernard Lyon 1, INSA Lyon, Villeurbanne, France

T.E. Kehl-Fie
Department of Microbiology, University of Illinois at Urbana-Champaign, Urbana, IL, United States

J.L. Kelliher
Department of Microbiology, University of Illinois at Urbana-Champaign, Urbana, IL, United States

C. Kruczek
Honors College, Texas Tech University, Lubbock, TX, United States

F. Lüthje
Department of Plant and Environmental Sciences, University of Copenhagen, Frederiksberg, Denmark

F. Meyer
Department of Biochemistry & Molecular Biology, Entomology & Plant Pathology, Mississippi State University, Starkville, MS, USA

G. Muskhelisvili
Department of Biology, University of Lyon, INSA-Lyon, Villeurbanne, Lyon, France

W. Nasser
Department of Biology, University of Lyon, INSA-Lyon, Villeurbanne, Lyon, France

R. Ravirala
Roche Molecular System, Pleasanton, CA, United States

C. Rensing
Department of Plant and Environmental Sciences, University of Copenhagen, Frederiksberg, Denmark

S. Reverchon
Department of Biology, University of Lyon, INSA-Lyon, Villeurbanne, Lyon, France

G. Rios-Sotelo
Department of Biology, University of Nevada, Reno, NV, United States

R. Rønn
Department of Biology, University of Copenhagen, Copenhagen, Denmark

E.B. Rosenblum
Department of Environmental Science Policy and Management, University of California, Berkeley, CA, United States

K.P. Rumbaugh
Department of Surgery, Texas Tech University Health Sciences Center, Lubbock, TX, United States; Department of Immunology and Molecular Microbiology, Texas Tech University Health Sciences Center, Lubbock, TX, United States

M. San Francisco
Department of Biological Sciences, Texas Tech University, Lubbock, TX, United States

A.A. Siddiqui
Department of Internal Medicine, Texas Tech University School of Medicine, Lubbock, TX, United States; Center of Tropical Medicine and Infectious Diseases, Texas Tech University School of Medicine, Lubbock, TX, United States

J. Thekkiniath
Department of Medicine, University of Massachusetts Medical School, Worcester, MA, United States

J. Voyles
Department of Biology, University of Nevada, Reno, NV, United States

C. Watters
Wound Infections Department, Naval Medical Research Center, Silver Spring, MD, United States

PREFACE

As advances in molecular biology, biochemstry, and genomics have furthered our understanding of biological systems, we are faced with new questions. These questions have become even more pressing in the study of cell–cell interactions, particularly those of pathogens with their hosts. Strategies utlized by microorganisms to acquire nutrition, evade host defenses, and gain a foothold in the host are varied and inventive. Host cells have, in turn, evloved mechanisms to supress pathogen processes, limit nutrient access and "seek and destroy" microbial invaders.

Nutritional immunity is at the center of the host–pathogen interaction, particularly with regard to metal acquistion. Two chapters address the acquisition of transition metals by pathogens. Work from the Kehl-Fie group discusses strategies for acquisition and sequestration of manganese by pathogen and host, respectively. Reduction of manganese avaliabilty can impair microbial spread and make them more susceptible to host defenses. German et al., describe how some other transition metals influence bacterial gene expression related to pathogenicity and virulence. They also highilght an interesting host strategy for pathogen elimination; termed "Brass Dagger" for its reliance on copper and zinc, the phagolysosomes of host macrophages accumulate metals to toxic levels to facilitate pathogen killing.

Bacterial pathogens of plants can cause great losses in agriculture; three chapters discuss various aspects of the genus *Dickeya* (formerly *Erwinia*), an important plant pathogen globally. Reverchon et al., focus their review on the complex regulatory networks that modulate early events of host adherence and virulence in the pathogen–plant interaction. The role that chromosomal superhelical density plays in regulating these interactions is of special note. Hogouvieux-Cotte-Pattat describes the dual role of plant cell wall-degrading enzymes as both nutritional providers and virulence factors. Pectate lyases in particular, which degrade the cementing pectin in plant cell walls, play important roles in modulating different phases of the infection process. Thekkiniath et al., discuss the role of multidrug efflux pumps in conferring *Dickeya* resistance to a powerful and varied arsenal of host-synthesized antimicrobial chemicals.

Pseudomonas aeruginosa is an opportunistic pathogen of plants and animals. This bacterium is a common of cause infection in the wounds of burn victims and in the lungs of cystic fibrosis patients. The highly nimble *Pseudomonas* expression platform permits facile adaptation to different environments, such as serum or mucus. Colmer-Hamood et al., describe work to mimic, in vitro, various host environments to study virulence gene expression in the bacterium. One mechanism employed by many successful pathogens, including *Pseudomonas*, is biofilm formation. Biofilms are important in adhesion, drug and toxin resistance, horizontal gene transfer, and long-term survival. Watters et al., reflect on the novel contribution of biofilms to immune evasion and supression of host immune responses.

Splicing of RNA to maximize coding potential in eukaryotic organisms is well known. RNA splicing in viral pathogens is likely the evloutionary ancestor of these systems. Meyer discusses mRNA biogenesis and stability in the context of RNA splicing in viruses and how these systems vary with different viruses.

Emerging diseases globally have risen many fold over the last decade. One of the most notable of these is the fungal chytrid pathogen of amphibians, *Batrachochytrium dendrobatidis*. Our understanding of this pathogen and its relationship to the host can be enhanced through effective use of genomic tools. Byrne et al., discuss the value of genomic tools with evolutionary, physiological, biochemical, epidemiological, immunological, and epidemiological approaches, to make important advances to guide in the conservation of these fragile hosts.

Ultimately, our understanding of microbes and the mechanisms they use to cause disease will allow us to devlop novel and useful strategies to prevent, diagnose, and treat infections. Gray et al., discuss strategies to develop treatments for neglected tropical diseases (NTDs). NTDs impact millions of individuals worldwide and yet are termed "neglected" in part because they have limited impact in western nations where funding is typically directed elsewhere. NTD research requires philanthropic and often multinational cooperation, hence the outgrowth of the Millinium Development Goals.

CHAPTER ONE

Competition for Manganese at the Host–Pathogen Interface

J.L. Kelliher, T.E. Kehl-Fie*
Department of Microbiology, University of Illinois at Urbana-Champaign, Urbana, IL, United States
* Corresponding author. E-mail address: kehlfie@illinois.edu

Contents

Abstract

Transition metals such as manganese are essential nutrients for both pathogen and host. Vertebrates exploit this necessity to combat invading microbes by restricting access to these critical nutrients, a defense known as nutritional immunity. During infection, the host uses several mechanisms to impose manganese limitation. These include removal of manganese from the phagolysosome, sequestration of extracellular manganese, and utilization of other metals to prevent bacterial acquisition of manganese. In order to cause disease, pathogens employ a variety of mechanisms that enable them to adapt to and counter nutritional immunity. These adaptations include, but are likely not limited to, manganese-sensing regulators and high-affinity manganese transporters. Even though successful pathogens can overcome host-imposed manganese starvation, this defense inhibits manganese-dependent processes, reducing the ability of these microbes to cause disease. While the full impact of host-imposed manganese starvation on bacteria is unknown, critical bacterial virulence factors such as superoxide dismutases are inhibited. This chapter will review the factors involved in the competition for manganese at the host–pathogen interface and discuss the impact that limiting the availability of this metal has on invading bacteria.

Progress in Molecular Biology and Translational Science, Volume 142
ISSN 1877-1173
http://dx.doi.org/10.1016/bs.pmbts.2016.05.002

1. INTRODUCTION

Transition metals such as iron (Fe), zinc (Zn), and manganese (Mn) are necessary for the proliferation of all organisms. Their importance is emphasized by analysis of protein databases, which predict that 30% of proteins utilize a metal cofactor.[1] Metals act as catalytic cofactors and structural components to perform a variety of tasks in the cell; metals including iron, zinc, and manganese also directly influence regulation of their own cellular homeostasis. Iron is utilized by almost every form of life and facilitates a variety of processes, such as respiration, metabolism, and macromolecule synthesis.[2] Iron is a cofactor in multiple types of catalytic centers, including mononuclear enzymes, such as superoxide dismutases; Fe–S cluster proteins, such as aconitase; and in heme-containing enzymes, such as cytochrome *c* oxidase. Zinc frequently functions as a structural cofactor, such as in the Fur and zinc-finger families of transcriptional regulators, and as catalytic cofactor. Zinc has a catalytic role in enzymes such as alcohol dehydrogenases, hydrolases, and kinases.[3] Manganese is an essential cofactor for a diverse set of processes, including in enzymes involved in nucleotide metabolism (ribonucleotide reductase), carbon metabolism (phosphoglycerate mutase), phosphorylation (serine/threonine kinase), and oxidative stress response (superoxide dismutase).[4–6]

To combat pathogens, vertebrates take advantage of the essential nature of transition metals by restricting their availability, a defense termed nutritional immunity. The most well characterized example of nutritional immunity is the iron-withholding response elaborated by the host. As a first line of defense, the availability of free iron in the absence of infection is kept very low throughout the body by multiple mechanisms. First, the majority of iron in the body is present in the form of heme, which is bound by hemoglobin within red blood cells.[2] Second, extracellular Fe^{2+} is rapidly oxidized to Fe^{3+}, which is insoluble at physiological pH.[7] Further restricting the availability of extracellular iron, scavenging molecules such as transferrin bind Fe^{3+}, and haptoglobin and hemopexin sequester hemoglobin and heme, respectively.[7,8] In response to infection, the host activates additional mechanisms to restrict the availability of iron.[2] Serum levels of the iron-oxidizing enzyme ceruloplasmin increase, presumably to increase the conversion of Fe^{2+} to Fe^{3+}, circulating levels of transferrin increase, and immune cells release lactoferrin, another protein capable of sequestering free iron, at sites of infection.[8–11] Despite the multiple tools used by the host to restrict iron availability, successful invaders possess mechanisms that enable them to circumvent this defense. To accomplish this task, bacteria utilize a variety of approaches, including expressing high-affinity iron-uptake systems such as siderophores.[2,12–15] The ongoing struggle for essential transition metals is highlighted by the observation that to combat bacterial siderophores, the host expresses siderocalin (lipocalin-2), which can bind enterobactin and prevent its uptake by bacteria.[7] In response, bacteria have evolved modified siderophores that can resist sequestration by siderocalin.[16] In addition to attempting to acquire free iron, some pathogens express receptors for heme, hemoglobin, and transferrin, allowing them to hijack iron-bound host molecules.[2,8,15] Additionally, in response to iron limitation, many bacteria activate an iron-sparing response, which functionally diverts this metal away from nonessential to essential enzymes.[17] Disrupting either the ability of the host to withhold iron or the ability of a pathogen to obtain this nutrient can significantly alter the outcome of infection in favor of one or the other party.[2,8,14] This experimental observation is manifested by the increased susceptibility of people with iron overload to infection by a diverse collection of pathogens, including *Yersinia enterocolitica, Listeria monocytogenes, Mycobacterium tuberculosis*, and *Plasmodium falciparum*.[8,12]

It has become apparent that in addition to restricting iron availability, the host also limits the availability of manganese and zinc. While the timing and mechanisms employed by the host to restrict the availability of these essential metals during infection has not been fully elucidated, it is clear that restricting them contributes to the ability of the host to combat invaders.[12,13,18,19] This chapter will focus on the mechanisms used by the host to limit the ability of pathogens to acquire manganese. It will also discuss how bacteria adapt and respond to host-imposed manganese limitation and the impact of this host defense on invaders. Finally, current questions in the field and broader impacts will be highlighted.

2. IMPOSITION OF MANGANESE STARVATION BY THE HOST

In contrast to iron, which is constantly and globally restricted, manganese availability appears only to be restricted in the presence of invading microbes and at the site of infection. Manganese can be sequestered from pathogens both intracellularly and extracellularly (Fig. 1).[18,20] A key cellular factor in preventing intracellular pathogens from obtaining manganese is the

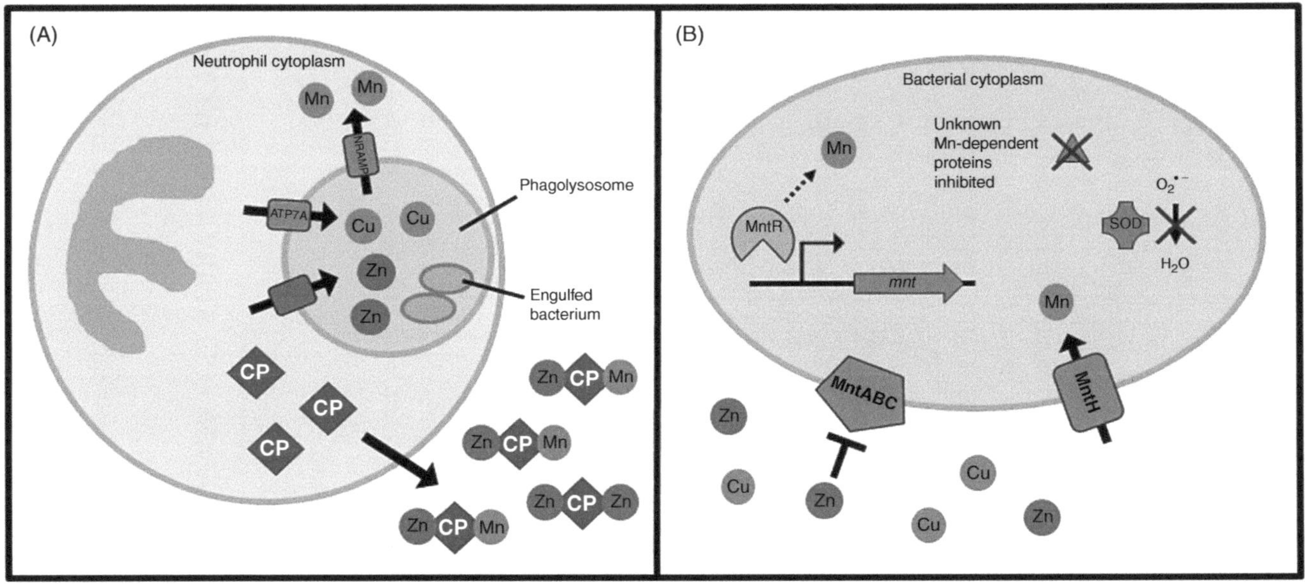

Figure 1 The host and pathogens compete for manganese during infection. (A) Diagram of the mechanisms utilized by the host to limit the ability of invaders to obtain manganese. In response to microbial invaders, host cells, primarily neutrophils, release calprotectin (CP), which limits the ability of extracellular pathogens to obtain manganese (and zinc). In the phagolysosomal membrane, NRAMP transporters remove manganese (Mn) from the phagolysosome. Additionally, zinc (Zn) and copper (Cu) are imported into this compartment, inhibiting the activity of bacterial manganese importers. (B) Diagram of how bacteria respond to manganese limitation and the processes that are disrupted by host-imposed manganese limitation. In response to manganese starvation, the MntR regulon is derepressed, and the expression of dedicated manganese importers such as MntH and MntABC increases. Despite the expression of high-affinity manganese importers by invading pathogens, the host remains capable of imposing manganese starvation, which inactivates manganese-dependent superoxide dismutase (SOD) and unknown essential enzymes.

divalent cation transporter NRAMP1 (natural resistance-associated macrophage protein-1), also known as DMT-2. The importance of this transporter to host defense was first revealed by the observation that macrophages lacking NRAMP1 are more susceptible to intracellular pathogens, including *Mycobacterium bovis* and *Salmonella enterica* Typhimurium.[20–25] NRAMP1 is constitutively expressed by macrophages and lymphocytes, where it associates with lysosomes, late endosomes, and maturing phagosomes.[26–30] In addition to *M. bovis* and *S. enterica* Typhimurium, loss of NRAMP1 in mice leads to increased susceptibility to a variety of pathogens, including *Toxoplasma gondii*, *Candida albicans*, *Mycobacterium lepraemurium*, and *Leishmania donovani*.[24,25,31–33] The contribution of NRAMP1 to restricting manganese availability during infection was revealed by investigations with *S. enterica* Typhimurium. Analysis of *Salmonella* mutants lacking high-affinity manganese importers revealed that they are less capable than wild type bacteria of surviving in primary peritoneal macrophages derived from $NRAMP1^{+/+}$ mice, but not those derived from $NRAMP1^{-/-}$ mice.[34] Consistent with this result, in an oral infection model, the *Salmonella* manganese uptake mutants are attenuated in $NRAMP1^{+/+}$ mice but not $NRAMP1^{-/-}$ mice.[34] The importance of NRAMP1 and intracellular manganese sequestration to host defense is emphasized by the identification of polymorphisms in humans that are associated with increased risk of developing tuberculosis, leishmaniasis, meningococcal disease, and others.[20,23,35,36]

NRAMP1 belongs to SLC11 family of solute transporters, members of which are present in all three domains of life.[29,37] Humans express two SLC11 family members, NRAMP1 and DMT-1.[29,37] The latter transporter facilitates absorption of iron, and potentially manganese, in the intestine.[29] This family of transporters contains 11–12 transmembrane segments that form a single channel pore.[37,38] NRAMP transporters symport a divalent cation and a proton, coupling metal transport with the energetically favorable flow of protons out of the phagolysosome.[37] In vitro, NRAMP1 can transport Mn^{2+}, Fe^{2+}, Zn^{2+}, Co^{2+}, Ni^{2+}, and Cd^{2+}, but not alkaline earth metals.[39,40] In vitro fluorescent probe-based assays and infection experiments, however, indicate NRAMP1 is important for the transport of Mn^{2+} and Fe^{2+} out of the phagolysosome.[20,29,41,42] Since NRAMP1 is an integral membrane protein, biophysical studies of the transporter have been challenging; however, a prokaryotic homolog from *Staphylococcus capitis* has been structurally characterized.[38] The metal-binding site is located within two short, unstructured regions, in between two sets of membrane-spanning alpha helices. The metal is coordinated by the side chains of N52 (both an oxygen and nitrogen ligand) and D49 (oxygen), the oxygen atom of the peptide bond linking residues 223 and 224, and the thioether of M226, resulting in a planar coordination.[38] These residues are conserved among the NRAMP family, and the human and *S. capitis* residues are identical.[38] The binding site is selective for Mn^{2+}, Fe^{2+}, Co^{2+}, Ni^{2+}, Cd^{2+}, and to some extent Cu^{2+} and Zn^{2+}. However, the latter two metals are coordinated by slightly different residues than Mn^{2+}, Fe^{2+}, Co^{2+}, Ni^{2+}, and Cd^{2+} and may not be effectively transported.[38] Cumulatively, the biophysical and infection studies suggest that NRAMP1 contributes to host defense by removing manganese and iron from the phagolysosome during infection.

In addition to restricting intracellular manganese availability, the extracellular availability of this metal is also limited by the host during infection.[43,44] This discovery was made possible by the application of advanced elemental imaging techniques, such as laser ablation inductively coupled plasma mass spectrometry (LA-ICP-MS), to the study of infection. LA-ICP-MS enables the assessment of the spatial distribution of metals within a tissue.[45] The prototypical example of the extracellular manganese-withholding response is the *Staphylococcus aureus* abscess, which is rendered virtually devoid of manganese by the host.[43,44] Notably, while the staphylococcal abscess is depleted of manganese, total tissue levels of this metal do not decrease.[44] Similar to manganese, zinc is also withheld from the staphylococcal abscess, and total tissue levels of this metal do not change.[43,44] These findings highlight both, the ability of the host to locally restrict metal availability in response to infection and the importance of assessing metal distribution within a tissue when evaluating the impact of nutritional immunity on invaders. A critical component of the manganese-withholding response is the manganese- and zinc-binding protein calprotectin (also known as S100A8/S100A9, calgranulin A/B, and MRP8/14). This innate immune effector is constitutively expressed by neutrophils, where it accounts for approximately 50% of the total cytosolic protein.[46] In addition to neutrophils, proinflammatory cytokines such as IL-17 and IL-22 can induce the production of calprotectin in other cell types, most notably epithelial cells.[47,48] At sites of infection where neutrophils release calprotectin, extracellular concentrations can be found in excess of 1 mg/mL.[49] Calprotectin-deficient ($S100A9^{-/-}$) mice have defects in manganese sequestration and are more susceptible to a variety of bacterial and fungal pathogens, including *S. aureus*, *Klebsiella pneumoniae*, *Acinetobacter baumanii*, *Aspergillus nidulans*, *Aspergillus fumigatus*, and *C. albicans*.[43,44,50–54] In vitro, the sequestration of transition metals by calprotectin inhibits the growth of a range of

Gram-positive, Gram-negative, and fungal pathogens, including *S. aureus*, *K. pneumoniae*, *A. baumanii*, *C. albicans*, and *A. nidulans*.[43,50–53,55] Analysis of metal distribution during staphylococcal infection revealed that while calprotectin-deficient mice do not remove manganese from staphylococcal liver abscess, they are still able to deplete kidney abscesses of manganese.[43,44] This finding indicates that the host possesses additional unknown mechanisms for restricting manganese availability at sites of infection. The importance of restricting extracellular manganese availability to host defense is emphasized by the observation that staphylococcal strains lacking high-affinity manganese uptake systems have a virulence defect in the livers of wild type mice but not calprotectin-deficient mice.[44]

Calprotectin is a member of the S100 family of calcium-binding EF-hand proteins. Unlike the other members of this family, which are homodimers, calprotectin is a heterodimer comprised of S100A8 and S100A9. Similar to calprotectin, a subset of the S100 family, including S100A7 (psoriasin), S100A12 (calgranulin C), and S100A15 (koebnerisin), are capable of binding transition metals.[56–59] These three proteins possess two identical transition metal binding sites located at the dimer interface. The canonical S100 protein transition metal binding site, possessed by S100A7 and S100A12, is composed of three histidines and an aspartic acid. Two of the histidines, arranged in an HXXXH motif, are contributed to the binding site by one of the monomers, while the third histidine and aspartic acid are contributed by the other monomer.[58,60] Unlike the other S100 family members, calprotectin has two non-identical transition metal binding sites. Based on homology with other S100 proteins, the first transition metal binding site possessed by calprotectin was originally thought to be comprised of H17 and H27 from S100A8 and H91 and H95 from S100A9.[61,62] However, subsequent investigations revealed that H103 and H105, contributed by a C-terminal extension of S100A9, also contribute to the ability of this site to bind manganese.[55,63] Crystallographic studies revealed that the six histidines bind manganese with a nearly perfect octahedral geometry.[55] This hexahistidine coordination, which had not previously been observed in proteins, has been confirmed by solution-based electron paramagnetic resonance.[63,64] The critical importance of the C-terminal extension, which only S100A9 possesses, and the observation that neither S100A7 nor S100A12 are capable of binding manganese suggests that among the S100 proteins, calprotectin is unique in its ability to bind this metal.[55,63] The second transition metal binding site is identical to the canonical transition metal binding site found in other S100 proteins, comprised of H83 and H87 from S100A8 and H20 and D30 from S100A9.[61,62]

Analysis of wild type calprotectin and variants lacking the two transition metal binding sites revealed that the first site is capable of binding both manganese and zinc tightly (subsequently referred to as the Mn/Zn site), while the second site is capable of binding only zinc tightly (subsequently referred to as the Zn site).[55,65,66] A combination of isothermal titration calorimetry (ITC) and dye competition studies revealed that the Mn/Zn site binds manganese with an affinity (K_d) of approximately 10 nM or less and zinc with an affinity of less than 240 pM.[55,62,65] Weaker affinities for manganese have been reported;[66] however, they are not consistent with the ability of calprotectin to prevent manganese acquisition by bacteria, which express manganese-importers with low nanomolar affinities.[67–69] Dye competition studies revealed that the Zn site binds zinc with an affinity of less than 10 pM.[65] Obtaining more precise binding affinities for manganese and zinc binding has been hampered by the lower limit of resolution for both ITC and dye-based studies.[65,69] Additional studies revealed that H103 and H105, which are located in the C-terminal extension of S100A9, are essential for the Mn/Zn site to bind manganese, but not zinc, tightly.[55,63] Similar to other S100 proteins, the ability of calprotectin to bind transition metals is influenced by the presence of calcium. In the absence of calcium, the K_d of calprotectin for manganese weakens to 5 μM, and the affinity of the Mn/Zn site and Zn site for Zn increase to 219 nM and 133 pM, respectively.[65,66] Modeling suggests that calcium binding to the EF-hand elongates two alpha helices that contain the transition metal binding sites.[70] Relative to the extracellular space, the cytoplasm is calcium-limited, leading to the suggestion that calcium binding serves as a switch to ensure that calprotectin does not bind manganese and zinc until released into the extracellular space.[65,66]

Activity assays utilizing the calprotectin binding site variants revealed that the Mn/Zn site is necessary for maximal antimicrobial activity against a wide range of Gram-positive and -negative pathogens, including *S. aureus*, *Staphylococcus epidermidis*, *A. baumanii*, *Escherichia coli*, *Enterococcus faecalis*, *Pseudomonasaeruginosa*, and *Shigellaflexeneri*.[55] The importance of the Mn/Zn site suggests that manganese binding is necessary for maximal broad-spectrum antimicrobial activity by calprotectin. However, not surprisingly, given the diversity of microbes, the Mn/Zn site and presumably manganese binding is not necessary for maximal antimicrobial activity in all cases.[71] In addition to Mn^{2+} and Zn^{2+}, calprotectin has been observed to bind Fe^{2+} in vitro via the Mn/Zn site in reducing environments.[71] Notably, this is in contrast to several prior studies in which iron binding by calprotectin was not observed.[43,55] In these studies the ionic state of iron was not controlled but due to the aerobic nature of these experiments, iron was most likely present as Fe^{3+}.[43,55] Based on the observation that calprotectin is capable of binding Fe^{2+}, it has been suggested that the antimicrobial activity of calprotectin is due to iron sequestration.[71] However, several observations argue against this proposal. First, manganese-dependent enzymes in *S. aureus* are inhibited by calprotectin, both in culture and during infection.[55,62] Second, staphylococcal mutants lacking high-affinity manganese importers are more sensitive to calprotectin in culture and have a virulence defect in wild type but not calprotectin-deficient mice.[44] Third, following growth in the presence of calprotectin, *A. baumanii* has reduced intracellular levels of manganese and zinc, but not iron. In fact, following exposure to calprotectin in *A. baumanii* iron levels actually increase.[51] Additionally, our current understanding of iron homeostasis suggests that due to the generally oxidizing nature of the extracellular space, outside of the cytoplasm iron should be present not as Fe^{2+} but as Fe^{3+}, a form of iron that calprotectin cannot bind.[8,43,55,71] Due to the oxidative burst of immune cells, sites of infection are likely to be more oxidizing than healthy tissues. This idea is supported by the observation that pathogens such as *S. enterica* Typhimurium have evolved to take advantage of metabolites that are generated by the oxidative burst as terminal electron receptors.[72] Additionally, Fe^{3+}-binding proteins such as transferrin and lactoferrin are critical to controlling infection.[2,7] While in select environments exceptions may exist, these observations strongly suggest that manganese and zinc sequestration are primarily responsible for the antimicrobial activity of calprotectin, both in culture and during infection.

Even though transition metals are necessary for life, they can also be toxic.[12,73] To prevent transition metal toxicity, intracellular levels of these metals are highly regulated through the coordinated expression of metal importers and exporters.[12,13] Transition metal efflux pumps for zinc and copper contribute to the ability of organisms such as *Streptococcus pyogenes*, *M. tuberculosis*, *Streptococcus pneumoniae*, *Neisseria meningitidis*, *Brucella abortus*, and *Helicobacterpylori* to cause disease,[74–80] suggesting that pathogens encounter toxic levels of these metals during infection. Further supporting this idea, zinc colocalizes with *S. pyogenes* in neutrophils, and chelation of this metal reduces the antimicrobial activity of the cells.[75] Supporting the use of the antimicrobial properties of copper by the host is the observation that this metal is transported into phagolysosome via ATP7A.[81]

The toxicity of transition metals is thought to be driven by the ability of a metal to bind inappropriately to (mismetalate) noncognate metalloenzymes or, in some cases, the ability of the metal to generate reactive oxygen species.

The general affinity of a metal for organic molecules is described by the Irving–Williams series, where $Mg^{2+}/Ca^{2+} < Mn^{2+} < Fe^{2+} < Co^{2+} < Ni^{2+} < Cu^{2+} > Zn^{2+}$.[82,83] Functionally, this means that an overabundance of metals such as copper and zinc can inhibit the activity of metalloproteins that use a weaker binding metal, such as manganese, as a cofactor.[82] Zinc and copper are capable of inhibiting a variety of intracellular enzymes and processes. For example, zinc can inhibit glycolytic enzymes such as phosphofructokinase, and copper can disrupt Fe–S cluster-containing enzymes.[73,84–88] While pathogens can attempt to regulate cytoplasmic levels of transition metals, they are unable to control their surrounding environment. As such, extracellular metalloproteins such as manganese-specific transporters (discussed subsequently) are particularly vulnerable to elevated extracellular concentrations of zinc and copper. In vitro, zinc binds irreversibly to PsaA, the solute-binding protein of the pneumococcal PsaABC manganese importer, preventing it from binding manganese.[67,89–91] In culture, a 30:1 ratio of zinc to manganese prevents *S. pneumoniae* from importing manganese and inhibits bacterial growth. Ratios of zinc to manganese in excess of this can be found in tissues during pneumococcal infection.[67] While the ability of copper to inhibit manganese uptake has not been directly evaluated, it also binds irreversibly to PsaA, suggesting that copper should also inhibit manganese acquisition.[89] Similar results have also been obtained with the staphylococcal solute-binding protein, suggesting that manganese specific ABC transporters are generally susceptible to zinc and copper poisoning.[68]

3. BACTERIAL ADAPTATION TO MANGANESE LIMITATION

In order to successfully cause disease, bacteria must adapt and respond to the ever-changing environment within the host, including the availability of manganese. Many bacteria sense manganese availability through a DtxR-family transcriptional repressor usually called MntR (Fig. 1). MntR homologs are present in a variety of Gram-positive and -negative bacteria, including *S. aureus*, *Bacillus subtilis*, *S. pneumoniae*, *M. tuberculosis*, *S. enterica* Typhimurium, *E. coli*, and *Treponema pallidum*.[92–98] A canonical repressor, manganese-bound MntR represses gene expression when manganese levels are sufficient. When manganese becomes scarce, apo-MntR releases from the DNA and allows transcription of targets to occur. In several species, including *S. aureus, S. pneumoniae, S. enterica* Typhimurium, *B. subtilis, S. flexneri*, and *Corynebacterium diphtheriae*, MntR represses the expression of high-affinity manganese importers when manganese is available.[92–94,99–101] Loss of the MntR homolog PsaR results in reduced virulence of *S. pneumoniae*,[102] suggesting that in addition to encountering manganese limitation, pathogens may be exposed to toxic levels of this metal. This idea is further supported by the identification of manganese efflux systems that contribute to virulence, such as MntE of *S. pneumoniae* and MntX of *N. meningitidis*.[77,78] While it is not clear when invaders would experience elevated levels of manganese, these observations highlight the diversity of metal environments encountered by pathogens within the host. Given the position of manganese in the Irving–Williams series, it is less apparent why elevated levels of this metal are toxic. It has been proposed that regulation of intracellular manganese levels is necessary to maintain an appropriate ratio of manganese to iron within the cell.[82,103] The importance of maintaining an appropriate balance between manganese and iron is emphasized by the observation that the manganese exporter MntX of *N. meningitidis* plays a part in regulating the intracellular ratio of these two metals.[78] Further supporting this idea, in response to manganese limitation *Bradyrhizobium japonicum* reduces the accumulation of iron.[104] In addition to MntR, several other metal-responsive/binding transcriptional regulators have been shown to bind and respond to manganese. These regulators include Fur, which canonically responds to iron availability, and the peroxide sensor PerR.[69,92,95,96,99,105,106] The full implications of the ability of manganese to bind these regulators are still being understood.

In addition to increasing the expression of manganese importers, bacteria frequently respond to manganese limitation by modifying the expression of numerous other cellular processes.[98,102,107,108] One example is the altered expression of genes involved in glucose utilization in *S. pneumoniae* when manganese availability is restricted.[109] However, analysis of the PsaR regulon indicates that this regulator does not control the expression of these genes.[102] This observation suggests that there are likely to be additional regulators that coordinate the bacterial response to host-imposed manganese limitation.

While the holistic response of bacteria to host-imposed manganese limitation and the cellular factors that control this response are still being elucidated, the expression of high-affinity manganese acquisition systems has emerged as a common theme (Fig. 1). The vast majority of bacteria express manganese importers belonging to either the NRAMP or ABC family of transporters.[110] NRAMP homologs are typically referred to as MntH, while ABC transporters are frequently named MntABC. There are, however, a few prominent examples of MntABC homologs with alternative names, including PsaABC in *S. pneumoniae* and SitABC in *Salmonella* species. Many bacteria such as *M. tuberculosis*, *Mycobacterium leprae*, and some strains of *E. coli* express only MntH, while others such as *Yersinia pestis*, *Porphyromonas gingivalis*, *S. pneumoniae*, *S. pyogenes*, *C. diphtheriae*, *Bacillus anthracis*, and *E. faecalis* express only MntABC.[101,111–119] Notably, genetic redundancy of manganese transport systems is common among pathogens, with many species, including *S. aureus*, *S. enterica* Typhimurium, and *S. flexneri*, encoding for homologs of both MntH and MntABC.[92,100,113,120] In addition to these conserved manganese importers, other less widely distributed and less characterized systems have also been identified in several organisms. Interestingly, the ubiquitous human pathogen *H. pylori*, which expresses putatively manganese-dependent enzymes, appears to lack homologs of any known manganese importers.[121] This observation suggests that additional unidentified manganese importers may exist. Emphasizing the general importance of these systems to bacterial virulence and resisting host-imposed manganese starvation, both *S. enterica* Typhimurium and *S. aureus* mutants that lack dedicated manganese importers have virulence defects relative to wild type bacteria in mice that are capable of restricting manganese availability, but not mice with defects in sequestering this metal.[34,44] In addition to these two species, loss of manganese importers results in reduced virulence of numerous pathogens, including *S. pneumoniae*, *S. pyogenes*, *Streptococcus mutans*, *Streptococcus suis*, *B. abortus*, *Yersinia pseudotuberculosis*, *Neisseria gonorrhoeae*, and certain strains of *E. coli*.[117,122–129]

The bacterial MntH family is evolutionarily related to the NRAMP1 family of transporters used by eukaryotes to remove divalent cations from the phagolysosome. MntH is essential for virulence of some pathogens, such as *B. abortus* and *Y. pseudotuberculosis*.[126,127] To date, the only structurally characterized NRAMP transporter is that of *S. capitis*, discussed in detail in the previous section. However, this family of transporters and the residues that coordinate the transported metal are highly conserved, suggesting that the structure and metal specificity should be similar across species.[38] Similar to the eukaryotic transporters, MntH has a preference in vitro for Mn^{2+} and Fe^{2+} but is capable of transporting other divalent cations, such as Cd^{2+} and Co^{2+}, as well.[110] While several metals can be transported by MntH homologs, in bacteria manganese is generally the physiological relevant substrate. However, in some cases iron import may also be relevant.[114] In bacteria, the expression of MntH is frequently induced by manganese limitation or increased cellular demand for manganese.[92,93,95,98,130] These latter situations include when iron availability is reduced and the presence of peroxide stress, which triggers the replacement of iron with manganese to prevent Fenton chemistry-induced damage.[131–133] Additionally, the relative selectivity of MntH for manganese tends to be greater than for other cognate metals, as observed with *S. enterica* Typhimurium and others.[113,134] In *S. enterica* Typhimurium, MntH has a pseudoaffinity (solute concentration at half-maximal transport, or $K_{0.5}$) for Mn^{2+} of ~100 nM, whereas its affinity for Fe^{2+} is approximately 25 μM.[113]

ABC transporters, the second primary family of manganese importers expressed by bacteria, possess a four-domain structure. These domains include two transmembrane proteins that facilitate substrate translocation and two nucleotide-binding proteins that power import via ATP hydrolysis.[135] In addition to these domains possessed by all ABC transporters, importers also possess a high-affinity solute-binding protein (SBP).[136] ABC-type manganese transporters are widespread among pathogens and can be found in *S. aureus*, *S. pneumoniae*, *S. enterica* Typhimurium, *S. flexneri*, *Y. pestis*, *P. gingivalis*, *E. faecalis*, and more.[44,69,110,115,119] In some bacteria, the MntABC system is critical for pathogenesis, including in *S. pneumoniae*, *S. pyogenes*, *S. mutans*, *S. suis*, *S. aureus*, and *N. gonorrhoeae*.[44,117,122–125,128]

Regardless of species, all manganese SBPs belong to the Cluster A-I group of SBPs of ABC transporters, along with iron- and zinc-specific SBPs.[137] MntABC transporters are capable of remarkably high affinities for their substrate; the K_d of PsaA for Mn^{2+} is 3.3 nM, and the SBP of *S. aureus* has a K_d for Mn^{2+} of 8 nM.[67,68] The metal-binding site of Cluster A-I SBPs consists of two nitrogen atoms from two conserved histidine residues, one carboxylate from either an aspartate or a glutamate, and a variable fourth ligand, which is thought to dictate specificity for the metal.[138] In manganese-specific SBPs, this ligand is another carboxylate group donated by a glutamate residue,[67,68] which, as with the conserved carboxylate, can donate two oxygen ligands. This allows for a total of six coordinating ligands, which is the preferred coordination for manganese. However, the physical constraints imposed by the protein result in imperfect octahedral coordination, which facilitates release of manganese to the translocation domain.[89] In addition to their respective cognate metal, manganese SBPs are capable of binding a range of noncognate divalent transition metals including zinc and copper.[68,89] Differing from manganese, zinc is capable of binding to PsaA with a near perfect tetrahedral coordination. This coordination results in an extremely stable complex, which prevents the release of zinc even when extensively dialyzed against strong chelating agents. This stability is thought to prevent the release of zinc from PsaA to the translocation domain, rendering the transporter nonfunctional.[67,90] Hence, the specificity of the ABC transporter is driven by release of the metal rather than initial binding to the SBP. These findings also provide a mechanistic explanation for the ability of zinc and copper to poison these transporters.[67,89,90]

In addition to the canonical NRAMP and ABC-family transporters, other types of manganese importers have been identified in *Borrelia burgdorferi, Lactobacillus plantarum*, and *Vibrio* species.[139–141] Notably, *B. burgdorferi* and *L. plantarum* accumulate extremely high levels of manganese (~30 mM in *L. plantarum*, for example) and are thought to have eliminated the need for iron.[142,143] *B. burgdorferi* lacks homologs of MntH and MntABC, and instead utilizes a ZIP family homolog named BmtA to acquire manganese. BmtA is thought to be responsible for the import of both manganese and zinc.[139] Loss of BmtA in *B. burgdorferi* results in decreased intracellular manganese levels and abrogates virulence.[139] Differing from *B. burgdorferi, L. plantarum* expresses homologs MntH and MntABC, as well as a P-type ATPase transporter, MntA, which has been implicated in the import of manganese.[140,144] The expression of MntA in *L. plantarum* is induced in manganese-depleted media, and deletion of the gene abrogates high-affinity manganese import.[144] A third novel class of putative manganese transporter has also been identified and is widely conserved among marine bacteria, including the human pathogen *Vibrio cholerae*.[141] The transporter, named MntX (unrelated to the manganese efflux system of *N. meningitidis*), appears to be repressed in manganese-replete conditions and enhances growth of other *Vibrio* species, which have been engineered to lack a manganese importer, in manganese-poor media.[141] Additional studies including metal accumulation and transport assays are necessary to determine if MntA from *L. plantarum* and MntX from *V. cholerae* are true manganese importers. In addition to transporting manganese across the inner membrane, Gram-negative bacteria must also transport this nutrient across the outer membrane. Originally, transition metals including manganese were thought to pass nonspecifically through outer membrane porins. However, this assumption has been challenged by the identification of dedicated outer membrane channels that facilitate acquisition of divalent cations, specifically by the characterization of a zinc-specific outer membrane receptor in *N. meningitidis*.[145,146] An analogous protein MnoP in *B. japonicum* facilitates the passage of manganese across the outer membrane.[147] While similar manganese-specific systems have not yet been described in other species, MnoP belongs to the OprB superfamily, many members of which await characterization.[147]

4. IMPACT OF MANGANESE LIMITATION ON INVADING MICROBES

The finding that mice with defects in restricting manganese availability are more susceptible to infection indicates that despite expressing high-affinity manganese acquisition systems, invading pathogens experience manganese starvation during infection.[43,50–53] While the breadth of biological processes to which manganese can contribute is significant,[4] the impact that host-imposed manganese starvation has on invading pathogens is only just beginning to be elucidated. This task is hampered by an incomplete understanding of manganese-dependent processes in bacteria and the observation that metal-dependent enzymes are frequently capable of using more than one metal as a cofactor. The latter challenge is highlighted by the observation that in response to oxidative stress, *E. coli* replaces iron in mononuclear enzymes with manganese in order to limit Fenton chemistry-induced damage.[131]

While the role of manganese in many cellular processes may be uncertain, it is clear that manganese is a critical contributor to resisting oxidative stress, serving as a cofactor for manganese-dependent superoxide dismutases.[130,148] In addition to enzymatic dismutase activity, manganese in complex with phosphate or cellular metabolites, such as lactate, has dismutase activity.[130,142,149–152] This chemical activity has led to the suggestion that these complexes may contribute to the ability of bacteria that accumulate high levels of manganese to resist oxidative stress.[150,152] However, relative to enzymatic dismutation, these complexes are much less efficient, leading to uncertainty regarding the contribution of manganese complexes to resisting oxidative stress during infection.[130] Given the established link between manganese and resisting oxidative stress, the ability of pathogens to resist oxidative stress under conditions of host-imposed manganese starvation has received significant attention. This area has primarily been investigated using *S. aureus* and *S. pneumoniae*, both of which utilize manganese-dependent superoxide dismutases. In *S. aureus*, calprotectin-induced manganese starvation reduces staphylococcal superoxide dismutase activity and increases bacterial sensitivity to paraquat-induced oxidative stress.[44,55,62] Calprotectin also renders *S. aureus* more sensitive to neutrophil-mediated killing.[62] Infection experiments employing staphylococcal superoxide dismutase mutants and calprotectin-deficient mice revealed that staphylococcal superoxide dismutase activity is inhibited by host-imposed manganese starvation during infection.[55,62] In *S. pneumoniae*, elevated zinc levels lead to a reduced accumulation of manganese and reduced superoxide dismutase activity.[67,90] Additionally, elevated zinc levels increase pneumococcal sensitivity to paraquat-induced oxidative stresses and killing by polymorphonuclear leukocytes.[67,90] The enhanced killing by immune cells indicates that not only does restricting manganese availability inhibit growth, but also renders invaders more susceptible to other immune effectors. As superoxide dismutase activity is not essential for viability of *S. aureus* or *S. pneumoniae*, it seems likely that other cellular factors are also inhibited by host-imposed manganese starvation. The complement of manganese-dependent enzymes in any microbe and the specific impact of host-imposed manganese starvation are likely to be as varied as the diversity of lifestyles adopted by pathogens. Potential processes that may be inhibited by manganese limitation during infection include enzymes involved in energy generation, nucleotide metabolism, and cell signaling.[4,153–159]

5. CONCLUSIONS AND BROADER IMPACTS

Nutritional immunity is a powerful defense employed by the host to control invading pathogens. While canonically associated with restricting iron from invading microbes, the concept of nutritional immunity has been expanded to include limiting the availability of other essential metals, including manganese, during infection.[12,18,20,43] Even though significant progress has been made elucidating how the host imposes manganese starvation, it is clear that additional unidentified host factors contribute to this defense. This gap in knowledge is highlighted by the ability of calprotectin-deficient mice to remove manganese from kidney but not liver abscesses.[43,44] Simultaneously, it has been revealed that the host not only physically removes manganese from sites of infection but also harnesses the toxic properties of zinc and copper to prevent acquisition of this metal.[13,73] However, LA-ICP-MS and the importance of zinc importers to bacterial pathogenesis indicate that pathogens also encounter zinc limitation during infection.[43,44,51,160–165] These two disparate observations raise the question of when the host utilizes the toxic properties of zinc to control infection versus when it restricts the availability of this metal. Adding even more complexity, more recent investigations utilizing *C. albicans* suggest that the host may also restrict copper availability.[166] Both the Centers for Disease Control and World Health Organization have stated that due to the emergence and spread of antibiotic resistance, there is a critical need for new approaches to treating infection.[167,168] Despite our nascent understanding of how noniron metal levels are manipulated in order to combat invaders, it is clear that preventing pathogens from acquiring manganese contributes to host defense. It is equally clear that successful pathogens, despite expressing high-affinity metal acquisition systems, experience metal starvation and are able to overcome this host defense.[62] However, the adaptations that enable this success are unknown. Therapeutics that augment nutritional immunity by manipulating metal levels during infection or prevent bacteria from adapting to this host defense represent a promising new approach for treating infection. However,our ability to successfully harness the full potential of these approaches will require a greater understanding of metal homeostasis during infection, how the host utilizes transition metals to combat infection, and how invading microbes circumvent nutritional immunity.

REFERENCES

1. Andreini C, et al. Metal ions in biological catalysis: from enzyme databases to general principles. *J Biol Inorg Chem.* 2008;13(8):1205–1218.
2. Cassat JE, Skaar EP. Iron in infection and immunity. *Cell Host Microbe.* 2013;13 (5):509–519.
3. Cerasi M, Ammendola S, Battistoni A. Competition for zinc binding in the host–pathogen interaction. *Front Cell Infect Microbiol.* 2013;3:108.
4. Kehres DG, Maguire ME. Emerging themes in manganese transport, biochemistry and pathogenesis in bacteria. *FEMS Microbiol Rev.* 2003;27(2–3):263–290.
5. Donat S, et al. Transcriptome and functional analysis of the eukaryotic-type serine/threonine kinase PknB in *Staphylococcus aureus*. *J Bacteriol.* 2009;191(13):4056–4069.
6. Srinivasan VB, et al. Functional characterization of a novel Mn2+ dependent protein serine/threonine kinase KpnK, produced by *Klebsiella pneumoniae* strain MGH78578. *FEBS Lett.* 2012;586(21):3778–3786.
7. Ganz T, Nemeth E. Iron homeostasis in host defence and inflammation. *Nat Rev Immunol.* 2015;15(8):500–510.
8. Soares MP, Weiss G. The Iron age of host-microbe interactions. *EMBO Rep.* 2015;16 (11):1482–1500.
9. Beaumier DL, Caldwell MA, Holbein BE. Inflammation triggers hypoferremia and de novo synthesis of serum transferrin and ceruloplasmin in mice. *Infect Immun.* 1984;46 (2):489–494.
10. Cernat RI, et al. Serum trace metal and ceruloplasmin variability in individuals treated for pulmonary tuberculosis. *Int J Tuberc Lung Dis.* 2011;15(9):1239–1245. i.
11. Letendre ED, Holbein BE. Ceruloplasmin and regulation of transferrin iron during *Neisseria meningitidis* infection in mice. *Infect Immun.* 1984;45(1):133–138.
12. Becker KW, Skaar EP. Metal limitation and toxicity at the interface between host and pathogen. *FEMS Microbiol Rev.* 2014;38(6):1235–1249.
13. Hood MI, Skaar EP. Nutritional immunity: transition metals at the pathogen-host interface. *Nat Rev Microbiol.* 2012;10(8):525–537.

14. Nairz M, et al. Iron at the interface of immunity and infection. *Front Pharmacol.* 2014;5:152.
15. Carpenter C, Payne SM. Regulation of iron transport systems in Enterobacteriaceae in response to oxygen and iron availability. *J Inorg Biochem.* 2014;133:110–117.
16. Abergel RJ, et al. Microbial evasion of the immune system: structural modifications of enterobactin impair siderocalin recognition. *J Am Chem Soc.* 2006;128(34): 10998–10999.
17. Jacques JF, et al. RyhB small RNA modulates the free intracellular iron pool and is essential for normal growth during iron limitation in *Escherichia coli*. *Mol Microbiol.* 2006;62(4):1181–1190.
18. Kehl-Fie TE, Skaar EP. Nutritional immunity beyond iron: a role for manganese and zinc. *Curr Opin Chem Biol.* 2010;14(2):218–224.
19. Diaz-Ochoa VE, et al. Transition metal ions at the crossroads of mucosal immunity and microbial pathogenesis. *Front Cell Infect Microbiol.* 2014;4:2.
20. Cellier MF, Courville P, Campion C. Nramp1 phagocyte intracellular metal withdrawal defense. *Microbes Infect.* 2007;9(14–15):1662–1670.
21. Vidal S, et al. The Ity/Lsh/Bcg locus: natural resistance to infection with intracellular parasites is abrogated by disruption of the Nramp1 gene. *J Exp Med.* 1995;182 (3):655–666.
22. Peracino B, et al. Function and mechanism of action of Dictyostelium Nramp1 (Slc11a1) in bacterial infection. *Traffic.* 2006;7(1):22–38.
23. Blackwell JM, et al. Divalent cation transport and susceptibility to infectious and autoimmune disease: continuation of the Ity/Lsh/Bcg/Nramp1/Slc11a1 gene story. *Immunol Lett.* 2003;85(2):197–203.
24. Bellamy R. The natural resistance-associated macrophage protein and susceptibility to intracellular pathogens. *Microbes Infect.* 1999;1(1):23–27.
25. Blackwell JM, et al. SLC11A1 (formerly NRAMP1) and disease resistance. *Cell Microbiol.* 2001;3(12):773–784.
26. Gruenheid S, et al. Natural resistance to infection with intracellular pathogens: the Nramp1 protein is recruited to the membrane of the phagosome. *J Exp Med.* 1997;185 (4):717–730.
27. Searle S, et al. Localisation of Nramp1 in macrophages: modulation with activation and infection. *J Cell Sci.* 1998;111(Pt 19):2855–2866.
28. Wessling-Resnick M. Nramp1 and other transporters involved in metal withholding during infection. *J Biol Chem.* 2015;290(31):18984–18990.
29. Forbes JR, Gros P. Divalent-metal transport by NRAMP proteins at the interface of host–pathogen interactions. *Trends Microbiol.* 2001;9(8):397–403.
30. Hedges JF, et al. Solute carrier 11A1 is expressed by innate lymphocytes and augments their activation. *J Immunol.* 2013;190(8):4263–4273.
31. Blackwell JM, et al. Influence of macrophage resistance gene Lsh/Ity/Bcg (candidate Nramp) on *Toxoplasma gondii* infection in mice. *Clin Exp Immunol.* 1994;97(1): 107–112.
32. Puliti M, et al. Influence of the Bcg locus on macrophage response to the dimorphic fungus *Candida albicans*. *Infect Immun.* 1995;63(10):4170–4173.
33. Skamene E, et al. Genetic regulation of resistance to intracellular pathogens. *Nature.* 1982;297(5866):506–509.
34. Zaharik ML, et al. The *Salmonella enterica* serovar typhimurium divalent cation transport systems MntH and SitABCD are essential for virulence in an Nramp1G169 murine typhoid model. *Infect Immun.* 2004;72(9):5522–5525.
35. Li X, et al. SLC11A1 (NRAMP1) polymorphisms and tuberculosis susceptibility: updated systematic review and meta-analysis. *PLoS One.* 2011;6(1):e15831.

36. Gallant CJ, et al. Reduced in vitro functional activity of human NRAMP1 (SLC11A1) allele that predisposes to increased risk of pediatric tuberculosis disease. *Genes Immun.* 2007;8(8):691–698.
37. Courville P, Chaloupka R, Cellier MF. Recent progress in structure-function analyses of Nramp proton-dependent metal-ion transporters. *Biochem Cell Biol.* 2006;84 (6):960–978.
38. Ehrnstorfer IA, et al. Crystal structure of a SLC11 (NRAMP) transporter reveals the basis for transition-metal ion transport. *Nat Struct Mol Biol.* 2014;21(11):990–996.
39. Mackenzie B, Hediger MA. SLC11 family of H+-coupled metal-ion transporters NRAMP1 and DMT1. *Pflugers Arch.* 2004;447(5):571–579.
40. Gunshin H, et al. Cloning and characterization of a mammalian proton-coupled metal-ion transporter. *Nature.* 1997;388(6641):482–488.
41. Jabado N, et al. Natural resistance to intracellular infections: natural resistance-associated macrophage protein 1 (Nramp1) functions as a pH-dependent manganese transporter at the phagosomal membrane. *J Exp Med.* 2000;192(9):1237–1248.
42. Forbes JR, Gros P. Iron, manganese, and cobalt transport by Nramp1 (Slc11a1) and Nramp2 (Slc11a2) expressed at the plasma membrane. *Blood.* 2003;102(5):1884–1892.
43. Corbin BD, et al. Metal chelation and inhibition of bacterial growth in tissue abscesses. *Science.* 2008;319(5865):962–965.
44. Kehl-Fie TE, et al. MntABC and MntH contribute to systemic *Staphylococcus aureus* infection by competing with calprotectin for nutrient manganese. *Infect Immun.* 2013;81(9):3395–3405.
45. Becker JS, Matusch A, Wu B. Bioimaging mass spectrometry of trace elements—recent advance and applications of LA-ICP-MS: a review. *Anal Chim Acta.* 2014;835:1–18.
46. Dale I, Fagerhol MK, Naesgaard I. Purification and partial characterization of a highly immunogenic human leukocyte protein, the L1 antigen. *Eur J Biochem.* 1983;134 (1):1–6.
47. Zindl CL, et al. IL-22-producing neutrophils contribute to antimicrobial defense and restitution of colonic epithelial integrity during colitis. *Proc Natl Acad Sci USA.* 2013;110 (31):12768–12773.
48. Liu JZ, et al. Zinc sequestration by the neutrophil protein calprotectin enhances *Salmonella* growth in the inflamed gut. *Cell Host Microbe.* 2012;11(3):227–239.
49. Gebhardt C, et al. S100A8 and S100A9 in inflammation and cancer. *Biochem Pharmacol.* 2006;72(11):1622–1631.
50. Achouiti A, et al. Myeloid-related protein-14 contributes to protective immunity in gram-negative pneumonia derived sepsis. *PLoS Pathog.* 2012;8(10):e1002987.
51. Hood MI, et al. Identification of an *Acinetobacter baumannii* zinc acquisition system that facilitates resistance to calprotectin-mediated zinc sequestration. *PLoS Pathog.* 2012;8 (12):e1003068.
52. Urban CF, et al. Neutrophil extracellular traps contain calprotectin, a cytosolic protein complex involved in host defense against *Candida albicans*. *PLoS Pathog.* 2009;5(10): e1000639.
53. Bianchi M, et al. Restoration of anti-Aspergillus defense by neutrophil extracellular traps in human chronic granulomatous disease after gene therapy is calprotectin-dependent. *J Allergy Clin Immunol.* 2011;127(5). 1243.e7–1252.e7.
54. Clark HL, et al. Zinc and manganese chelation by neutrophil S100A8/A9 (calprotectin) limits extracellular *Aspergillus fumigatus* hyphal growth and corneal infection. *J Immunol.* 2015;196(1):336–344.
55. Damo SM, et al. Molecular basis for manganese sequestration by calprotectin and roles in the innate immune response to invading bacterial pathogens. *Proc Natl Acad Sci USA.* 2013;110(10):3841–3846.

56. Buchau AS, et al. S100A15, an antimicrobial protein of the skin: regulation by *E. coli* through Toll-like receptor 4. *J Invest Dermatol.* 2007;127(11):2596–2604.
57. Glaser R, et al. Antimicrobial psoriasin (S100A7) protects human skin from *Escherichia coli* infection. *Nat Immunol.* 2005;6(1):57–64.
58. Moroz OV, et al. Structure of the human S100A12-copper complex: implications for host-parasite defence. *Acta Crystallogr D Biol Crystallogr.* 2003;59(Pt 5):859–867.
59. Moroz OV, et al. Both Ca2+ and Zn2+ are essential for S100A12 protein oligomerization and function. *BMC Biochem.* 2009;10:11.
60. Brodersen DE, Nyborg J, Kjeldgaard M. Zinc-binding site of an S100 protein revealed. Two crystal structures of Ca2+-bound human psoriasin (S100A7) in the Zn2+-loaded and Zn2+-free states. *Biochemistry.* 1999;38(6):1695–1704.
61. Korndorfer IP, Brueckner F, Skerra A. The crystal structure of the human (S100A8/S100A9)2 heterotetramer, calprotectin, illustrates how conformational changes of interacting alpha-helices can determine specific association of two EF-hand proteins. *J Mol Biol.* 2007;370(5):887–898.
62. Kehl-Fie TE, et al. Nutrient metal sequestration by calprotectin inhibits bacterial superoxide defense, enhancing neutrophil killing of *Staphylococcus aureus*. *Cell Host Microbe.* 2011;10(2):158–164.
63. Brophy MB, et al. Contributions of the S100A9 C-terminal tail to high-affinity Mn(II) chelation by the host-defense protein human calprotectin. *J Am Chem Soc.* 2013;135(47):17804–17817.
64. Gagnon DM, et al. Manganese binding properties of human calprotectin under conditions of high and low calcium: X-ray crystallographic and advanced electron paramagnetic resonance spectroscopic analysis. *J Am Chem Soc.* 2015;137(8):3004–3016.
65. Brophy MB, Hayden JA, Nolan EM. Calcium ion gradients modulate the zinc affinity and antibacterial activity of human calprotectin. *J Am Chem Soc.* 2012;134(43):18089–18100.
66. Hayden JA, et al. High-affinity manganese coordination by human calprotectin is calcium-dependent and requires the histidine-rich site formed at the dimer interface. *J Am Chem Soc.* 2013;135(2):775–787.
67. McDevitt CA, et al. A molecular mechanism for bacterial susceptibility to zinc. *PLoS Pathog.* 2011;7(11):e1002357.
68. Gribenko A, et al. Three-dimensional structure and biophysical characterization of *Staphylococcus aureus* cell surface antigen-manganese transporter MntC. *J Mol Biol.* 2013;425(18):3429–3445.
69. Lisher JP, Giedroc DP. Manganese acquisition and homeostasis at the host–pathogen interface. *Front Cell Infect Microbiol.* 2013;3:91.
70. Hsu K, et al. Anti-infective protective properties of s100 calgranulins. *Antiinflamm Antiallergy Agents Med Chem.* 2009;8(4):290–305.
71. Nakashige TG, et al. Human calprotectin is an iron-sequestering host-defense protein. *Nat Chem Biol.* 2015;11(10):765–771.
72. Winter SE, et al. Gut inflammation provides a respiratory electron acceptor for *Salmonella*. *Nature.* 2010;467(7314):426–429.
73. Djoko KY, et al. The role of copper and zinc toxicity in innate immune defense against bacterial pathogens. *J Biol Chem.* 2015;290(31):18954–18961.
74. Botella H, et al. Mycobacterial p(1)-type ATPases mediate resistance to zinc poisoning in human macrophages. *Cell Host Microbe.* 2011;10(3):248–259.
75. Ong CL, et al. An antimicrobial role for zinc in innate immune defense against group A streptococcus. *J Infect Dis.* 2014;209(10):1500–1508.
76. Johnson MDL, et al. Role of copper efflux in pneumococcal pathogenesis and resistance to macrophage-mediated immune clearance. *Infection and Immunity.* 2015;83(4):1684–1694.

77. Rosch JW, et al. Role of the manganese efflux system mntE for signalling and pathogenesis in *Streptococcus pneumoniae*. *Mol Microbiol.* 2009;72(1):12–25.
78. Veyrier FJ, et al. A novel metal transporter mediating manganese export (MntX) regulates the Mn to Fe intracellular ratio and *Neisseria meningitidis* virulence. *PLoS Pathog.* 2011;7(9):e1002261.
79. Sheehan LM, et al. Coordinated zinc homeostasis is essential for the wild-type virulence of *Brucella abortus*. *J Bacteriol.* 2015;197(9):1582–1591.
80. Stahler FN, et al. The novel *Helicobacter pylori* CznABC metal efflux pump is required for cadmium, zinc, and nickel resistance, urease modulation, and gastric colonization. *Infect Immun.* 2006;74(7):3845–3852.
81. White C, et al. A role for the ATP7A copper-transporting ATPase in macrophage bactericidal activity. *J Biol Chem.* 2009;284(49):33949–33956.
82. Waldron KJ, et al. Metalloproteins and metal sensing. *Nature.* 2009;460(7257):823–830.
83. Irving H, Williams RJP. The stability of transition-metal complexes. *Journal of the Chemical Society.* 1953;3:3192–3210.
84. Maret W. Inhibitory zinc sites in enzymes. *Biometals.* 2013;26(2):197–204.
85. Scheie AA, Pearce EI. The effect of mineral-derived zinc ions on in vitro glucose metabolism of *Streptococcus mutans* NCTC 10449. *Caries Res.* 1994;28(5):329–334.
86. Macomber L, Imlay JA. The iron-sulfur clusters of dehydratases are primary intracellular targets of copper toxicity. *Proc Natl Acad Sci USA.* 2009;106(20):8344–8349.
87. Ong CL, Walker MJ, McEwan AG. Zinc disrupts central carbon metabolism and capsule biosynthesis in *Streptococcus pyogenes*. *Sci Rep.* 2015;5:10799.
88. Johnson MDL, Kehl-Fie TE, Rosch JW. Copper intoxication inhibits aerobic nucleotide synthesis in *Streptococcus pneumoniae*. *Metallomics.* 2015;7(5):786–794.
89. Counago RM, et al. Imperfect coordination chemistry facilitates metal ion release in the Psa permease. *Nat Chem Biol.* 2014;10(1):35–41.
90. Eijkelkamp BA, et al. Extracellular zinc competitively inhibits manganese uptake and compromises oxidative stress management in *Streptococcus pneumoniae*. *PLoS One.* 2014;9(2):e89427.
91. Li N, et al. Varied metal-binding properties of lipoprotein PsaA in *Streptococcus pneumoniae*. *Journal of Biological Inorganic Chemistry.* 2014;19(6):829–838.
92. Horsburgh MJ, et al. MntR modulates expression of the PerR regulon and superoxide resistance in *Staphylococcus aureus* through control of manganese uptake. *Mol Microbiol.* 2002;44(5):1269–1286.
93. Que Q, Helmann JD. Manganese homeostasis in *Bacillus subtilis* is regulated by MntR, a bifunctional regulator related to the diphtheria toxin repressor family of proteins. *Mol Microbiol.* 2000;35(6):1454–1468.
94. Hendriksen WT, et al. Strain-specific impact of PsaR of *Streptococcus pneumoniae* on global gene expression and virulence. *Microbiology.* 2009;155(Pt 5):1569–1579.
95. Kehres DG, et al. Regulation of *Salmonella enterica* serovar Typhimurium mntH transcription by H(2)O(2), Fe(2+), and Mn(2+). *J Bacteriol.* 2002;184(12):3151–3158.
96. Patzer SI, Hantke K. Dual repression by Fe(2+)-Fur and Mn(2+)-MntR of the mntH gene, encoding an NRAMP-like Mn(2+) transporter in *Escherichia coli*. *J Bacteriol.* 2001;183(16):4806–4813.
97. Posey JE, et al. Characterization of a manganese-dependent regulatory protein, TroR, from *Treponema pallidum*. *Proc Natl Acad Sci USA.* 1999;96(19):10887–10892.
98. Pandey R, et al. MntR(Rv2788): a transcriptional regulator that controls manganese homeostasis in *Mycobacterium tuberculosis*. *Mol Microbiol.* 2015;98(6):1168–1183.
99. Ikeda JS, et al. Transcriptional regulation of sitABCD of *Salmonella enterica* serovar Typhimurium by MntR and Fur. *J Bacteriol.* 2005;187(3):912–922.

100. Runyen-Janecky L, et al. Role and regulation of the *Shigella flexneri* sit and MntH systems. *Infect Immun.* 2006;74(8):4666–4672.
101. Schmitt MP. Analysis of a DtxR-like metalloregulatory protein, MntR, from *Corynebacterium diphtheriae* that controls expression of an ABC metal transporter by an Mn(2+)-dependent mechanism. *J Bacteriol.* 2002;184(24):6882–6892.
102. Johnston JW, et al. Mn2+-dependent regulation of multiple genes in *Streptococcus pneumoniae* through PsaR and the resultant impact on virulence. *Infect Immun.* 2006;74 (2):1171–1180.
103. Foster AW, Osman D, Robinson NJ. Metal preferences and metallation. *J Biol Chem.* 2014;289(41):28095–28103.
104. Puri S, Hohle TH, O'Brian MR. Control of bacterial iron homeostasis by manganese. *Proc Natl Acad Sci USA.* 2010;107(23):10691–10695.
105. Lee JW, Helmann JD. The PerR transcription factor senses H2O2 by metal-catalysed histidine oxidation. *Nature.* 2006;440(7082):363–367.
106. Makthal N, et al. Crystal structure of peroxide stress regulator from *Streptococcus pyogenes* provides functional insights into the mechanism of oxidative stress sensing. *J Biol Chem.* 2013;288(25):18311–18324.
107. Guedon E, et al. The global transcriptional response of *Bacillus subtilis* to manganese involves the MntR, Fur, TnrA and sigmaB regulons. *Mol Microbiol.* 2003;49(6):1477–1491.
108. Wu HJ, et al. PerR controls Mn-dependent resistance to oxidative stress in *Neisseria gonorrhoeae. Mol Microbiol.* 2006;60(2):401–416.
109. Ogunniyi AD, et al. Central role of manganese in regulation of stress responses, physiology, and metabolism in *Streptococcus pneumoniae. J Bacteriol.* 2010;192(17):4489–4497.
110. Papp-Wallace KM, Maguire ME. Manganese transport and the role of manganese in virulence. *Annu Rev Microbiol.* 2006;60:187–209.
111. Agranoff D, et al. Mycobacterium tuberculosis expresses a novel pH-dependent divalent cation transporter belonging to the Nramp family. *J Exp Med.* 1999;190(5):717–724.
112. Reeve I, et al. Overexpression, purification, and site-directed spin labeling of the Nramp metal transporter from *Mycobacterium leprae. Proc Natl Acad Sci USA.* 2002;99 (13):8608–8613.
113. Kehres DG, et al. The NRAMP proteins of *Salmonella typhimurium* and *Escherichia coli* are selective manganese transporters involved in the response to reactive oxygen. *Mol Microbiol.* 2000;36(5):1085–1100.
114. Bearden SW, Perry RD. The Yfe system of *Yersinia pestis* transports iron and manganese and is required for full virulence of plague. *Mol Microbiol.* 1999;32(2):403–414.
115. Dashper SG, et al. A novel *Porphyromonas gingivalis* FeoB plays a role in manganese accumulation. *J Biol Chem.* 2005;280(30):28095–28102.
116. McAllister LJ, et al. Molecular analysis of the psa permease complex of *Streptococcus pneumoniae. Mol Microbiol.* 2004;53(3):889–901.
117. Janulczyk R, Ricci S, Bjorck L. MtsABC is important for manganese and iron transport, oxidative stress resistance, and virulence of *Streptococcus pyogenes. Infect Immun.* 2003;71 (5):2656–2664.
118. Gat O, et al. The solute-binding component of a putative Mn(II) ABC transporter (MntA) is a novel *Bacillus anthracis* virulence determinant. *Mol Microbiol.* 2005;58 (2):533–551.
119. Low YL, et al. Manganese-dependent regulation of the endocarditis-associated virulence factor EfaA of *Enterococcus faecalis. J Med Microbiol.* 2003;52(Pt 2):113–119.
120. Kehres DG, et al. SitABCD is the alkaline Mn(2+) transporter of *Salmonella enterica* serovar Typhimurium. *J Bacteriol.* 2002;184(12):3159–3166.
121. Lee MJ, et al. Identification and biochemical characterization of a unique Mn2+ -dependent UMP kinase from *Helicobacter pylori. Arch Microbiol.* 2010;192(9): 739–746.

122. Berry AM, Paton JC. Sequence heterogeneity of PsaA, a 37-kilodalton putative adhesin essential for virulence of *Streptococcus pneumoniae*. *Infect Immun.* 1996;64(12):5255–5262.
123. Dintilhac A, et al. Competence and virulence of *Streptococcus pneumoniae*: Adc and PsaA mutants exhibit a requirement for Zn and Mn resulting from inactivation of putative ABC metal permeases. *Mol Microbiol.* 1997;25(4):727–739.
124. Paik S, et al. The sloABCR operon of *Streptococcus mutans* encodes an Mn and Fe transport system required for endocarditis virulence and its Mn-dependent repressor. *J Bacteriol.* 2003;185(20):5967–5975.
125. Wichgers Schreur PJ, et al. TroA of *Streptococcus suis* is required for manganese acquisition and full virulence. *J Bacteriol.* 2011;193(19):5073–5080.
126. Anderson ES, et al. The manganese transporter MntH is a critical virulence determinant for *Brucella abortus* 2308 in experimentally infected mice. *Infect Immun.* 2009;77(8): 3466–3474.
127. Champion OL, et al. *Yersinia pseudotuberculosis* mntH functions in intracellular manganese accumulation, which is essential for virulence and survival in cells expressing functional Nramp1. *Microbiology.* 2011;157(Pt 4):1115–1122.
128. Lim KH, et al. Metal binding specificity of the MntABC permease of *Neisseria gonorrhoeae* and its influence on bacterial growth and interaction with cervical epithelial cells. *Infect Immun.* 2008;76(8):3569–3576.
129. Porcheron G, et al. Iron, copper, zinc, and manganese transport and regulation in pathogenic Enterobacteria: correlations between strains, site of infection and the relative importance of the different metal transport systems for virulence. *Front Cell Infect Microbiol.* 2013;3:90.
130. Imlay JA. Cellular defenses against superoxide and hydrogen peroxide. *Annu Rev Biochem.* 2008;77:755–776.
131. Anjem A, Varghese S, Imlay JA. Manganese import is a key element of the OxyR response to hydrogen peroxide in *Escherichia coli*. *Mol Microbiol.* 2009;72(4):844–858.
132. Anjem A, Imlay JA. Mononuclear iron enzymes are primary targets of hydrogen peroxide stress. *J Biol Chem.* 2012;287(19):15544–15556.
133. Imlay JA. The molecular mechanisms and physiological consequences of oxidative stress: lessons from a model bacterium. *Nat Rev Microbiol.* 2013;11(7):443–454.
134. Makui H, et al. Identification of the *Escherichia coli* K-12 Nramp orthologue (MntH) as a selective divalent metal ion transporter. *Mol Microbiol.* 2000;35(5):1065–1078.
135. Cui J, Davidson AL. ABC solute importers in bacteria. *Essays Biochem.* 2011;50(1):85–99.
136. Maqbool A, et al. The substrate-binding protein in bacterial ABC transporters: dissecting roles in the evolution of substrate specificity. *Biochem Soc Trans.* 2015;43 (5):1011–1017.
137. Berntsson RP, et al. A structural classification of substrate-binding proteins. *FEBS Lett.* 2010;584(12):2606–2617.
138. Morey JR, McDevitt CA, Kehl-Fie TE. Host-imposed manganese starvation of invading pathogens: two routes to the same destination. *Biometals.* 2015;28(3):509–519.
139. Ouyang Z, et al. A manganese transporter, BB0219 (BmtA), is required for virulence by the Lyme disease spirochete, *Borrelia burgdorferi*. *Proc Natl Acad Sci USA.* 2009;106 (9):3449–3454.
140. Groot MN, et al. Genome-based in silico detection of putative manganese transport systems in *Lactobacillus plantarum* and their genetic analysis. *Microbiology.* 2005;151 (Pt 4):1229–1238.
141. Green RT, Todd JD, Johnston AW. Manganese uptake in marine bacteria; the novel MntX transporter is widespread in Roseobacters, Vibrios, Alteromonadales and the SAR11 and SAR116 clades. *ISME J.* 2013;7(3):581–591.
142. Archibald FS, Fridovich I. Manganese and defenses against oxygen toxicity in *Lactobacillus plantarum*. *J Bacteriol.* 1981;145(1):442–451.

143. Posey JE, Gherardini FC. Lack of a role for iron in the Lyme disease pathogen. *Science.* 2000;288(5471):1651–1653.
144. Hao Z, Chen S, Wilson DB. Cloning, expression, and characterization of cadmium and manganese uptake genes from *Lactobacillus plantarum*. *Appl Environ Microbiol.* 1999;65 (11):4746–4752.
145. Stork M, et al. An outer membrane receptor of *Neisseria meningitidis* involved in zinc acquisition with vaccine potential. *PLoS Pathog.* 2010;6:e1000969.
146. Calmettes C, et al. The molecular mechanism of Zinc acquisition by the neisserial outer-membrane transporter ZnuD. *Nat Commun.* 2015;6:7996.
147. Hohle TH, et al. Bacterial outer membrane channel for divalent metal ion acquisition. *Proc Natl Acad Sci USA.* 2011;108(37):15390–15395.
148. Miller AF. Superoxide dismutases: ancient enzymes and new insights. *FEBS Lett.* 2012;586(5):585–595.
149. Archibald FS, Fridovich I. The scavenging of superoxide radical by manganous complexes: in vitro. *Arch Biochem Biophys.* 1982;214(2):452–463.
150. Horsburgh MJ, et al. Manganese: elemental defence for a life with oxygen. *Trends Microbiol.* 2002;10(11):496–501.
151. Aguirre JD, Culotta VC. Battles with iron: manganese in oxidative stress protection. *J Biol Chem.* 2012;287(17):13541–13548.
152. Culotta VC, Daly MJ. Manganese complexes: diverse metabolic routes to oxidative stress resistance in prokaryotes and yeast. *Antioxid Redox Signal.* 2013;19(9): 933–944.
153. Juttukonda LJ, Skaar EP. Manganese homeostasis and utilization in pathogenic bacteria. *Mol Microbiol.* 2015;97(2):216–228.
154. Gajadeera CS, et al. Structure of inorganic pyrophosphatase from *Staphylococcus aureus* reveals conformational flexibility of the active site. *J Struct Biol.* 2015;189(2):81–86.
155. Fraser HI, Kvaratskhelia M, White MF. The two analogous phosphoglycerate mutases of *Escherichia coli*. *FEBS Lett.* 1999;455(3):344–348.
156. Mohamed SF, et al. Ribonucleotide reductase in *Bacillus subtilis*—evidence for a Mn-dependent enzyme. *Biofactors.* 1998;7(4):337–344.
157. Reddy SK, et al. Eukaryotic-like adenylyl cyclases in *Mycobacterium tuberculosis* H37Rv: cloning and characterization. *J Biol Chem.* 2001;276(37):35141–35149.
158. Johnson GS, et al. Role of the spoT gene product and manganese ion in the metabolism of guanosine 5′-diphosphate 3′-diphosphate in *Escherichia coli*. *J Biol Chem.* 1979;254 (12):5483–5487.
159. Ohtani N, et al. Identification of the genes encoding Mn2+-dependent RNase HII and Mg2+-dependent RNase HIII from *Bacillus subtilis*: classification of RNases H into three families. *Biochemistry.* 1999;38(2):605–618.
160. Ammendola S, et al. High-affinity Zn2+ uptake system ZnuABC is required for bacterial zinc homeostasis in intracellular environments and contributes to the virulence of *Salmonella enterica*. *Infect Immun.* 2007;75(12):5867–5876.
161. Campoy S, et al. Role of the high-affinity zinc uptake znuABC system in *Salmonella enterica* serovar Typhimurium virulence. *Infect Immun.* 2002;70(8):4721–4725.
162. Davis LM, Kakuda T, DiRita VJ. A *Campylobacter jejuni* znuA orthologue is essential for growth in low-zinc environments and chick colonization. *J Bacteriol.* 2009;191 (5):1631–1640.
163. Rosadini CV, et al. A novel zinc binding system, ZevAB, is critical for survival of nontypeable *Haemophilus influenzae* in a murine lung infection model. *Infect Immun.* 2011;79(8):3366–3376.
164. Corbett D, et al. Two zinc uptake systems contribute to the full virulence of *Listeria monocytogenes* during growth in vitro and in vivo. *Infect Immun.* 2012;80(1): 14–21.

165. Bayle L, et al. Zinc uptake by *Streptococcus pneumoniae* depends on both AdcA and AdcAII and is essential for normal bacterial morphology and virulence. *Mol Microbiol.* 2011;82(4):904–916.
166. Li CX, et al. *Candida albicans* adapts to host copper during infection by swapping metal cofactors for superoxide dismutase. *Proc Natl Acad Sci USA.* 2015;112(38): E5336–E5342.
167. CDC, *Antibiotic Resistance Threats in the United States, 2013*. 2013, U.S. Centers for Disease Control and Prevention: Online
168. WHO. *Antimicrobial Resistance Global Report on Surveillance*. Geneva, Switzerland: World Health Organization; 2014.

CHAPTER TWO

Microbial Virulence and Interactions With Metals

N. German*, F. Lüthje[†], X. Hao[‡], R. Rønn[§], C. Rensing[†,1]

[*] Department of Pharmaceutical Sciences, Texas Tech University Health Sciences Center, Amarillo, TX, United States
[†] Department of Plant and Environmental Sciences, University of Copenhagen, Frederiksberg, Denmark
[‡] Institute of Urban Environment, Chinese Academy of Sciences, Xiamen, China
[§] Department of Biology, University of Copenhagen, Copenhagen, Denmark
[1] Corresponding author. E-mail address: rensing@iue.ac.cn

Contents

Abstract

Transition metals, such as iron, copper, zinc, and manganese play an important role in many bacterial biological processes that add to an overall evolutional fitness of bacteria. They are often involved in regulation of bacterial virulence as a mechanism of host invasion. However, the same transition metals are known to play an important role in host-defense mechanisms against bacteria through Fenton chemistry evoked toxicity as an example. Copper and zinc are used as a mechanism to poison bacteria whereas other metals, such as, iron and manganese are withheld by the predator to prevent reconstruction of Fe–S clusters and the use of Mn as a protectant against reactive oxygen species. Therefore, tight regulation of transition metal distribution in bacteria and hosts is a vital part of host–pathogen interactions.

The ability of pathogens to withstand selective pressure evoked in the presence of the host-immune system, microbial competitors, or antimicrobial agents, and survive, reproduce, or spread defines their evolutionary fitness. The genes that encode the elements responsible for such fitness will be passed

Progress in Molecular Biology and Translational Science, Volume 142
ISSN 1877-1173
http://dx.doi.org/10.1016/bs.pmbts.2016.05.010

on by successful pathogens to subsequent generations, providing the basis for positive directional selection. Dawkins et al.,[1] summarized this concept in 1979, proposing that the presence of a continuous cyclic evolutionary arms race between the host and pathogen should lead to a fixation of beneficial alleles in a population. However, this model does not explain the plasticity of bacterial genomes in which different bacterial alleles can be beneficial in one environment and auxiliary, at best, in another. Therefore, another model was proposed in which polymorphic diversity within a population was explained by a frequency-dependent selection capable of preserving rare alleles.[2] It is well accepted today that both of these selection types influence microbial populations.

Opportunistic pathogens are organisms that can become pathogenic only after certain perturbations to the host, such as, disease, wounding, prior infection, alteration of immune system, and aging.[3] There are examples of normally commensal bacteria, such as, *Staphylococcus aureus* and *Streptococcus pneumoniae*, which become opportunistic pathogens, as well as, bacteria acquired from the environment (*Pseudomonas aeruginosa* and *Burkholderia cepacia*). It is safe to suggest that opportunistic pathogens must have well-developed plastic responses to succeed in different environments. Bacteria rely on tight regulation to ensure that specific genes are turned on only under certain conditions. For example, activation of flagella expression in *Salmonella enterica* serovar Typhi is required to obtain nutrients from the environment. But this feature is downregulated when these bacteria invade the host to avoid the host-derived immune response.[4] Another example is the ability of *P. aeruginosa* to switch to a mucoid phenotype in vivo to ensure biofilm formation, thereby increasing survival of the bacteria in the lungs.[5]

The process by which pathogens harm their host, through invasion and activation of pathways that cause tissue damage, is defined as microbial virulence.[6] Virulence varies significantly among pathogens from being lethal to nearly asymptomatic. Microbial virulence is regulated by virulence factors that are classically defined as components that influence virulence but do not affect viability when deleted.[6] Such factors may or may not affect the growth rate of bacteria within the host. Virulence factors may be involved in many processes, including bacterial cell adhesion, motility, quorum sensing, iron (Fe) uptake, antiphagocytosis mechanisms (capsulation), and biofilm formation. Certain virulence factors are known to specific determinants of pathogenicity, such as, the production of a polysaccharide capsule by *S. pneumoniae*,[7] whereas other factors contribute broadly to the overall pathogenicity of a bacterium or fungus, altering the magnitude of the disease state, but not defining it.[6]

Several theories have been proposed to explain the evolution of virulence.[8] The virulence hypothesis predicts that pathogens would evolve to become less virulent, as killing the host would also lead to the pathogen's death. Lethal pathogens are therefore considered maladapted. This hypothesis was later largely replaced by the trade-off hypothesis, which states that a trait that increases virulence can be favored if it, at the same time, increases the chances of transmission. The original evolution hypothesis explains a selection for virulence factors, which increase the pathogen's fitness within the host, despite the fact that it decreases the chances for transmission to new hosts. The coincidental evolution hypothesis states that virulence factors have evolved as a response to selection pressures in other ecological niches. Therefore, it explains why we find many pathogens for which humans are an evolutionary dead-end, but which still possesses strong virulence factors. For example, despite the fact that the soil bacteria *Clostridium tetani* and *Clostridium botulinum* do not naturally inhabit human hosts, are not transmitted between humans and gain nothing from killing their host, are still able to grow inside human tissue and produce neurotoxins that serve as extremely potent virulence factors.[6] Most likely, these neurotoxins give them an advantage in their usual soil environment and they have coincidentally allowed the Clostridia species that produce them to become lethal pathogens.[9] Another example is *Legionella pneumophila*, the causative agent of Legionnaires' disease. *L. pneumophila* is not transmitted from person-to-person but can survive and multiply inside some species of free-living amoebae[10]; thus, the amoebae constitute a reservoir and are likely the natural host of this bacterium. The ability of *L. pneumophila* to infect human macrophages is likely a consequence of its evolved ability to infect amoeba. These and similar examples provide the basis for the emerging paradigm of environmental-based preselection for virulence factors.[3] This concept highlights how the evolution of bacterial resistance to predation by protists in the environment has been preselected for adaptation mechanisms that allow for enhanced bacterial survival in human macrophages.[11] Interestingly, it has been observed that many genes associated with bacterial virulence factors are located on genomic islands (or pathogenicity islands), which are gene clusters presumably originating from horizontal gene transfer events ensuring efficient "sharing" of this information in a microbial community, if needed.[12]

Overall, it is clear that bacterial virulence has emerged as a response to selective pressure presented by the host or the environment. Hence, in-depth exploration of the mechanisms underlying pathogen–host interactions is essential for understanding virulence-factor activation and defining therapeutic targets for potential antivirulence therapies.

Metal ions, such as, Fe, copper (Cu), zinc (Zn), and manganese (Mn), are required for many crucial biological processes.[13] These transition metals are key micronutrients for eukaryotic and prokaryotic cells and act as inorganic cofactors for up to 25% of all cellular proteins.[14] At the same time, metal ions are known to cause cell toxicity by displacement of metal cofactors from various enzymes, rendering them inactive, as well as, through production of reactive oxygen species (ROS) (Fenton chemistry). The mechanism of the latter includes reaction of reduced form of transitions metals, such as, Fe, Cu, and Ni, with dissolved oxygen to produce superoxide. Superoxide, in turn, can react with another molecule of the reduced metal to produce hydrogen peroxide (H_2O_2) that promotes production of ROS, such as, hydroxyl radicals and oxidized forms of metals. (Fig. 1)

It is known that metals from the upper end of Irving–Williams series, such as, Mn (II) $<$ Fe (II) $<$ Co (II) $<$ Ni (II) $<$ Cu (II) $>$ Zn (II), can displace metals from the lower end. Cu (II) complexes are known to display the highest stability in the range of these transition metals. Bacteria use various strategies to regulate intracellular concentrations of metals and prevent cell toxicity. These mechanisms include transport of ions by means of metal-efflux pumps, channels, cation-specific metalloregulatory proteins, small noncoding RNAs, and two-component signal-transduction systems.

This important aspect of host–pathogen interaction is based on the regulation of metal distribution in both the bacteria and the host. Transition metals, in particular, Fe and Zn, are required for bacterial survival and growth. Therefore, the first line of immune defense on bacterial invasion is to starve the pathogen of these metals. This host-derived strategy, termed "nutritional immunity,"[15] is achieved via activation of metal-sequestering mechanisms. In turn, bacterial survival "in host" is dependent on the upregulation of siderophore production (to scavenge Fe) and upregulation of metal ion-transport systems, which can be considered as examples of virulence factors.

$$\text{1. Me (II)} + O_2 \longrightarrow \text{Me (III)} + {}^{\bullet}O_2^-$$

$$\text{2. Me (III)} + {}^{\bullet}O_2^- + 2\,H^+ \longrightarrow \text{Me (III)} + H_2O_2$$

$$\text{3. Me (III)} + H_2O_2 \longrightarrow \text{Me (III)} + {}^{\bullet}OH + OH^-$$

Figure 1 Production of ROS via Fenton-like reactions.

The other aspect of metal-based host-immune defense is to increase concentrations of transition metals in the phagosome of macrophages to levels that are toxic to the bacteria. This strategy has been termed *brass dagger* due to its reliance on Cu and Zn. In response to host-mediated metal poisoning, bacteria have developed ion-efflux defense mechanisms.

In this chapter we summarize reported evidence of bacterial virulence modulation by various transition metals.

1. IRON

Although Fe is the most abundant element on Earth, by mass, its biological availability after the Great Oxidation Event became greatly limited due to decreased solubility under physiological conditions. Despite this fact, Fe is required for a plethora of biological processes in eukaryotic and prokaryotic cells. In vertebrates, heme, a porphyrin-complexed form of Fe, is required for oxygen transport and storage. Heme is also the cytochrome cofactor, important in myriad oxidation–reduction reactions in vertebrates and bacteria. In addition, bacteria require Fe to maintain Fe–S clusters involved in electron transfer and protein stabilization.[16] However, excess levels of Fe promote formation of ROS that cause DNA damage and inactivation of enzymes.[17] Therefore, Fe concentrations are highly regulated by organisms to both minimize metal-induced toxicity and ensure the presence of Fe concentrations necessary for bacterial growth. In response to host-mediated Fe starvation during infection, bacteria activate several virulence factors, including upregulation of Fe-uptake systems and production of siderophores (Fig. 2).

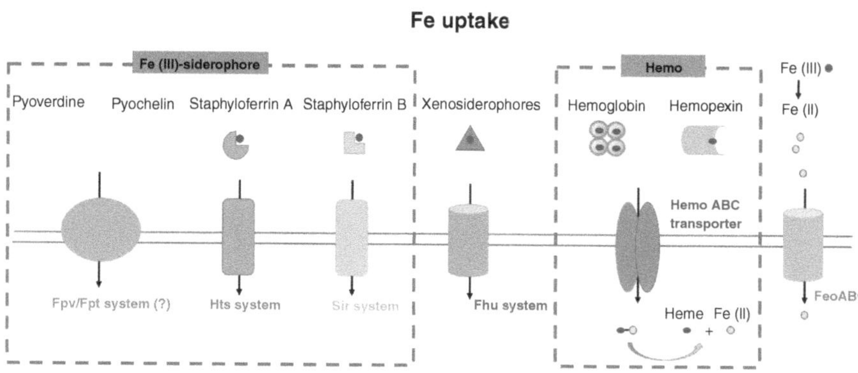

Figure 2 Fe sources and Fe-uptake systems employed in bacteria.

1.1 Sequestering of Iron by Bacteria

Due to the essential role of Fe in many metabolic processes, bacteria have developed several mechanisms to regulate intracellular Fe concentration. The ability to sequester Fe from the environment determines how successful a bacterial invasion will be and what niche the bacterium will be able to occupy relative to other bacterial species present at the site of infection. Therefore, mechanisms of Fe sequestering, including production of siderophores, are considered to be components of bacterial virulence.

Siderophores (*Greek* for Fe carriers) (Fig. 3) are small metal-chelating compounds that are secreted by many pathogenic and nonpathogenic bacteria.[18] Siderophore production is a well-studied characteristic for the vast majority of microbes with almost over 500 different molecules described to date.[18] Evolution of siderophore biosynthesis is thought to be a response to the appearance of oxygen in the early atmosphere, an event that threatened to convert soluble ferrous Fe (Fe II) to its insoluble ferric form (Fe III). Biosynthetic pathways for siderophore production may occur

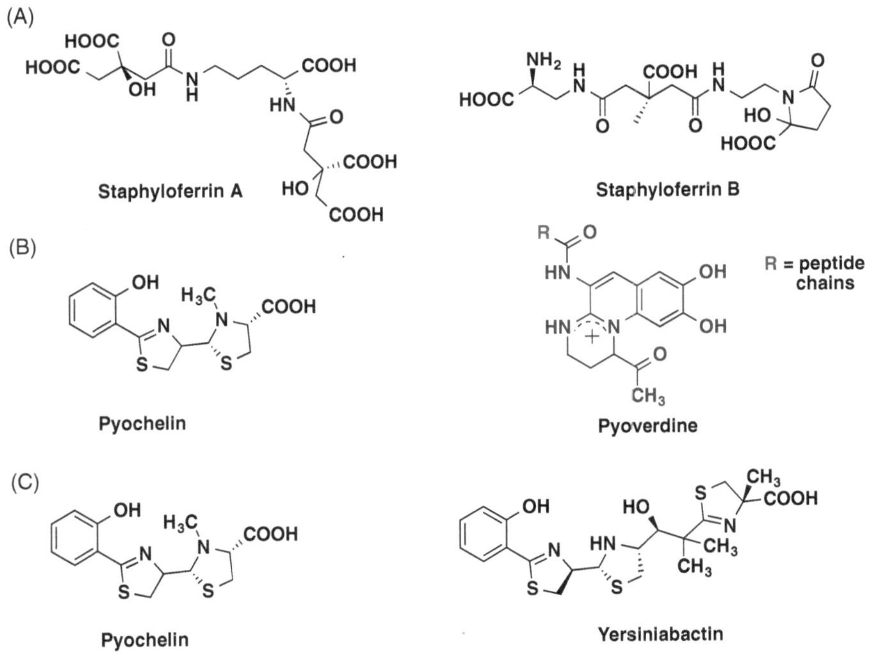

Figure 3 Representative structures of siderophores discussed in this chapter. (A) Siderophores produced by *S. aureus*. (B) Siderophores produced by *P. aeruginosa*. (C) Representative siderophores that capable of interacting with Cu.

via nonribosomal peptide synthase/polyketide synthase (NPPS/PKS)[2]-dependent or -independent pathways.[18] Siderophores are characterized by their high affinity to oxidized Fe.

S. aureus, a Gram-positive nonmotile cocci, is known to produce two types of siderophores: staphyloferrin A and staphyloferrin B. Both of these compounds belong to the carboxylate family and are capable of scavenging Fe from different host proteins.[19] The production of siderophores is regulated by the Fe-dependent ferric uptake regulator (fur), which activates corresponding genes in response to an Fe-deprived environment.[20] In addition to initiating biosynthesis of siderophores, *fur* is known to regulate expression of virulence factors required for cell adhesion and biofilm formation.[21] Moreover, it activates the release of immunomodulatory toxins causing suppression of the host-immune system.[22] On binding Fe, siderophore–Fe complexes are recognized by membrane lipoprotein receptors: HtsA (staphyloferrin A) and SirA (staphyloferrin B). This ligand–receptor interaction leads to a conformational change in the corresponding receptor and entrapment of the siderophore–Fe complex.[23] ABC permeases, HtcBC, and SirBC, facilitate transport of staphyloferrin A and B into the cytoplasm.

In addition, *S. aureus* is capable of utilizing xenosiderophores—siderophores produced by other bacteria.[19] Such a capability provides an additional advantage for *S. aureus* survival and allows it to outcompete other species. Xenosiderophores are recognized and transported using the *fhu*CBG-encoded system; receptors FhuD1 and FhuD2 undergo less of a conformational change on ligand binding than SirA and HtsA.[24] A weaker affinity for ligands allows those receptors to achieve a broad spectrum, promiscuous binding for many xenosiderophores.

Heme is another attractive source of Fe and many bacteria have developed mechanisms of utilizing host-derived heme.[25] Due to its high hydrophobicity, heme can easily interact with cell membranes where it can promote nonenzymatic redox reactions. Therefore, heme Fe exists in a bound form, for example, hemoglobin or hemopexin, and bacteria require additional mechanisms for isolating Fe from these complex molecules. *S. aureus*, for example, utilizes the IsdH system containing three near-Fe transporter (NEAT) domains.[26] By passing the hemoglobin complex through these binding domains bacteria "strip" heme from its auxiliary components. The complete mechanism of heme transport into the cytoplasm is not yet fully understood. It is known that *S. aureus* uses the IsdDEF system in which the permease IsdF utilizes energy from ATP hydrolysis by IsdD to deliver heme into the cytoplasm, where Fe is released from the complex on enzymatic degradation. However, genetic deletion of *isd*DEF had only a slight effect on the growth of *S. aureus* on heme as the sole Fe source, suggesting the presence of additional unknown pathways for utilizing heme-based Fe.[19]

P. aeruginosa is a Gram-negative bacterium that is one of the most common causes of opportunistic infections in humans. In part, the high degree of pathogenicity can be explained by the high adaptability of this organism and its ability to occupy many different niches. *P. aeruginosa* has three major strategies to obtain Fe from the environment, including production of siderophores/utilization of xenosiderophores, heme uptake, and extracellular reduction of Fe (III) to Fe (II).

There are two types of siderophore molecules produced by *P. aeruginosa*: pyoverdine and pyochelin. Interestingly, these molecules have different affinities for Fe and they are utilized at different stages of the infection. Pyoverdine is a high affinity mixed siderophore consisting of a peptide chain (6–12 aa) and a chrompohore.[27] The chrompohore moiety is highly conserved, whereas the composition of the peptide chain varies between different strains, even within the same species.[28] It has been shown that the production of pyoverdine is required to initiate infection in a burned mouse model or in a pulmonary infection mouse model.[29] Moreover, the production of pyoverdine triggers release of other virulence factors essential for bacterial invasion of host tissue.[30] At the same time, pyochelin, a low affinity siderophore,[31] has been shown to play an important role in establishing chronic infections, such as, cystic fibrosis (CF) and high concentrations of pyochelin were detected in CF-sputum medium.[32] Production of this compound results in chronic inflammation of the host tissue, causing damage and promoting subsequent release of hemoproteins and other Fe-containing proteins into the environment.

During the chronic disease stage, *P. aeruginosa* also employs extracellular reduction of Fe (III) to its soluble Fe (II) form, which can diffuse through the outer membrane and be transported into the cytoplasm using the FeOABC system.[33] This type of Fe uptake is characteristic in microaerobic or anaerobic environments, such as in CF-lung mucus, and promotes biofilm formation. Reduction of Fe (III) occurs via phenazine-induced redox reactions, therefore phenazines, secondary metabolites *of P. aeruginosa*, are also considered to be virulence factors.

From these examples, it is clear that the uptake of Fe from the environment is essential for bacterial survival and the establishment of an acute/chronic infection in the host tissue and this paradigm extends to many species of bacteria. For example, Modarresi et al. have shown that Fe-limited environments promote the formation of biofilms by *Acinetobacter baumannii*, possibly through the induction of acyl homoserine lactone (AHL) signaling.[34] Similarly, Fe-limiting conditions led to production of virulence factors in *Escherichia coli*, *Vibrio cholerae*, and *Corynebacterium diphtheriae*.[35] Interestingly, higher concentrations of Fe have been shown to lead to increased pathogenicity for numerous bacteria; a study in African children demonstrated that Fe-fortified foods can lead to a higher number of pathogenic bacteria in the microflora.[36] Similar results were obtained in the other studies, where Fe enrichment resulted in slightly higher diarrhea cases among children.[37] Under these conditions, pathogenic bacteria were shown to have increased growth rates and elevated production of virulence factors to promote cell adhesion and cell invasion. Therefore, both Fe-enriched and Fe-deprived conditions can promote virulence and result in increased pathogenicity.

2. COPPER

The bactericidal properties of Cu have been known and taken advantage of for thousands of years. Due to a very low bacterial tolerance for intracellular Cu, host organisms have developed several Cu-accumulation strategies to initiate metal-induced toxicity.[38] Although the exact mechanism of Cu toxicity to the bacterial cell is still not fully understood, certain pathways that are believed to mediate this toxicity, including ROS-associated Cu response, are well studied. ROS production leads to the damage of bacterial DNA, proteins, and lipids. Cu has been shown to induce damage of Fe–S clusters essential for bacterial survival and growth. As discussed earlier, Cu toxicity is employed by macrophages to eliminate pathogenic bacteria from the host organism. It was proposed that such mechanisms of bacterial killing originated in protozoa and evolved only later in eukaryotic phagocytes. Many interactions between prokaryotes and eukaryotes are governed by differences in response to different metals and there is evidence that free-living protozoa employ various metal-poisoning strategies to kill bacteria in their phagosome.[39] These observations have led to the hypothesis that metal-induced virulence evolved as a resistance mechanism to grazing by free-living protozoan predators.[40] Therefore, coevolution of bacteria and free-living unicellular eukaryote predators has predisposed bacteria to become pathogenic (opportunistically) in multicellular host organisms.

The dual role of Cu as an essential micronutrient and as a toxic compound requires tight control of the intracellular concentration and bacteria make use of several mechanisms to handle this challenge. This text will focus on *E. coli* since its Cu homeostasis mechanisms are relatively well described. No Cu-specific uptake proteins have been identified for *E. coli* but several systems to remove free Cu are known.[41] In general three strategies are used to handle Cu toxicity.

- Cu efflux—excess Cu is removed by pumping it across cell membranes.
- Cu sequestration—Cu-chelating agents bind and inactivate Cu ions either outside the cell to prevent them from entering the cell or inside the periplasm where the chelator delivers the ion to an export system.
- Cu oxidation—oxidizes Cu (I) to Cu (II). Cu (I) is more toxic than Cu (II) because, as a strong soft metal, it can displace Fe in Fe–S clusters, in addition to being able to generate ROS through a Fenton-like reaction. Furthermore, both Cu (I) and Cu (II) can cross the outer membrane of *E. coli* but only Cu (I) can permeate the inner membrane and reach the cytoplasm by an unknown mechanism.

E. coli encodes two Cu tolerance systems called *cue* (*Cu e*fflux) and *cus* (*Cu s*ensing). The *cue* system is the main system under aerobic conditions, while the *cus* system is necessary for proper Cu tolerance in anaerobic environments.[42]

CueR, a Cu-responsive metalloregulator, which controls the expression of *copA* and *cueO*, regulates the *cue* system. CueR has a sensitivity to free Cu (I) in the zeptomolar (10^{-21} M) range; corresponding to its activation at a concentration 11 orders of magnitude lower than what corresponds to 1 free Cu atom per cell.[43] This fine-tuned regulation of the intracellular Cu concentration highlights how important proper Cu homeostasis is for the cell. CopA is a Cu exporting P_{1B}-type ATPase, which transports Cu (I) from the cytoplasm to the periplasm. CueO is a multicopper oxidase localized in the periplasm. It oxidizes Cu (I) to the less toxic Cu (II) and reduces dioxygen to water. As the reaction catalyzed by this enzyme requires oxygen, CueO only works under aerobic conditions.[44a]

The *cus* system encodes a multiunit pump spanning the periplasmic space, transporting Cu (I) from the periplasm out of the cell.[45] This pump called CusCBA comprises three distinct components: an inner membrane proton antiporter (CusA), an outer membrane channel (CusC), and a periplasmic membrane fusion protein (CusB) connecting CusA and CusC. In addition, the *cus* determinant also encodes CusF, a periplasmic metallochaperone that binds Cu (I) and can deliver it to CusB, thus enhancing periplasmic Cu efflux from the cell. The *cus* operon has two regulatory genes, *cusS* and *cusR*. Membrane-bound CusS senses Cu (I) in the periplasm and can activate CusR, which in turn regulates the expression of *cusCFBA*.[44]

2.1 Plasmid-Borne Resistance

Many strains, able to survive in environments with high Cu concentrations, contain additional plasmid-encoded determinants that confer Cu resistance. The *pco* (*p*lasmid-borne *Cu* resistance) system is an example of this. This gene cluster was initially thought to comprise seven genes, *pcoABCDRSE*, arranged in two operons, *pcoABCD* and *pcoSR*, plus *pcoE* further downstream. It has now been shown that the original 7-gene *pco* operon is part of a larger 19-gene cluster. The Pco system does not confer higher Cu tolerance in *ΔcopA* mutants, indicating a link between chromosomally-encoded and plasmid-borne resistance. Presumably the *pco* gene cluster plays a role in sequestering the Cu transported to the periplasm from the cytoplasm by CopA. *PcoE* is controlled by CusRS, whereas, the expression of *PcoABCD* is primarily regulated by PcoRS. PcoE does not confer Cu-resistance by itself but is strongly induced by Cu, and allows *E. coli* to recover more quickly from Cu stress.[46] This suggests that PcoE sequesters Cu in the periplasm, giving time for the remaining *pco* genes to be fully induced and the encoded proteins to function effectively. PcoA is a multicopper oxidase distantly related to CueO; PcoC is a periplasmic chaperone that binds Cu (I); PcoB is predicted to be an outer membrane protein; PcoD is involved in Cu uptake across the inner membrane.[44a]

2.2 Copper-Binding Siderophores

It is well accepted that Cu-binding molecules in bacteria have evolved as defense tools, allowing the acquisition and/or neutralization of host-derived Cu. The virulence of some pathogenic Enterobacteriaceae, including uropathogenic *E. coli*, is strongly linked to the expression of yersiniabactin (Ybt), a salicylate-based siderophore. In addition to scavenging Fe, Ybt can also sequester Cu extracellularly, preventing it from entering the bacterial cell. Ybt forms a stable complex with Cu (II) that, in addition to shielding against the toxicity of Cu, also protects against ROS-mediated damage. The protection mechanism is based on its superoxide dismutase (SOD)-like activity, catalyzing the dismutation of superoxide into oxygen and H_2O_2.[41] Ybt–Cu (II) catalyzes the following reactions:

$$O_{2\bullet}^{-} + Cu^{2+} - Ybt \rightarrow Cu^{+} - Ybt + O_2$$

$$O_{2\bullet}^{-} + Cu^{2+} - Ybt \rightarrow 2H^{+} \rightarrow Cu^{2+} - Ybt + H_2O_2$$

The Ybt–Cu (II) complex can thus decrease the concentration of the reductant, superoxide, and increase the concentration of H_2O_2 and may indirectly reduce the concentration of Cu (I).

Another new class of Cu-chelating siderophores, methanobactins, was described very recently for Gram-negative species. This class of molecules results from the biosynthesis of the Cu-containing particulate methane monooxygenase.[47]

3. MANGANESE

Mn also plays an essential role in bacterial cellular processes; it is present as a cofactor in some hydrolytic enzymes and certain enzymes involved in metabolic processes.[13] In addition, Mn is required for bacteria to neutralize host-mediated oxidative stress; thus, Mn increases the virulence of pathogenic bacteria.[48] It has been shown that ROS stress activates Mn import and that deletion of Mn-import systems from bacteria causes reduced cell viability due to high sensitivity to superoxide-anion radicals or H_2O_2. It was also shown that low concentrations of H_2O_2 in *E. coli* induced the OxyR regulon that includes genes encoding MntH, a Mn-import pump.[49] Similar correlations between the level of H_2O_2 and the upregulation of the PerR regulon, an analog of OxyR present in Gram-positive bacteria, have been noted.[50]

The inorganic "fingerprint" of an unstressed bacterial cell, defined as the total concentration of all cell-associated metals over all cellular fractions (membrane, cytoplasm, etc.) can be used to segregate Fe-centered bacteria from Mn-centered bacteria based on the Mn:Fe ratio.[48] The first group of bacteria is characterized by lower accumulation of Mn in the cell when compared to the intracellular Fe level. The difference in ratio can range from 10-fold (*E. coli*) to 1000-fold (*Pseudomonas putida*).[51] A second group of organisms, such as, the lactic acid bacteria *S. pneumoniae*[52] and *Lactobacillus planatarum*, pathogenic *Neisseria meningitis*[53] and the UV-resistant *Deinococcus radiodurans*,[54] have a Mn:Fe ratio at the level of ≥1. It has been proposed that such high levels of Mn are required for survival of these species due to their ability to neutralize damage caused by γ-radiation (*D. radiodurans*) or by the high concentration of the H_2O_2. H_2O_2 is produced in millimolar concentrations by lactic acid bacteria under aerobic conditions; therefore, effective mechanisms of detoxification, possibly involving Mn, are required for cell survival.

There are several Mn-induced antioxidant mechanisms in bacterial cells (Fig. 4). One involves the Mn–SOD catalyzed transformation of the superoxide anion radical to H_2O_2 and oxygen, followed by the further removal of H_2O_2 by catalase and peroxidase enzymes.[55] However, it has been shown that the presence of free or complexed Mn (II) can protect SOD-mutants from oxidative stress, indicating that there are other detoxification pathways involving Mn.[56] At the same time, another pathway was discovered that requires simple Mn compounds, such as $Mn_3(PO_4)_2$, to ensure nonenzymatic disproportionation of peroxide in vivo.[57] This mechanism was shown to play a complimentary role to major the Mn–SOD pathway for many types of bacteria including *Bacillis subtilis*, *E. coli* and *D. radiodurans*. However, Tseng et al. have reported that Mn-induced resistance to oxidative stress in *Neisseria gonorrhoeae* is independent from the SOD-catalyzed process.[58] A third mechanism of Mn-induced resistance to oxidative stress is based on similarities in the coordination parameters of Mn (II) and Fe (II). It is well known that Fe, as a cofactor in an enzyme, undergoes reversible inactivation in the presence of lower H_2O_2 concentrations, followed by irreversible changes if the oxidative stress persists.[59] Recent studies have suggested a cofactor substitution of Fe (II) for Mn (II) as a protective mechanism preventing inactivation of these enzymes in vivo. For example, Anhem and Imlay have shown that under oxidative stress conditions *E. coli* can switch from a Fe-based metabolism to a Mn-based one.[59b] Overall, Mn is an essential player in providing the bacterial cell with multiple ways to protect against oxidative stress and increasing the fitness of bacterial species during the host immune-mediated response. Therefore, the ability of bacteria to maintain adequate Mn levels significantly contributes to virulence and pathogenicity.

Currently there is no evidence that bacteria employ chelating molecules, similar to siderophores, to obtain Mn form the environment. Therefore, it is hypothesized that Mn-import systems are directly involved in the capture and transport of metal. There are two major classes of Mn (II) transporters in prokaryotes, Nramp H^+–Mn^{2+} transporters and ATP-binding cassette (ABC) Mn permeases, with many bacteria expressing both types.[60]

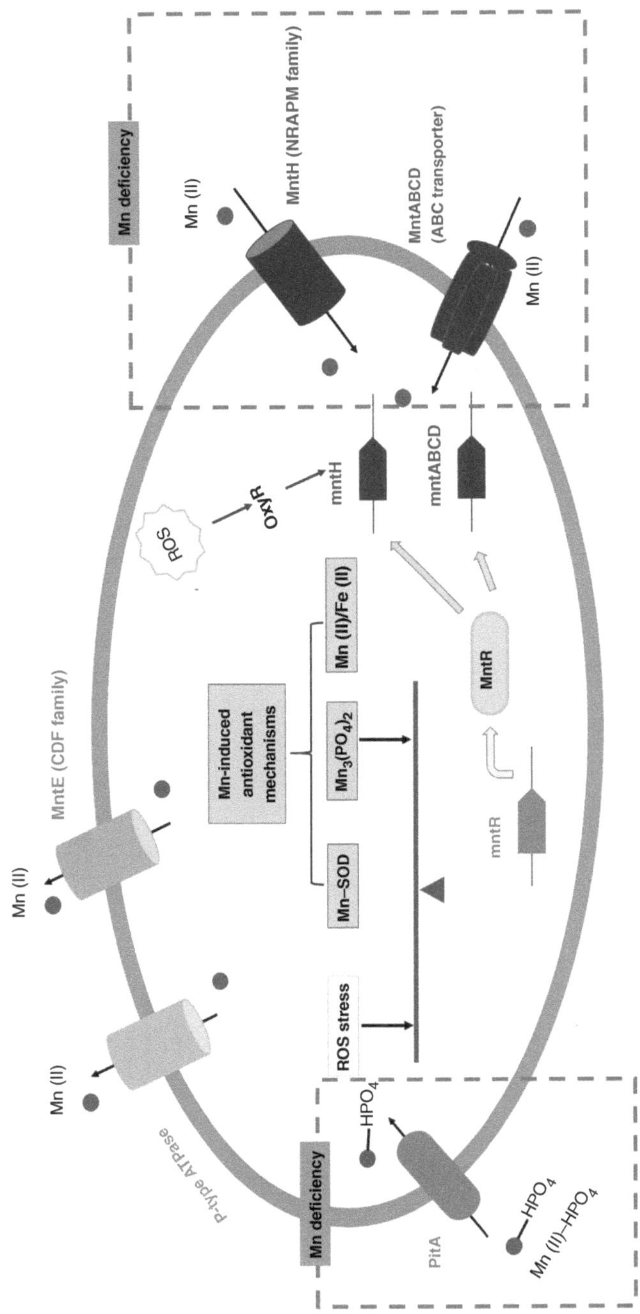

Figure 4 Mn-induced antioxidant mechanisms in bacterial cells.

In addition, *Lactobacillus* species were shown to have a third type of transporter, a Mn (II) transporting P-type ATPase.

In 2000, Que and Helmann identified a bacterial transcription factor of the DtxR family involved in regulation of Mn metabolism in *B. subtilis*, MntR.[61] Deletion of this factor led to increased sensitivity to Mn (II) and Cd (II). Further studies identified *mntH* (Nramp analogue) and *mntABCD* (ABC transporter) as the targets for regulation by MntR in *E. coli*, *S. aureus*, and *Salmonella typhimurium*.[60] Interestingly, these Nramp transporters originally were defined as Fe transporters; however, current data have shown that these transporters exhibit higher selectivity for Mn (II) when compared to Zn (II), Fe (II), or Cd (II).[62]

ABC transporters for Mn (II) in Gram-positive bacteria are composed of the extracellular lipoprotein that binds Mn from the environment, the cytoplasmic ATP-binding protein, and an integral membrane protein that is responsible for ion influx[63] Gram-negative bacteria employ metal transporters that have a similar structure to that of the ABC transporters for Mn (II) in Gram-positive species, with the soluble periplasmic cation-binding protein substituting for the extracellular lipoprotein of the Gram-positive species. In both types of organisms the ABC transporters are capable of transporting other transition metals, such as, Zn (II), Fe (II), and Cd (II), but with much lower affinity than for the Mn ion.[60]

The major mechanism of Mn (II)-transport regulation occurs via MntR, which functions as a repressor in the presence of bound metal ions. At high metal concentrations these transport pathways are downregulated and further accumulation of Mn is mediated through other systems, for example PitA, an inorganic phosphate transporter with high affinity for Mn–HPO_4 complexes.

The efflux of Mn occurs through MntE (CDF), which is found in many different bacteria. Interestingly, MntE mutants accumulate 3 times higher inner concentration of the metal resulting in higher resistance to oxidative stress and lower pathogenicity of these mutants. This finding indicates that efflux of Mn may play a role in virulence of bacteria.[64]

Another pathway for transporting Mn out of bacterial cells involves the P-type ATPase CtpC. This type of efflux was identified in *Mycobacterium tuberculosis* and *Mycobacterium smegmatis*. Interestingly, deletion of this pump did not lead to Mn accumulation in cells, suggesting that this transporter may play a role in providing Mn for incorporation into secreted enzymes.[65]

Overall, Mn-induced virulence has been shown to play an important role in establishing pathogenicity in many types of bacteria in vivo. For example,

PsaA, a Mn transporter in *S. pneumoniae*, was shown to control bacterial adhesion and a deletion of *psaA* resulted in avirulent strains in a mouse model of respiratory tract disease.[66] Another example includes the decreased pathogenicity of *Yersenia pestis*, the causative agent of bubonic and pneumonic plague in humans, in the absence of the *yfeABCD* transporter.[67] It was shown that intravenous infection of NIH/Swiss Webster mice (Nramp $1^{+/+}$) with a *yfeABCD* mutant results in a 10^6-fold decrease in virulence when compared to the wild type

4. ZINC

Compared to other transition metals, Zn has the largest involvement as a structural or catalytic cofactor in a range of proteins. This is due to the fully filled *d* orbital of Zn (II) that causes redox stability of the ion. In addition, the ability of Zn to act as a Lewis acid makes it useful in various catalytic reactions. Moreover, Zn is capable of forming stable chemical bonds with oxygen, nitrogen, and sulfur atoms, and accepts various numbers of coordinating bonds. All these properties secure an essential role of Zn ion in the life cycle of many bacterial species and its presence is required for bacterial survival and fitness. However, the same useful characteristics of Zn can create the potential problem of unspecific binding to cell proteins, causing cell toxicity at higher concentrations. Therefore, bacterial cells tightly regulate the amount of available Zn (II) by keeping it in the "bound" form. It has been shown that the overall amount of Zn in prokaryotic cells is in the submillimolar range,[68] and an in vivo study has identified unbound "free" form of Zn at the 20 pM level via the use of ratiometric biosensors.[69] Due to this dual effect of Zn on bacterial growth and survival, hosts may employ two opposite strategies to combat and defeat bacteria: sequester Zn ions to prevent bacterial growth (innate immunity) or sharply increase concentration of Zn ions to initiate toxicity. As a result bacteria have evolved elaborate systems to control Zn efflux and uptake to account for both host strategies (Fig. 5).

The bacterial outer membrane is relatively permeable to Zn ions; therefore, Zn concentration in the periplasmic space directly correlates with the ion availability in the extracellular environment. It has been established that Gram-negative bacteria use low affinity-transport systems, such as ZupT, for Zn uptake when Zn is freely available. ZupT, a constitutively expressed as a low affinity transporter (ZIP family), has a distinct preference for Zn ions

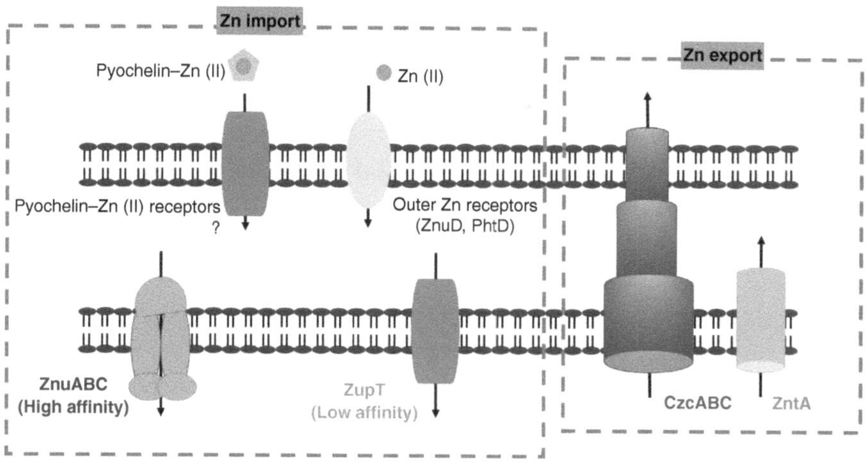

Figure 5 Zn-influx and -efflux systems in bacteria discussed in this chapter.

over other transition metals but can also transport Cu when overexpressed.[70] Under conditions of low Zn availability Gram-negative bacteria activate expression of *znuABC*, enconding a high affinity-import system. ZnuABC consists of three proteins: ZnuB, a protein channel; ZnuC, an ATPase responsible for providing energy for the ion transport; and ZnuA, soluble protein that binds Zn (II) in the periplasmic space and delivers it to ZnuB. Recent studies have identified several Zn receptors, such as ZnuD (*N. meningitis*)[71] and PhtD (*Pneumococcus*),[72] which are capable of binding the ion directly from extracellular space. However, no other Gram-negative bacteria were reported to have similar Zn receptors. There is no in vivo evidence that pathogenic bacteria produce Zn (II)-chelating molecules, called tsinkophores; however, the possibility for this biosynthesis still remains. Indirectly, the existence of tsinkophores is supported by the fact that pyochelin, a low affinity-Fe siderophore produced by *P. aeruginosa*, has been shown to chelate Zn (II) in vitro.

The role of Zn transport in bacterial virulence and pathogenicity has been well studied. For example, the ZnuABC uptake system is required for the adhesion of *E. coli* O157:H7 to epithelial cells[73] and for the overall fitness of uropathogenic *E. coli* during urinary tract infection.[74] Deletion of the ZnuABC system was shown to critically affect virulence of *A. baumanii*, *Brucella abortus*, *C. jejuni*, *Haemophilus ducreyi*, *S. enterica*, *Yersenia ruckeri*, *Moraxella catarrhalis*, and *Pasterurella multocida*. Absence of ZnA and ZnT components in *S. typhimurium* led to impaired growth of bacteria and decreased virulence in infected mice.[75] Both ZurAM and ZinABC

Zn-uptake systems are present in *Listeria monocytogenes* and their deletion led to a loss of virulence in an oral mouse model.[76]

Detoxification of the bacterial cell can occur using two major systems: P-type ATPases or RND transporters. It was shown that ZntA, P_{1B}-type Zn transporter, plays a critical role in Zn tolerance in *E. coli*[77] and in maintaining the optimum intracellular level of Zn.[78] Interestingly, it was reported that Zn-efflux transporters are upregulated in Mn-limited conditions, suggesting that Zn toxicity is more potent when Mn concentrations in environment are lowered.

In conclusion, understanding the role of metals at the interface of host–pathogen interaction is essential for determining metal-induced pathogenicity of bacteria. As has been shown in numerous studies, activation of virulence mechanisms results from the shift in the delicate balance of environmental metal concentrations. Elucidating these triggering points will allow the development of novel therapeutic strategies for the treatment of infectious diseases.

REFERENCES

1. Dawkins R, Krebs JR. Arms races between and within species. *Proc R Soc Lond B Biol Sci.* 1979;205(1161):489–511.
2. Woolhouse ME, Webster JP, Domingo E, Charlesworth B, Levin BR. Biological and biomedical implications of the co-evolution of pathogens and their hosts. *Nat Genet.* 2002;32(4):569–577.
3. Brown SP, Cornforth DM, Mideo N. Evolution of virulence in opportunistic pathogens: generalism, plasticity, and control. *Trends Microbiol.* 2012;20(7):336–342.
4. Salazar-Gonzalez RM, Srinivasan A, Griffin A, Muralimohan G, Ertelt JM, Ravindran R, Vella AT, McSorley SJ. *Salmonella* flagellin induces bystander activation of splenic dendritic cells and hinders bacterial replication in vivo. *J Immunol.* 2007;179 (9):6169–6175.
5. Wu W, Badrane H, Arora S, Baker HV, Jin S. MucA-mediated coordination of type III secretion and alginate synthesis in *Pseudomonas aeruginosa. J Bacteriol.* 2004;186 (22):7575–7585.
6. Finlay BB, Falkow S. Common themes in microbial pathogenicity revisited. *Microbiol Mol Biol Rev.* 1997;61(2):136–169.
7. Russmann H. Inverted pathogenicity: the use of pathogen-specific molecular mechanisms for prevention or therapy of disease. *Int J Med Microbiol.* 2004;293(7–8): 565–569.
8. Freemann SH, Herron JC. Evolution and human health. In: Gilfillan A, ed. *Evolutionary Analysis 4th ed. San Francisco, USA: Pearson; 2007:* 529–574.
9. Levin BR. The evolution and maintenance of virulence in microparasites. *Emerg Infect Dis.* 1996;2(2):93–102.
10. Steinert M. Pathogen-host interactions in *Dictyostelium, Legionella, Mycobacterium* and other pathogens. *Semin Cell Dev Biol.* 2011;22(1):70–76.
11. (a) Anisimova M, Bielawski J, Dunn K, Yang Z. Phylogenomic analysis of natural selection pressure in *Streptococcus* genomes. *BMC Evol Biol.* 2007;7:154;

(b) Dozois CM, Daigle F, Curtiss 3rd R. Identification of pathogen-specific and conserved genes expressed in vivo by an avian pathogenic *Escherichia coli* strain. *Proc Natl Acad Sci USA*. 2003;100(1):247–252;
(c) Stabler RA, Marsden GL, Witney AA, Li Y, Bentley SD, Tang CM, Hinds J. Identification of pathogen-specific genes through microarray analysis of pathogenic and commensal *Neisseria* species. *Microbiology*. 2005;151(Pt 9):2907–2922;
(d) Dunning Hotopp JC, Grifantini R, Kumar N, Tzeng YL, Fouts D, Frigimelica E, Draghi M, Giuliani MM, Rappuoli R, Stephens DS, Grandi G, Tettelin H. Comparative genomics of *Neisseria meningitidis*: core genome, islands of horizontal transfer and pathogen-specific genes. *Microbiology*. 2006;152(Pt 12):3733–3749;
(e) Champion OL, Gaunt MW, Gundogdu O, Elmi A, Witney AA, Hinds J, Dorrell N, Wren BW. Comparative phylogenomics of the food-borne pathogen *Campylobacter jejuni* reveals genetic markers predictive of infection source. *Proc Natl Acad Sci USA*. 2005;102(44):16043–16048.

12. Ho Sui SJ, Fedynak A, Hsiao WW, Langille MG, Brinkman FS. The association of virulence factors with genomic islands. *PLoS One*. 2009;4(12):e8094.
13. Maret W. Metalloproteomics, metalloproteomes, and the annotation of metalloproteins. *Metallomics*. 2010;2(2):117–125.
14. Waldron KJ, Robinson NJ. How do bacterial cells ensure that metalloproteins get the correct metal? *Nat Rev Microbiol*. 2009;7(1):25–35.
15. Hood MI, Skaar EP. Nutritional immunity: transition metals at the pathogen-host interface. *Nat Rev Microbiol*. 2012;10(8):525–537.
16. Beinert H, Holm RH, Munck E. Iron-sulfur clusters: nature's modular, multipurpose structures. *Science*. 1997;277(5326):653–659.
17. Braun V. Avoidance of iron toxicity through regulation of bacterial iron transport. *Biol Chem*. 1997;378(8):779–786.
18. Holden VI, Bachman MA. Diverging roles of bacterial siderophores during infection. *Metallomics*. 2015;7(6):986–995.
19. Hammer ND, Skaar EP. Molecular mechanisms of *Staphylococcus aureus* iron acquisition. *Annu Rev Microbiol*. 2011;65:129–147.
20. Friedman DB, Stauff DL, Pishchany G, Whitwell CW, Torres VJ, Skaar EP. *Staphylococcus aureus* redirects central metabolism to increase iron availability. *PLoS Pathog*. 2006;2(8):e87.
21. Johnson M, Cockayne A, Morrissey JA. Iron-regulated biofilm formation in *Staphylococcus aureus* Newman requires ica and the secreted protein Emp. *Infect Immun*. 2008;76(4):1756–1765.
22. Nizet V. Understanding how leading bacterial pathogens subvert innate immunity to reveal novel therapeutic targets. *J Allergy Clin Immunol*. 2007;120(1):13–22.
23. (a) Grigg JC, Cooper JD, Cheung J, Heinrichs DE, Murphy ME. The *Staphylococcus aureus* siderophore receptor HtsA undergoes localized conformational changes to enclose staphyloferrin A in an arginine-rich binding pocket. *J Biol Chem*. 2010;285(15):11162–11171;
(b) Grigg JC, Cheung J, Heinrichs DE, Murphy ME. Specificity of Staphyloferrin B recognition by the SirA receptor from *Staphylococcus aureus*. *J Biol Chem*. 2010;285(45):34579–34588.
24. (a) Sebulsky MT, Shilton BH, Speziali CD, Heinrichs DE. The role of FhuD2 in iron(III)-hydroxamate transport in *Staphylococcus aureus*. Demonstration that FhuD2 binds iron(III)-hydroxamates but with minimal conformational change and implication of mutations on transport. *J Biol Chem*. 2003;278(50):49890–49900;
(b) Sebulsky MT, Speziali CD, Shilton BH, Edgell DR, Heinrichs DE. FhuD1, a ferric hydroxamate-binding lipoprotein in *Staphylococcus aureus*: a case of gene duplication and lateral transfer. *J Biol Chem*. 2004;279(51):53152–53159.

25. Drabkin DL. Metabolism of the hemin chromoproteins. *Physiol Rev.* 1951;31 (4):345–431.
26. Mazmanian SK, Skaar EP, Gaspar AH, Humayun M, Gornicki P, Jelenska J, Joachmiak A, Missiakas DM, Schneewind O. Passage of heme-iron across the envelope of *Staphylococcus aureus*. *Science.* 2003;299(5608):906–909.
27. Meyer JM. Pyoverdines: pigments, siderophores and potential taxonomic markers of fluorescent *Pseudomonas* species. *Arch Microbiol.* 2000;174(3):135–142.
28. Meyer JM, Gruffaz C, Raharinosy V, Bezverbnaya I, Schafer M, Budzikiewicz H. Siderotyping of fluorescent *Pseudomonas*: molecular mass determination by mass spectrometry as a powerful pyoverdine siderotyping method. *Biometals.* 2008;21(3):259–271.
29. (a) Meyer JM, Neely A, Stintzi A, Georges C, Holder IA. Pyoverdin is essential for virulence of *Pseudomonas aeruginosa*. *Infect Immun.* 1996;64(2):518–523;
(b) Takase H, Nitanai H, Hoshino K, Otani T. Impact of siderophore production on *Pseudomonas aeruginosa* infections in immunosuppressed mice. *Infect Immun.* 2000;68 (4):1834–1839.
30. Lamont IL, Beare PA, Ochsner U, Vasil AI, Vasil ML. Siderophore-mediated signaling regulates virulence factor production in *Pseudomonas aeruginosa*. *Proc Natl Acad Sci USA.* 2002;99(10):7072–7077.
31. Cox CD, Rinehart Jr KL, Moore ML, Cook Jr JC. Pyochelin: novel structure of an iron-chelating growth promoter for *Pseudomonas aeruginosa*. *Proc Natl Acad Sci USA.* 1981;78 (7):4256–4260.
32. Hare NJ, Soe CZ, Rose B, Harbour C, Codd R, Manos J, Cordwell SJ. Proteomics of *Pseudomonas aeruginosa* Australian epidemic strain 1 (AES-1) cultured under conditions mimicking the cystic fibrosis lung reveals increased iron acquisition via the siderophore pyochelin. *J Proteome Res.* 2012;11(2):776–795.
33. Cartron ML, Maddocks S, Gillingham P, Craven CJ, Andrews SC. Feo-transport of ferrous iron into bacteria. *Biometals.* 2006;19(2):143–157.
34. Modarresi F, Azizi O, Shakibaie MR, Motamedifar M, Mosadegh E, Mansouri S. Iron limitation enhances acyl homoserine lactone (AHL) production and biofilm formation in clinical isolates of *Acinetobacter baumannii*. *Virulence.* 2015;6(2):152–161.
35. Payne SM. Regulation of bacterial toxin synthesis by iron. In: Rappuoli R, ed. *Bacterial Protein Toxins. Washington, DC: ASM Press; 2003:* 25–38.
36. Zimmermann MB, Chassard C, Rohner F, N'Goran EK, Nindjin C, Dostal A, Utzinger J, Ghattas H, Lacroix C, Hurrell RF. The effects of iron fortification on the gut microbiota in African children: a randomized controlled trial in Cote d'Ivoire. *Am J Clin Nutr.* 2010;92(6):1406–1415.
37. Gera T, Sachdev HP. Effect of iron supplementation on incidence of infectious illness in children: systematic review. *BMJ.* 2002;325(7373):1142.
38. Waldron KJ, Rutherford JC, Ford D, Robinson NJ. Metalloproteins and metal sensing. *Nature.* 2009;460(7257):823–830.
39. Hao X, Luthje FL, Qin Y, McDevitt SF, Lutay N, Hobman JL, Asiani K, Soncini FC, German N, Zhang S, Zhu YG, Rensing C. Survival in amoeba-a major selection pressure on the presence of bacterial copper and zinc resistance determinants? Identification of a "copper pathogenicity island". *Appl Microbiol Biotechnol.* 2015;99(14):5817–5824.
40. Adiba S, Nizak C, van Baalen M, Denamur E, Depaulis F. From grazing resistance to pathogenesis: the coincidental evolution of virulence factors. *PLoS One.* 2010;5(8):e11882.
41. Chaturvedi KS, Hung CS, Giblin DE, Urushidani S, Austin AM, Dinauer MC, Henderson JP. Cupric yersiniabactin is a virulence-associated superoxide dismutase mimic. *ACS Chem Biol.* 2014;9(2):551–561.
42. Outten FW, Huffman DL, Hale JA, O'Halloran TV. The independent cue and cus systems confer copper tolerance during aerobic and anaerobic growth in *Escherichia coli*. *J Biol Chem.* 2001;276(33):30670–30677.

43. Changela A, Chen K, Xue Y, Holschen J, Outten CE, O'Halloran TV, Mondragon A. Molecular basis of metal-ion selectivity and zeptomolar sensitivity by CueR. *Science.* 2003;301(5638):1383–1387.
44. (a) Bondarczuk K, Piotrowska-Seget Z. Molecular basis of active copper resistance mechanisms in Gram-negative bacteria. *Cell Biol Toxicol.* 2013;29(6):397–405;
(b) Chaturvedi KS, Henderson JP. Pathogenic adaptations to host-derived antibacterial copper. *Front Cell Infect Microbiol.* 2014;4:3.
45. Kim EH, Nies DH, McEvoy MM, Rensing C. Switch or funnel: how RND-type transport systems control periplasmic metal homeostasis. *J Bacteriol.* 2011;193(10):2381–2387.
46. Zimmermann M, Udagedara SR, Sze CM, Ryan TM, Howlett GJ, Xiao Z, Wedd AG. PcoE-a metal sponge expressed to the periplasm of copper resistance *Escherichia coli.* Implication of its function role in copper resistance. *J Inorg Biochem.* 2012;115:186–197.
47. Knapp CW, Fowle DA, Kulczycki E, Roberts JA, Graham DW. Methane monooxygenase gene expression mediated by methanobactin in the presence of mineral copper sources. *Proc Natl Acad Sci USA.* 2007;104(29):12040–12045.
48. Lisher JP, Giedroc DP. Manganese acquisition and homeostasis at the host-pathogen interface. *Front Cell Infect Microbiol.* 2013;3:91.
49. Zheng B, Zhang Q, Gao J, Han H, Li M, Zhang J, Qi J, Yan J, Gao GF. Insight into the interaction of metal ions with TroA from *Streptococcus suis. PLoS One.* 2011;6(5):e19510.
50. Gu M, Imlay JA. The SoxRS response of *Escherichia coli* is directly activated by redox-cycling drugs rather than by superoxide. *Mol Microbiol.* 2011;79(5):1136–1150.
51. Outten CE, O'Halloran TV. Femtomolar sensitivity of metalloregulatory proteins controlling zinc homeostasis. *Science.* 2001;292(5526):2488–2492.
52. Jacobsen FE, Kazmierczak KM, Lisher JP, Winkler ME, Giedroc DP. Interplay between manganese and zinc homeostasis in the human pathogen *Streptococcus pneumoniae. Metallomics.* 2011;3(1):38–41.
53. Veyrier FJ, Boneca IG, Cellier MF, Taha MK. A novel metal transporter mediating manganese export (MntX) regulates the Mn to Fe intracellular ratio and *Neisseria meningitidis* virulence. *PLoS Pathog.* 2011;7(9):e1002261.
54. Daly MJ, Gaidamakova EK, Matrosova VY, Vasilenko A, Zhai M, Venkateswaran A, Hess M, Omelchenko MV, Kostandarithes HM, Makarova KS, Wackett LP, Fredrickson JK, Ghosal D. Accumulation of Mn(II) in *Deinococcus radiodurans* facilitates gamma-radiation resistance. *Science.* 2004;306(5698):1025–1028.
55. (a) May BP, Dennis PP. Evolution and regulation of the gene encoding superoxide dismutase from the archaebacterium *Halobacterium cutirubrum. J Biol Chem.* 1989;264(21):12253–12258;
(b) Wintjens R, Noel C, May AC, Gerbod D, Dufernez F, Capron M, Viscogliosi E, Rooman M. Specificity and phenetic relationships of iron- and manganese-containing superoxide dismutases on the basis of structure and sequence comparisons. *J Biol Chem.* 2004;279(10):9248–9254.
56. Chang EC, Kosman DJ. Intracellular Mn (II)-associated superoxide scavenging activity protects Cu, Zn superoxide dismutase-deficient *Saccharomyces cerevisiae* against dioxygen stress. *J Biol Chem.* 1989;264(21):12172–12178.
57. (a) Archibald FS, Fridovich I. Manganese and defenses against oxygen toxicity in *Lactobacillus plantarum. J Bacteriol.* 1981;145(1):442–451;
(b) Barnese K, Gralla EB, Valentine JS, Cabelli DE. Biologically relevant mechanism for catalytic superoxide removal by simple manganese compounds. *Proc Natl Acad Sci USA.* 2012;109(18):6892–6897.
58. Tseng HJ, Srikhanta Y, McEwan AG, Jennings MP. Accumulation of manganese in *Neisseria gonorrhoeae* correlates with resistance to oxidative killing by superoxide anion and is independent of superoxide dismutase activity. *Mol Microbiol.* 2001;40(5):1175–1186.

59. (a) Sobota JM, Imlay JA. Iron enzyme ribulose-5-phosphate 3-epimerase in *Escherichia coli* is rapidly damaged by hydrogen peroxide but can be protected by manganese. *Proc Natl Acad Sci USA*. 2011;108(13):5402–5407;
(b) Anjem A, Imlay JA. Mononuclear iron enzymes are primary targets of hydrogen peroxide stress. *J Biol Chem*. 2012;287(19):15544–15556.
60. Papp-Wallace KM, Maguire ME. Manganese transport and the role of manganese in virulence. *Annu Rev Microbiol*. 2006;60:187–209.
61. Que Q, Helmann JD. Manganese homeostasis in *Bacillus subtilis* is regulated by MntR, a bifunctional regulator related to the diphtheria toxin repressor family of proteins. *Mol Microbiol*. 2000;35(6):1454–1468.
62. (a) Portnoy ME, Jensen LT, Culotta VC. The distinct methods by which manganese and iron regulate the Nramp transporters in yeast. *Biochem J*. 2002;362(Pt 1):119–124;
(b) Kehres DG, Zaharik ML, Finlay BB, Maguire ME. The NRAMP proteins of *Salmonella typhimurium* and *Escherichia coli* are selective manganese transporters involved in the response to reactive oxygen. *Mol Microbiol*. 2000;36(5):1085–1100.
63. Kitten T, Munro CL, Michalek SM, Macrina FL. Genetic characterization of a *Streptococcus mutans* LraI family operon and role in virulence. *Infect Immun*. 2000;68 (8):4441–4451.
64. Rosch JW, Gao G, Ridout G, Wang YD, Tuomanen EI. Role of the manganese efflux system mntE for signalling and pathogenesis in *Streptococcus pneumoniae*. *Mol Microbiol*. 2009;72(1):12–25.
65. Padilla-Benavides T, McCann CJ, Arguello JM. The mechanism of Cu^{+} transport ATPases: interaction with Cu^{+} chaperones and the role of transient metal-binding sites. *J Biol Chem*. 2013;288(1):69–78.
66. Marra A, Lawson S, Asundi JS, Brigham D, Hromockyj AE. In vivo characterization of the psa genes from *Streptococcus pneumoniae* in multiple models of infection. *Microbiology*. 2002;148(Pt 5):1483–1491.
67. Bearden SW, Perry RD. The Yfe system of *Yersinia pestis* transports iron and manganese and is required for full virulence of plague. *Mol Microbiol*. 1999;32(2):403–414.
68. Eide DJ. Zinc transporters and the cellular trafficking of zinc. *Biochim Biophys Acta*. 2006;1763(7):711–722.
69. Wang D, Hurst TK, Thompson RB, Fierke CA. Genetically encoded ratiometric biosensors to measure intracellular exchangeable zinc in *Escherichia coli*. *J Biomed Opt*. 2011;16(8):087011.
70. (a) Grass G, Wong MD, Rosen BP, Smith RL, Rensing C. ZupT is a Zn(II) uptake system in *Escherichia coli*. *J Bacteriol*. 2002;184(3):864–866;
(b) Grass G, Franke S, Taudte N, Nies DH, Kucharski LM, Maguire ME, Rensing C. The metal permease ZupT from *Escherichia coli* is a transporter with a broad substrate spectrum. *J Bacteriol*. 2005;187(5):1604–1611.
71. Stork M, Bos MP, Jongerius I, de Kok N, Schilders I, Weynants VE, Poolman JT, Tommassen J. An outer membrane receptor of *Neisseria meningitidis* involved in zinc acquisition with vaccine potential. *PLoS Pathog*. 2010;6:e1000969.
72. Loisel E, Chimalapati S, Bougault C, Imberty A, Gallet B, Di Guilmi AM, Brown J, Vernet T, Durmort C. Biochemical characterization of the histidine triad protein PhtD as a cell surface zinc-binding protein of *Pneumococcus*. *Biochemistry*. 2011;50 (17):3551–3558.
73. Gabbianelli R, Scotti R, Ammendola S, Petrarca P, Nicolini L, Battistoni A. Role of ZnuABC and ZinT in *Escherichia coli* O157:H7 zinc acquisition and interaction with epithelial cells. *BMC Microbiol*. 2011;11:36.
74. Sabri M, Houle S, Dozois CM. Roles of the extraintestinal pathogenic *Escherichia coli* ZnuACB and ZupT zinc transporters during urinary tract infection. *Infect Immun*. 2009;77(3):1155–1164.

75. Ammendola S, Pasquali P, Pistoia C, Petrucci P, Petrarca P, Rotilio G, Battistoni A. High-affinity Zn^{2+} uptake system ZnuABC is required for bacterial zinc homeostasis in intracellular environments and contributes to the virulence of *Salmonella enterica. Infect Immun.* 2007;75(12):5867–5876.
76. Corbett D, Wang J, Schuler S, Lopez-Castejon G, Glenn S, Brough D, Andrew PW, Cavet JS, Roberts IS. Two zinc uptake systems contribute to the full virulence of *Listeria monocytogenes* during growth in vitro and in vivo. *Infect Immun.* 2012;80(1):14–21.
77. (a) Beard SJ, Hashim R, Membrillo-Hernandez J, Hughes MN, Poole RK. Zinc(II) tolerance in *Escherichia coli* K-12: evidence that the zntA gene (o732) encodes a cation transport ATPase. *Mol Microbiol.* 1997;25(5):883–891;
(b) Rensing C, Mitra B, Rosen BP. The *zntA* gene of *Escherichia coli* encodes a Zn(II)-translocating P-type ATPase. *Proc Natl Acad Sci USA.* 1997;94(26):14326–14331.
78. Wang D, Hosteen O, Fierke CA. ZntR-mediated transcription of *zntA* responds to nanomolar intracellular free zinc. *J Inorg Biochem.* 2012;111:173–181.

CHAPTER THREE

Virulence Program of a Bacterial Plant Pathogen: The *Dickeya* Model

S. Reverchon*, G. Muskhelisvili, W. Nasser
Department of Biology, University of Lyon, INSA-Lyon, Villeurbanne, Lyon, France
* Corresponding author. E-mail address: Sylvie.reverchon-pescheux@insa-lyon.fr

Contents

Abstract

The pectinolytic *Dickeya* spp. are Gram-negative bacteria causing severe disease in a wide range of plant species. Although the *Dickeya* genus was initially restricted to tropical and subtropical areas, two *Dickeya* species (*D. dianthicola* and *D. solani*) emerged recently in potato cultures in Europe. Soft-rot, the visible symptoms, is caused by plant cell wall degrading enzymes, mainly pectate lyases (Pels) that cleave

Progress in Molecular Biology and Translational Science, Volume 142
ISSN 1877-1173
http://dx.doi.org/10.1016/bs.pmbts.2016.05.005

the pectin polymer. However, an efficient colonization of the host requires many additional elements including early factors (eg, flagella, lipopolysaccharide, and exopolysaccharide) that allow adhesion of the bacteria and intermediate factors involved in adaptation to new growth conditions encountered in the host (eg, oxidative stress, iron starvation, and toxic compounds). To facilitate this adaptation, *Dickeya* have developed complex regulatory networks ensuring appropriate expression of virulence genes. This review presents recent advances in our understanding of the signals and genetic circuits impacting the expression of virulence determinants. Special attention is paid to integrated control of virulence functions by variations in the superhelical density of chromosomal DNA, and the global and specific regulators, making the regulation of *Dickeya* virulence an especially attractive model for those interested in relationships between the chromosomal dynamics and gene regulatory networks.

1. INTRODUCTION

At the beginning of this 21st century, one of the main challenges of agricultural research is the reinforcement of global food security in the context of continuous human population growth. Indeed, global food production must increase by 50% to meet the projected demand of the world's population by 2050.[1] Many plant pathogens are factors limiting food production throughout the world.[2] Especially, bacterial plant pathogens are a major worldwide problem for agriculture because they are particularly difficult to control due to the scarcity of chemical control agents for bacteria with the exception of antibiotics. However, the use of antibiotics is restricted in many countries due to the high potential for evolution of antibiotic resistance and its eventual transmission to human pathogens. In addition, bacteria show an incredible ability to disseminate and adapt to changing environment (colonization of a new host, geographic invasion of new areas, etc.).[3] Climate change is likely to accelerate the evolution of pathogens and increase their diversity. To migrate or evolve are two options for plant pathogens under conditions of climate change.[4] Therefore, the ways for combating bacterial plant diseases are limited. In this context, there is a critical need for comprehensively understanding the process of bacterial pathogenesis in plants in order to develop novel knowledge-based strategies to control plant bacterial diseases.

This review focuses on the plant pathogenic bacteria of the genus *Dickeya*, which are among the top 10 most important bacterial pathogens in agriculture limiting the crop yield and quality.[5] Like many plant pathogenic bacteria, *Dickeya* spend most of their parasitic life in the intercellular space of the plants—the apoplast. The apoplast is a nutrient-limited environment that is guarded by plant defenses, so *Dickeya* have evolved with an intricate strategy to successfully colonize this niche. In this review, we will present the multifaceted process of plant–*Dickeya* interaction and the molecular mechanisms used by these bacteria to coordinate their virulence program during plant invasion.

2. *DICKEYA* ECOLOGY

The soft rot enterobacterial plant pathogens *Pectobacterium* and *Dickeya* species (formerly classified as pectinolytic *Erwinia* spp.) are an important group of bacteria responsible for disease in a wide range of angiosperm plant species in geographically diverse regions.[6] Whereas *Pectobacterium* is found both in temperate and tropical regions, *Dickeya* populations were, initially, considered to be restricted to tropical and subtropical plant hosts and areas. The *Dickeya* genus was first described by Samson et al.,[7] and six species were initially discriminated as *D. dadantii, D. dieffenbachiae, D. chrysanthemi, D. paradisiaca, D. zeae*, and *D. dianthicola*. Subsequently, *D. dieffenbachiae* has been reported to belong to the same species as *D. dadantii* and has been reclassified as *D. dadantii* subspecies *dieffenbachiae*.[8] In the past two or three decades, "cold-tolerant" *Dickeya dianthicola* strains have been isolated from potato plants in Western Europe.[9] Since 2005, a new genetic clade, probably representing a highly virulent *Dickeya* species, has been detected in Europe.[10,11] Strains belonging to this clade were isolated from seed potatoes in France, Finland, Poland, the Netherlands, and Israel.[12] In many of these countries, the pathogen was introduced via the international trade of seed potatoes. The same genetic clade has also been found in hyacinth, suggesting a transfer from hyacinth to potato in the recent past, possibly via the contaminated irrigation water.[11] The successive emergence of two *Dickeya* species (*D. dianthicola* and *D. solani*) in potato cultures in Europe highlights the adaptation capacity and dissemination of *Dickeya* to new geographic areas and to new hosts.[9,13] More recently, *Dickeya* strains isolated from freshwater rivers were characterized as a new species designated as *Dickeya aquatica*.[14] Accordingly, the *Dickeya* genus now comprises seven species (*D. chrysanthemi, D. dadantii, D. dianthicola, D. zeae, D. paradisiaca, D. solani*, and *D. aquatica*). Genome sequences are available for each species except *D. aquatica*. However, several sequenced *Dickeya* strains have also been reclassified and public databases often failed to be updated accordingly: Ech586 was moved from *D. dadantii* to *D. zeae*, Ech703 from *D. dadantii* to *D. paradisiaca* and Ech1591 from *D. zeae* to *D. chrysanthemi*.[15]

Dickeya species were reported to cause diseases both in monocotyledons (rice, maize, pineapple, banana) and dicotyledons (potato, Chicory, etc.). Typical symptoms of the soft rot disease elicited by *Dickeya* in different hosts are described in Reverchon and Nasser.[16] It has been suggested that plant families with a high cell wall calcium content, or low intercellular pH, are more resistant to soft rot enterobacteria.[6] The epidemiological and aetiological aspects of potato soft rot disease caused by *Dickeya* have been reviewed.[12] These bacteria are devastating phytopathogens that exert significant negative impacts on agricultural production, causing losses in fields and in storage. They cause both stem rot (blackleg), which affects the growing plant, and tuber soft rot of potato. The most characteristic symptom of potato blackleg caused by *Dickeya* species is a slimy, wet, black rot lesion spreading from the rotting mother tuber up the stems, especially under wet conditions. Tuber soft rot is initiated at lenticels, the stolon end and/or in wounds under wet conditions. The major source for potato infection is the latently infected seed (mother) tuber. During spring, bacteria from infected seed spread into young stems and roots. Bacteria invade intercellular spaces, dissolve the plant cell walls, and liquefy the mother tuber. Subsequently, bacteria are released into the soil and transmitted by soil water to contaminate neighboring progeny tubers. Czajkowski et al.,[17] showed that the bacteria in soil could also colonize potato roots and subsequently move via the vascular system into progeny tubers. In addition, *Dickeya* could be isolated from the roots of healthy weeds in agricultural fields.[18] These infested healthy weeds may represent an environmental reservoir responsible for pathogen persistence and its spread to potato crops. *Dickeya* have also been identified on leaf surfaces, where they can live as epiphytes. Wounds caused during potato harvest and handling provide multiple points of entry into the tuber as infection may spread to the whole tuber and thence to neighboring tubers in storage. During winter, *Dickeya* survive in association with crop residue in soil. Moreover, several cases of contamination of surface and irrigation waters by *Dickeya* have been reported, whereby the bacterial contamination may persist as assessed by repeated annual samplings (reviewed in Toth et al.[19]). Furthermore, *Dickeya*, encode the butanediol pathway, which results in the production of the potent insect attractant acetoin, suggesting that these bacteria may attract insects.[20] During the whole disease cycle, insects could be involved in disease development either because they cause wounds on plant organs on which they feed thus allowing bacteria to enter the plant, or when stopping over an infected plant, they become contaminated with bacteria and could transfer them to other plants.[21] The life cycle of *Dickeya* during infection of potatoes is described in Ref. [16]. Investigation of the migration of the bacteria within *Arabidopsis* plant tissues revealed the existence of two distinct phases in the infection process[22] one of which is asymptomatic, when the bacteria progress intercellularly with no substantial multiplication, whereas the other is symptomatic and associated with strongly increased bacterial fitness. Globally, the four main steps of plant infection by *Dickeya* are: (1) adherence to the plant surface and entry into the plant tissues, either via wound sites or through natural openings such as stomata; (2) colonization of the apoplastic spaces between plant cells; (3) suppression of the host defense response; and (4) plant cell wall degradation resulting in development of disease symptoms. Each of these disease stages and life-history transitions require the detection of signals from the environment, the host, and the pathogen population.

3. THE VIRULENCE DETERMINANTS OF *D. DADANTII*

Plant–pathogen interaction is a multifaceted process, mediated by the pathogen- and plant-derived molecules. Secreted molecules, derived from the pathogens, are the key factors, which determine their pathogenicity and allow successful colonization of the host. On the other hand, plant-derived molecules are involved in the recognition of pathogen and triggering of the defense response.

3.1 The Effectors of Type-I and Type-II Secretion Systems

In the *Dickeya* genus, most genetic and virulence studies have been performed in the *D. dadantii* 3937 strain, isolated from African violet (*Saintpaulia ionantha)*, that was chosen as a model by the *Dickeya* international community because of its amenability to genetic manipulation. *D. dadantii* 3937 is able to infect a wide range of plant species including potato and can cause tuber rotting and blackleg symptoms. The plant cell wall is a complex and dynamic meshwork of polymers (cellulose, hemicellulose, pectin) and represents a barrier that *Dickeya* need to breach to colonize the plant tissue (Fig. 1A). Therefore, the virulence of *Dickeya* is correlated with its ability to synthesize and secrete plant cell wall degrading enzymes (CWDEs), including a full set

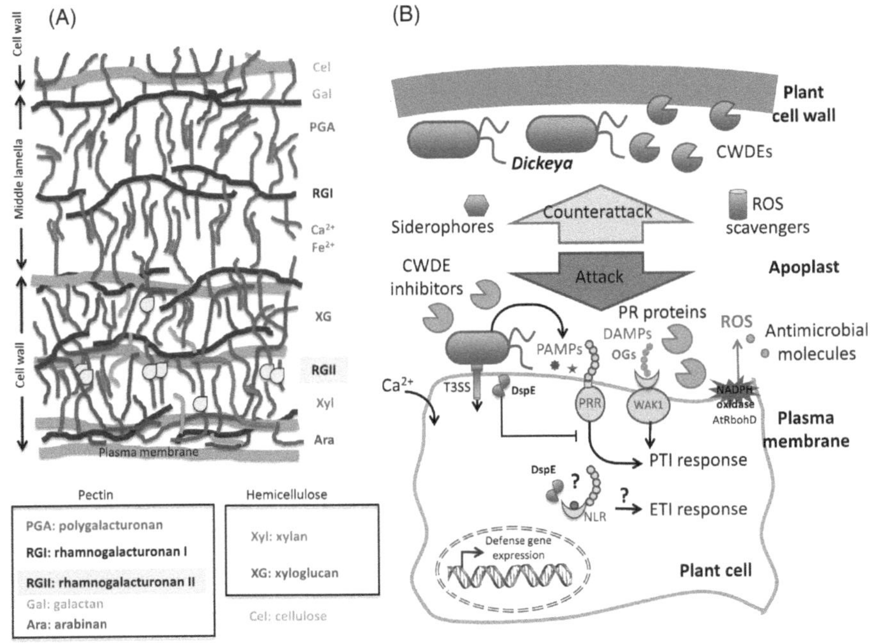

Figure 1 Structure of the plant cell wall (A) and the sketch of Plant–*Dickeya* interaction (B). *Dickeya* propagate in the extracellular spaces of plant tissues. The molecules released from bacteria (PAMPs) into the apoplast and oligogalacturonides (DAMPs) resulting from pectin cleavage by bacterial pectinases are recognized, respectively, by plant receptors (PRRs) and by the wall-associated kinase (WAK1) leading to elicitation of the PAMP/DAMP-triggered immunity (PTI). Plant produces ROS, antimicrobial molecules, and pathogenesis-related (PR) proteins, some of which are CWDE inhibitors. To cope with ROS, bacteria produce the indigoidine ROS scavenger. *Dickeya* deliver the DspE effector in the host cell by the T3SS secretion pilus. DspE is supposed to suppress PTI and facilitates bacterial propagation *in planta*. Inside the apoplast, *D. dadantii* produces siderophores, which allow the bacterium to cope with the restricted iron bioavailability in the plant. These siderophores also manipulate plant immunity.

of pectinases,[23] a glucuronoxylanase XynA,[24] a periplasmic endogalactanase GanA,[25] a cellulase CelZ,[26] and four proteases PrtA, PrtB, PrtC, PrtG.[27] Pectinases and the cellulase are secreted by the Out type-II secretion system[28] whereas proteases are secreted by the PrtDEF type-I secretion system.[29] Mutation of the Out type-II secretion system inhibits plant tissue maceration while mutants in the PrtDEF type-I secretion system are delayed in symptom progression illustrating the importance of CWDEs in plant pathogenesis.[30,31] The soft rot symptom results from the general disorganization of plant tissues following the degradation of pectin by the pectinolytic equipment of *D. dadantii*. Pectin is the major component of the middle lamella that cements adjacent plant cells and provides cohesion to the plant tissue. Accordingly, soft rot symptoms can be obtained in vitro using purified pectate lyases.[32] Pectin includes both linear regions composed of polygalacturonan and ramified regions (RGI and RGII) (Fig. 1A). RGI contains a rhamnogalacturonan backbone and various lateral chains such as galactan, arabinan, and arabinogalactan.[33] RGII contains a short galacturonan backbone, carrying four side chains, with a diversity of rare monosaccharides.[34] The carboxylic groups of D-galacturonate residues are methyl-esterified to various degrees (up to 80%) and these residues are, to a lesser extent, acetylated at the C2 and/or C3 positions. Feruloyl esters are a type of modification commonly found in arabinan and galactan chains of pectin hairy regions.[35] The ferulic acid esters can esterify polysaccharides and permit their covalent cross-linking by forming dehydrodimers. In addition, these phenolic compounds present in the cell walls chelate high amounts of iron.[36] Efficient degradation of pectin by *D. dadantii* results from the cooperative action of depolymerases that have either an *endo*- or an *exo*-cleaving mode and esterases that remove methyl, acetyl and feruloyl groups from pectin. The *D. dadantii* genome contains twenty-one genes encoding pectinases including nine *endo*-pectate lyases (PelA, PelB, PelC, PelD, PelE, PelI, PelL, PelN, PelZ) that cleave pectin by a ß-elimination mechanism and require a divalent cation as cofactor (calcium for most of them, except PelN which requires Fe^{2+}, suggesting a possible evolutionary adaptation of PelN to the presence of this element in the close environment of its substrate), two *exo*-pectate lyases (PelX and PelW), five polygalacturonases that cleave pectin by a classic hydrolysis mechanism (PehV, PehW, PehX, PehK, PehN), one rhamnogalacturonate lyase RhiE able to cleave the rhamnogalacturonan chains,[37] two pectin methylesterases (PemA, PemB), two pectin acetylesterases (PaeX, PaeY), and two feruloyl esterases (FaeD, FaeT).[38] In addition, two genes encoding proteins (PnlH and PnlG = Dda3937_03551) showing similarities with pectin lyases supposed to depolymerize methylated pectin are present in *Dickeya* genome; however, no enzymatic activities have been identified so far for these proteins. It has been suggested that PnlH, a protein anchored at the outer face of the outer membrane, could allow the bacteria to bind the plant pectic polysaccharides.[39] In contrast to other extracellular pectinases that are secreted by the Out system, PnlH is secreted by the Stt specific type-II secretion system.[39] The different steps of pectin degradation are described in detail by Hugouvieux-Cotte-Pattat in the Chapter 8.

The oligogalacturonides (OGs) released from the degraded pectic component of the cell wall by the action of pectinases produced by the pathogens activate the plant defenses that restrict invasion and colonization.[40] These endogenous plant molecules released during pathogen attack are designated as damage-associated molecular patterns (DAMPs) (Fig. 1B). The biological activity of OGs is dependent on their degree of polymerization (DP). OGs with DP between 10 and 15 are required to trigger an immune response.[41] This size is optimal for the formation of Ca^{2+}-mediated intermolecular cross-links resulting in structures called "eggboxes,"[42,43] that are thought to be necessary for OG activity. The different endopectate lyases from *D. dadantii* release oligogalacturonides of different sizes.[44] Whereas PelB and PelD display their highest activities on hexagalacturonate and tetragalacturonate, respectively, PelA, PelI, and PelL are most active on the octamer, the largest tested substrate in vitro. The preferential products formed from the oligogalacturonates are unsaturated dimers for PelD, unsaturated trimers for PelB, and unsaturated tetramers for PelI and PelL. For PelA, preferential products are dependent on the size of the initial oligogalacturonate.[44] However, these results obtained in vitro at pH 8.5 (optimum pH for Pels) are not relevant to the initial stage of infection when apoplastic fluid is slightly acidic. Therefore, the low activity of Pels under acidic pH could lead to production of long OGs that can induce the plant defense reaction. The OGs and polygalacturonate are recognized by the wall-associated kinase 1 (WAK1) which carries a N-terminal pectin-binding domain that interacts with nonmethylesterified polygalacturonate and OGs in a Ca^{2+}-dependent manner[45] (Fig. 1B). A chimeric receptor approach revealed that WAK1 acts as a receptor of OGs and activates the plant innate immunity.[46] Therefore, when pathogens start degrading the plant cell wall components, plants are capable of perceiving the loss of wall integrity and subsequently activate the defense signaling pathways.[47] Thus for an efficient pathogenesis strategy, it is thus crucial to prevent production of pectinases in the initial stage of infection to avoid release of cell wall components that are sensed as damage signals by plant.

PelA, PelD, and PelE are among the most important Pels for virulence on Saintpaulia plants.[48] However, the contribution of each individual Pel to virulence depends on the host plants. For example, on pea, only the *pelA* mutant exhibited a reduced virulence.[49] Therefore, the multiplicity of pectinases probably reflects the evolutionary adaptation of the bacterium to use a wide range of plant hosts.

Most pectinase genes are strongly induced by pectin during stationary phase except *pelL*, *pelN*, and *pehV*, which may act more specifically on other particular pectic glycans.[49a] Regarding response to stress, most genes encoding pectate lyases are repressed under acidic stress in keeping with both their optimum activity at alkaline pH (pH 8.0–9.5)[23] and their requirement at the late stage of infection, when the pH is slightly basic.[49a] One exception is the *pelA* gene, which is induced under acidic stress.[50] PelA has a very low pectate lyase activity, is unable to macerate potato tubers but paradoxically, the virulence of *pelA* mutant is strongly affected suggesting an important role of PelA in initiating pectin degradation at the early stage of infection.[48] In contrast to pectinases that are downregulated under acidic conditions characteristic of early stages of infection, the genes of polygalacturonases PehV, W, X are upregulated under acidic stress[49a] in keeping with their activity optimum at acidic pH.[51] Notably, deletion of the *pehVWX* cluster only affects the initiation of pectin degradation.[52] Polygalacturonases, which liberate dimers, could hasten the formation of intracellular inducers of pectate lyase synthesis. The acid-specific induction of polygalacturonase genes *pehV*, *pehW*, and *pehX* could in turn be dependent on the two-component system PhoP-PhoQ, especially since in *Pectobacterium*, the PehR-PehS (PhoP-PhoQ) system was shown to specifically control the production of the polygalacturonase PehA without affecting other plant cell wall degrading enzymes.[53] However, in *Dickeya*, the PhoP-PhoQ system was shown to control pectate lyase production in response to organic acids and to impact virulence.[54] Also the glucuronoxylanase *xynA* gene is upregulated under acidic stress, as well as ABF-0018996 gene encoding a protein conserved in various plant pathogenic microorganisms.[49a] This latter gene was also found to be induced during the asymptomatic phase of *Arabidopsis* infection.[55] The rationale for downregulation of most of the genes encoding plant cell wall degrading enzymes under acidic conditions mimicking the early stage of infection is to prevent an early detection of plant breakdown products by the host, particularly oligogalacturonides, that would reveal the presence of the pathogen before it reaches a population density appropriate for successful infection.[56] In keeping with this notion, not only the expression of plant cell wall degrading enzymes but also that of the operon encoding oligogalacturonide transporter (*togMNAB*) is downregulated under acidic stress, so that the intracellular inducer of *pel* genes, 2-keto-3-deoxy-gluconate (KDG), would not be present in the cytoplasm.[49a] Downregulation of *togMNAB* operon was also observed during the asymptomatic phase of *Arabidopsis* infection.[55] Interestingly, the *pelA*, *pelD*, and *pelE* genes originating from a common ancestor by duplication events display distinct responses to various kinds of stress. Expression of *pelA* is induced by acidic stress, *pelD* is expressed under oxidative stress whereas *pelE* is expressed under osmotic stress. This order probably reflects the temporal regulation of these genes during infection.[49a] Furthermore, like the *pel* genes, the *gan* genes required for degradation of the galactan ramification of pectin are repressed under acidic stress at the beginning of infection, suggesting that pectin must be probably first cleaved into smaller oligonucleotide fragments at the more advanced stage of infection, before the Gan enzymes can degrade the galactan ramifications. Indeed, the *gan* genes are induced under oxidative stress concomitantly with *pelD*.[49a] As expected, most of the genes of plant cell wall degrading enzymes including those for proteases PrtABCG, cellulase CelZ, and glucuronoxylanase XynA are induced in stationary phase (corresponding to late stage of infection) and in addition, all of these genes are subject to quorum sensing control[57] (see Section 5.1).

In addition to CWDEs, *Dickeya* utilize several other factors to colonize plant tissue and enhance disease progression. Such factors include the extracellular necrosis inducing protein NipE, the two proteins AvrL-AvrM (similar to avirulence protein in *Xanthomonas*), and two proteins ABF-0018996 and ABF-0014625 (similar to proteins in the rice pathogenic fungus *Magnaporthe grisea*). Secretome analysis revealed that secretion of these additional proteins is mediated by the type-II Out system.[58]

The mode of pathogenicity shared by *Pectobacterium* and *Dickeya* has been termed "brute force," as the CWDEs physically attack the plant cell walls and their surrounding apoplast.[59] This appears to contrast with the mode of action of other bacterial plant pathogens, such as *Pseudomonas syringae*, which instead use "stealth" mechanisms to attack plants. These mechanisms include type-III secreted effectors that enter plant cells and assist in suppressing plant defenses.[60] Unexpectedly, genome annotations of both *Pectobacterium atrosepticum* SCRI1043 and *D. dadantii* 3937 strains[61,62] revealed the presence of a Hrp type-III secretion system and some associated substrates leading to reconsider the overall strategy used by these pathogens in their associations with plants.

3.2 The Hrp Type-III Secretion System and Its Substrates

To cope with microbial attacks, plants have evolved a sophisticated network of responses, including a rapid influx of calcium ions from external store, a burst of reactive oxygen species, activation of MAPK signaling cascades resulting in induction of defense-related genes, cell wall reinforcement by callose deposition, and often localized cell death.[63,64] Plant innate immunity is activated by the perception of pathogen-associated molecular patterns (PAMPs) such as flagellin, lipopolysaccharide, peptidoglycan, and EF-Tu, which are recognized by transmembrane pattern-recognition receptors (PRR) eliciting a basal resistance response called PAMP-triggered immunity (PTI)[65] (Fig. 1B). To counteract basal plant defenses, Gram-negative pathogenic bacteria deliver virulence effector proteins directly into host cells through the type-III secretion system (T3SS). These effectors suppress the PTI immune response and promote pathogenesis. In turn, plants have evolved nucleotide-binding/leucine-rich-repeat (NLR) resistance proteins that recognize the injected effectors and induce a second layer of robust resistance response called effector-triggered immunity (ETI).[63] ETI reinstates and amplifies PTI basal transcriptional programs and antimicrobial defenses and is often associated with localized cell death referred to as the hypersensitive response, killing both the pathogen and infected plant cells.[66,67] Thus, effector proteins appear to play at least one of the two roles in virulence—inhibition of PAMP-triggered immunity (PTI) or induction of plant cell death.

Plants use vacuoles and vacuolar contents for programmed cell death (PCD) in two different—destructive and nondestructive—ways. The destructive way implicates vacuolar membrane collapse, followed by the release of vacuolar hydrolytic enzymes into the cytosol and resulting in rapid and direct cell death. The nondestructive way involves fusion of the vacuolar and the plasma membranes, which allows vacuolar defense proteins to be discharged into the extracellular space containing proliferating bacteria. A recent study showed that culture filtrates of *Pectobacterium* induce simplification of vacuole structure during plant cell death.[68] The vacuole shape simplifies and endoplasmic actin filaments disassemble prior to cell death. However, disintegration of the plasma membrane precedes vacuolar rupture suggesting that the simplification of vacuole structure might be a hallmark of ongoing autophagy.

In contrast to *P. syringae*, which secrete multiple effectors via the T3SS, only one effector, DspE, has been identified in the genome of *D. dadantii* alongside the two harpins HrpN and HrpW.[62] A third gene ID18999 located close to the T3SS genes could encode a potential harpin on the basis of its amino acid composition (35% serine, 0% cysteine) and its pI around 4, which are features typical of harpins.[69] These harpins are secreted through the T3SS but are targeted to the extracellular space of plant tissues, unlike the DspE effector protein that acts inside the plant cells. HrpN and HrpW are thought to act as translocators facilitating the injection of DspE effector into plant cells.[69] However, it was also speculated that harpins have membrane-pore-forming activity promoting cellular dysfunction and death. It was demonstrated that harpins form amyloid fibers in apoplastic fluids. Such amyloid fibrils may cause cell membrane breakage.[70] In *D. dadantii*, HrpN and the T3SS are reported to play a key role in the initial infection of plant hosts.[71,72] In addition, HrpN mediates cell–cell aggregation in *Dickeya*.[73] Such aggregation activity could be related to HrpN amyloid fiber formation. HrpW carries a pectate lyase domain in its C-terminus and can bind to calcium pectate but is devoid of pectate lyase activity.[69] Therefore, HrpW could anchor bacteria to pectin and promote bacterial-plant cell contact. Recently, a microarray containing over 5000 genes related to plant defense in potato and *Arabidopsis* was used to identify genes that were differentially expressed in potato in the presence and absence of HrpW during the early stages of infection using *P. atrosepticum* SCRI1043 as pathogen. One such gene encoding a putative DNA binding protein is highly expressed on exposure to the *hrpW* mutant. Therefore, HrpW contributes, probably indirectly, to the repression of this plant defense gene. Moreover, when this gene was constitutively expressed in a potato plant, the plant exhibited high levels of resistance to the wild type pathogen.[74]

The effector DspE belongs to the AvrE superfamily of type-III effectors (T3Es).[75] The AvrE family is the only family of T3Es present in all type-III-dependent, agriculturally important phytobacteria (enterobacteria, xanthomonads, pseudomonads, and *Ralstonia* spp.) indicating an ancestral acquisition of these effectors in bacterial plant pathogens.[76] Effectors of this superfamily inhibit salicylic acid-mediated plant defenses, interfere with vesicular trafficking, and promote bacterial growth *in planta*.[75] Recent studies indicated that AvrE-type effectors alter the sphingolipid pathway *in planta* by inhibiting the serine palmitoyl transferase.[77] The sphingolipid biosynthetic pathway is induced during the hypersensitive response that blocks pathogen attack at the site of infection.[78] Therefore, inhibition of this pathway could be seen as a mechanism that delays hypersensitive response cell death and allows bacterial development *in planta*[75] (Fig. 1B). Perturbation of the sphingolipid pathway *in planta* could also explain why effectors of the AvrE family induce a slow plant cell death, because sphingolipids represent structural lipids of plant membranes and prolonged inhibition of the sphingolipid pathway ultimately leads to plant cell death.[79] Recent studies revealed that AvrE is localized in the plant cell plasma membrane and downregulates the expression of the *NHL13* gene in *Arabidopsis*, which is required for antibacterial plant immunity.[80] Although the function of *D. dadantii* DspE has not been analyzed in detail, we assume that most of the characteristics that have emerged from studies of the different members of the AvrE family can be extrapolated to this effector. Interestingly, Peng et al.[81] by analyzing the expression of green fluorescent protein (GFP) fused to the *dspE* promoter of *D. dadantii* observed that a fraction of *D. dadantii* cells never expressed *dspE* in various media or host plants. This raises the possibility that individual bacterial cells may play different roles dividing the labor and working together as a community effort to ensure efficient disease development.

3.3 The Siderophores and Iron Metabolism

The siderophores function as potent iron-chelating compounds implicated in iron acquisition from the surroundings. *D. dadantii* produces two siderophores, the hydroxycarboxylate achromobactin, which is produced when iron becomes limiting,[82] and the catecholate chrysobactin,[83] which prevails under severe iron deficiency.[84] These siderophores are required for the systemic progression of maceration symptoms in the hosts (Fig. 1B). Mutants defective in chrysobactin- or achromobactin-mediated iron transport produce symptoms that remain localized to the inoculated leaf, indicating that these siderophores are required for bacteria to spread throughout the plant.[85,86] In addition to its nutritional role, chrysobactin appears to be involved in manipulation of the plant defense system.[86,87] Indeed, *Arabidopsis* leaf infiltration with chrysobactin causes activation of the plant defense pathways.[87] Such activation does not occur when chrysobactin is loaded with iron before leaf treatment. Therefore, plant immunity activation depends on the iron-scavenging activity of the siderophore. Consistently, chrysobactin infiltration in *Arabidopsis* leaves causes modification of the iron localization in plant tissues with an accumulation of iron in plant cell wall and depletion in plastids.[88] Such rapid change in iron status in the vascular system is followed by cell wall rigidification reflected in the callose deposition along the vascular system.[88] Moreover, siderophore treatment by mimicking iron deficiency induces perturbation of metal homeostasis in plant cell with upregulation of the gene encoding the major iron plant transporter *IRT1*.[87,88] In addition, in an *irt1* mutant affected in the uptake of several metals (Zn^{2+}, Mn^{2+}, Fe^{2+}) including iron, siderophores are unable to trigger the immune responses observed in the wild type plant.[88] Therefore, the iron-starvation effect caused by scavenging alone is not sufficient to trigger plant immunity. Instead, the metal homeostasis disturbance by the bacterial iron scavengers is probably a signal that triggers defense responses in plants. Modification in metal bound to metalloproteins could be the danger-signal sensed by plants.[89] Recently, Aznar et al.[90] showed that progression of *D. dadantii* in the leaf tissues causes loss of iron from plant cell walls and cellular compartments such as plastids that normally contain high amounts of iron. Loss of iron from the cell walls coincides with cell wall degradation by pectinases, whereas in macerated tissues, bacteria are loaded with iron suggesting that they benefit from release of high amount of this element for growth. At the same time, the release of free iron in the intercellular space amplifies the oxidative stress generated by the host defense mechanism through the Fenton reaction, producing the highly toxic and reactive hydroxyl radical $OH^{\bullet}$. Bacterial siderophores can interfere with this reaction by sequestering Fe (III). Therefore, for bacteria it is of crucial importance to be equipped with strong iron scavengers during progression in host tissues to cope with excess of free iron released by the host cells (Fig. 1B). *D. dadantii* produces several iron storage proteins, FntA, Bfr, and Dps, that contribute to its virulence, underscoring the importance of a perfect control of iron homeostasis during infection.[91] Iron also plays an important role as a regulatory signal in *Dickeya* pathogenicity via the Fur repressor. Indeed, Fur not only represses genes involved in iron metabolism using ferrous ions as cofactor but also acts as a repressor of pectinase genes. In the presence of Fe (II), Fur binds to the *pel* gene regulatory regions counteracting their activation by CRP.[92] Furthermore, under conditions of pectinolysis, when the pectinase genes are turned on, the achromobactin and chrysobactin iron transport systems are constitutively expressed. This coordinated regulation and metabolic coupling between pectinolysis and iron assimilation makes sense in the light of the modification of cellular iron localization during pathogenesis. From these observations, it could be postulated that during infection, bacteria are confronted with alternating phases of iron-poor and iron-rich environments. During the early phase of infection, plant tissues are intact, so little iron is available and bacteria secrete siderophores and a low amount of pectinases. The secreted pectinases are inactive under the acidic pH of the apoplast but during the interaction pH gradually increases[20,50] leading to pectin degradation by Pels and release of free iron in the intercellular space. This in turn results in repression of pectinase and siderophore genes by Fur in the macerated tissues. Bacteria multiply in the macerated tissues and afterward move to colonize intact plant tissues where they are again confronted with iron starvation. A new round of pectinase and siderophore production is reinitiated until the release of iron due to degradation of pectin. Such successive rounds of Pel and siderophore production are fully consistent with the kinetics of pectinase and chrysobactin gene expression observed *in planta*.[93,94]

4. PLANT–PATHOGEN INTERACTION PROCESS

4.1 Adhesion to Plant Surface and Entry Into Apoplast

At initial stage of infection, *D. dadantii* produce cellulose fibrils, which enable the development of aggregates on the plant surface.[95,96] These aggregates are embedded within an extracellular polymeric substance (EPS) that maintains a hydrated surface around the bacteria and, thus, helps them to survive under conditions of desiccation.[97] The CdiA/HexA type-V secreted protein, comprising a large hemagglutinin repeat region, also contributes to the adherence of bacteria to leaves.[98] An operon encoding a multirepeat adhesin (ABF-0016865) and a type-I secretion pathway is found in the genome of *D. dadantii* but the role of this putative adhesin in adherence is not yet established. However, this protein contains multiple cadherin-homologous domains likely to bind calcium ions strongly suggesting a role in adhesion. In *P. atrosepticum*, such a multirepeat adhesin secreted by a type-I pathway was shown to be required for binding to the host plant.[99] In addition, *D. dadantii* secretes a biosurfactant that increases wettability, probably resulting in increased plant surface colonization.[31] No fimbria genes were detected in the *D. dadantii* genome.[62] However, the genome of closely related *Dickeya chrysanthemi* strain Ech1591 was shown to contain a Flp/Tad pilus that could be involved in plant surface adherence.[100]

A role for flagella-mediated adherence has been demonstrated in many different plant and animal infection models.[101] Therefore, the *D. dadantii* flagella could also contribute to plant adhesion. Motility and chemotaxis are essential for *D. dadantii* when searching for favorable sites to enter into the plant apoplast. Indeed, mutants affected in the chemotaxis transduction system (*cheW*, *B*, *Y*, and *Z*) and in the structure of the flagellar motor (*motA*) are unable to enter into *Arabidopsis* leaves.[102] In particular, *D. dadantii* has a strong chemotactic response to jasmonic acid, which is produced by wounded tissue and may enable the bacterial cells to move toward plant wounds and facilitate systemic invasion.[102,103] A recent in silico analysis of the 47 methyl-accepting chemotaxis proteins (MCPs) present in *D. dadantii* revealed that 83% of these

MCPs have a periplasmic ligand-binding region (LBR) and could be involved in sensing the plant-specific compounds.[104] Among these, 13 LBRs are conserved in other plant-pathogenic bacteria beyond the *Dickeya* genus, whereas 12 LBRs are exclusively present in the *Dickeya*. These in silico data implicate these chemoreceptors in sensing the plant-derived compounds. Mutagenesis of these chemoreceptors and analysis of the chemotactic response of the corresponding mutants toward jasmonic acid and xylose revealed the involvement of at least two of them (ABF-0020167 and ABF-0046680, candidate receptors of jasmonic acid and xylose, respectively) in the chemotaxis-driven entry process through plant wounds.[104] Importantly, the global transcriptional response of *D. dadantii* to environmental stimuli relevant to plant infection revealed substantial variations of various MCPs in response to conditions encountered *inplanta*.[105] The most strongly repressed MCP in acidic, oxidative and osmotic stress conditions is ABF-0017419. Many MCPs (except ABF-0014722, ABF-0014726, ABF-0014722, ABF-0015513, ABF-0015603, ABF-0019306, ABF0019858) are repressed by osmotic stress, whereas three (ABF-0017665, ABF-0017668, and ABF-0014726) are strongly induced under acidic stress. Four MCPs (ABF-0019855, ABF-0016380, ABF-0019858, and ABF-0018541) are strongly induced under oxidative stress in exponential phase, and seven (*chmX*, ABF-0016436, ABF-0017090, ABF18765, ABF-0018892, ABF-0019050, ABF-0020252) are particularly inducible during PGA catabolism in stationary phase. These last seven MCPs are distributed between two distinct groups on the basis of their topology.[104] ChmX, ABF18765, ABF-0018892, ABF-0019050, and ABF-0020252 possess two transmembrane segments and a periplasmic LBR domain, whereas ABF-0016436 and ABF-0017090 contain a cytoplasmic LBR domain and no transmembrane segment. Therefore, from these seven MCPs, only ChmX, ABF18765, ABF-0018892, ABF-0019050, and ABF-0020252 could be involved in chemotaxis towards pectic oligomers.

4.2 Role of Multiple Efflux Pumps for Surviving in the Plant Apoplast

After penetration into the plant intercellular spaces, *D. dadantii* have to face many plant secondary metabolites, such as phytoalexins and alkaloids, which are likely to play a role in protecting plants against pathogens (Fig. 1B). To counteract the presence of these toxic substances, *Dickeya* uses multiple efflux pumps that contribute to virulence showing some degree of host specificity.[106–108] For example, the *D. dadantii* outer-membrane efflux pump component TolC participates in extrusion of the antimicrobial compound berberine and the *tolC* mutation severely impairs *in planta* multiplication and infection of plant tissue.[107] Similarly, both the Acr1AB and the Acr2AB systems were found to be required for full virulence on chicory leaves.[108] However, in Saintpaulia plants only the Acr1AB system contributed to virulence, whereas Acr2AB had no effect. Moreover, efflux pumps may also have a role in intermicrobial competition as several epiphytes synthesize antimicrobials. For example, the *D. dadantii* YbiT ABC transporter might play such a role. Indeed, the *D. dadantii ybiT* mutant retained virulence in potato tubers and chicory leaves, but was less infectious than the wild-type strain in coculture experiments with saprophytic bacteria such as *Pseudomonas putida* or *Pseudomonas fluorescens*.[106] This phenotype suggests that YbiT is capable of extruding toxic compounds produced by *P. putida* and *P. fluorescens*, whereas in its absence *D. dadantii* is outcompeted by the *Pseudomonas* strains.

Plant secondary metabolites are not the only compounds capable of inducing the expression of *D. dadantii* efflux pumps. For example, salicylic acid, an important signaling molecule triggering resistance to phytopathogens, also induces the expression of genes encoding the AcrAB and EmrAB efflux pumps.[109] The ability of bacteria to utilize defense-signaling compounds to induce the expression of efflux systems reveals their efficient adaptation to the plant defense system. However, when *Dickeya* enters in plant apoplast, it in addition encounters an acidic environment. The acidic pH specifically induces the MfbAB multidrug efflux by modulating DNA-binding activity of the MfbR transcriptional regulator.[110] Although MfbR represses the MfbAB multidrug efflux, it activates the genes encoding the CWDEs. Therefore, MfbR participates in temporal coordination of the factors needed for survival and pathogenesis. Indeed, the significant reduction of MfbR activity under acidic conditions of the apoplast leads to both an increase in the expression of the MfbAB efflux system and a failure to activate the expression of the CWDE genes at the early stage of infection.[110]

4.3 Adaptation to the Acidic Environment of Apoplast and to Plant Antimicrobial Peptides

When *D. dadantii* infects a plant, it colonizes intercellular apoplastic fluid, the pH value of which lies between 5.0 and 6.5.[111] Upon colonization, bacteria induce cell lysis, which results in variation of the pH value of the intercellular environment. Accordingly, it was shown that pH value of plant tissue varies

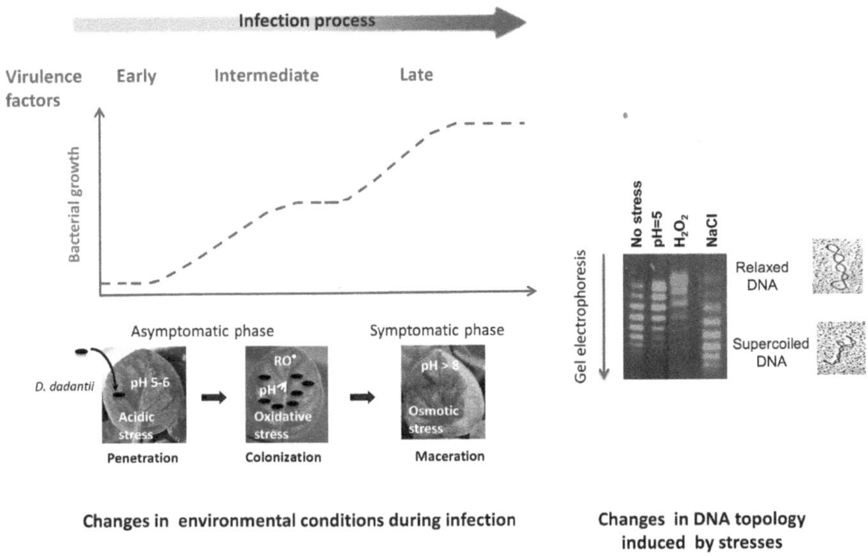

Figure 2 Infection process by *D. dadantii*. The *Dickeya* pathogenicity program comprises asymptomatic and symptomatic phases. In the asymptomatic phase during plant penetration and colonization, the bacteria experience acidic and oxidative stress. During the symptomatic phase, bacteria experience osmotic stress. In macerated tissue, bacteria multiply further.[131] The various stresses distinctly affect the DNA topology.

from acidic to basic during infection by *D. dadantii*[20,50] (Fig. 2). To cope with the acidic environment of plant apoplast, *D. dadantii* induces the production of indole-3-acetic acid (IAA),[49a] which is known to stimulate bacterial adaptation to stress, such as UV irradiation, changing salinity and acidity, thus improving the survival and persistence in the plant environment.[112] In addition, this auxin has been shown to promote stomatal opening thereby enhancing disease progression.[113,114] Concomitantly, *D. dadantii* promotes reactions consuming protons.[49a] For example, it uses the *oxlT-oxdD* oxalate decarboxylation pathway to remove protons from the cytoplasm as well as hydrogenases, which interconvert protons with hydrogen gas and generate H2 from formate through association with formate dehydrogenase. At the same time, *D. dadantii* induces metabolic switches, minimizing the acidification of the cytoplasm. For example, carbohydrate metabolism is down-regulated at the expense of fermentation pathway converting pyruvate into neutral end products acetoin or 2, 3-butanediol. As a second strategy, *Dickeya* modulates activities of proton transporters by up-regulating transporters that pump out protons and down-regulating transporters responsible for their entry. The transcriptional response of *D. dadantii* to acidic stress also includes envelop remodeling with covalent modifications of lipopolysaccharide (LPS) neutralizing the net negative charge on lipid A and decreasing the interaction between positively charged components and the bacterial surface, alteration of the outer-membrane protein composition, as well as of the phospholipid composition of the inner membrane.[49a] All these concurrent modifications modulate permeability properties of the cell envelope facilitating bacterial adaptation to acidic stress. Accordingly, this stress induces the expression of genes encoding two component-regulatory systems (TCS) responsible for cell envelope modification such as PhoP-PhoQ, EnvZ-OmpR, and PmrA-PmrB.[49a] Importantly, the *D. dadantii* PhoP-PhoQ system is both involved in survival at acid pH and also regulates several virulence determinants including some pectinolytic enzymes and iron-related metabolic processes.[115–117] Low magnesium concentrations also serve as a signal to activate the PhoP-PhoQ system. The actual magnesium concentration that surrounds *D. dadantii* in the apoplast is difficult to measure, although it has been hypothesized to be low during the first stage of infection.[115] Interestingly, a low concentration of magnesium was also suggested to serve as a signal to activate SlyA, a regulator of the MarR family that directly binds to the promoter region of PhoP and regulates virulence by controlling multiple pathogenicity-related genes.[118] The CpxAR system is also involved in the protection of *D. dadantii* against plant defenses (oxidative stress and antimicrobial peptides) and in the regulation of both the CWDEs and motility.[119] In *Pectobacterium carotovorum*, the PmrB-PmrA two-component system was also shown to be involved in regulation of virulence.[120] Indeed, disruption of *pmrB* resulted in decreased production of plant cell wall degrading enzymes and concomitant reduction of virulence both in potato and *Arabidopsis*. The role of PmrB has not been studied in *Dickeya* but the results obtained in *Pectobacterium* could probably be extrapolated to *Dickeya*. Regarding the EnvZ-OmpR system its role in virulence has not yet been established, but recent study revealed that the EnvZ-OmpR system is involved in survival at low pH[49a] and also controls the alternative expression of the two *D. dadantii* oligogalacturonide porins, KdgM, and KdgN.[121]

Notably, both PhoP-PhoQ and PmrA-PmrB systems are also involved in sensitivity to plant antimicrobial peptides.[120,122] Production of the latter contributes to plant immunity and the virulence of *D. dadantii* mutants with increased sensitivity to antimicrobial peptides is greatly affected.[123] Therefore, the mechanisms of resistance to antimicrobial peptides play a key role during infection. These mechanisms include proteolytic degradation of antimicrobial peptides, extrusion of these compounds by efflux systems, or modifications of the antimicrobial target (ie, the microbial cell envelope). *D. dadantii* uses each of these three mechanisms to resist the toxic effects of antimicrobial peptides. Indeed, mutants in the Sap transport system, which removes the antimicrobial peptides from the membrane and directs them to degradation pathways, are strongly affected for virulence.[123] Similarly, mutants in *ybjX* and *arnT* genes, which are linked to LPS modification, display compromised virulence in various plant hosts and compromised resistance to thionins and snakin-2, two widely distributed plant antimicrobial peptides.[122]

4.4 Adaptation to Anaerobiosis Conditions in Plant Tissue

In plant tissue, *D. dadantii* also experiences low oxygen availability, which is one of the key factors in the development of soft-rot diseases. Indeed, pectate lyase synthesis is induced under anaerobic conditions.[124] A transcriptome analysis revealed that more than 7% of the genes of this phytopathogen are differentially expressed under anaerobic conditions.[125] In addition to genes involved in anaerobic metabolism, the O_2 responsive stimulon contains a variety of virulence and pathogenicity relevant processes including secretion, response to environmental stress, metal homeostasis, and taxis.[125] Sensing of anaerobiosis is achieved by one transcription factor (TF), FNR, and the two TCS systems, NarXL and ArcAB.[125] FNR is an iron–sulfur cluster-containing transcriptional regulator that dimerizes in the absence of O_2. NarXL is responsible for activation of genes involved in nitrate/nitrite catabolism under anaerobic conditions and repression of genes involved in other anaerobic respiratory and fermentative pathways. The ArcAB system detects the oxidation state of ubiquinones from the aerobic respiratory chain and under anaerobic conditions represses genes required for aerobic metabolism. The involvement of this system in the virulence of *D. dadantii* was recently questioned since in the reference strain 3937, *arcB* contains a nonsense mutation that is supposed to inactivate ArcA-mediated regulatory function. However, this mutation does not impact the virulence of strain 3937.[125] Counter intuitively, in *D. dadantii*, the *ahpC*, *sodC*, and *dps* genes involved in oxidative stress are upregulated under anaerobic conditions. Babujee et al.[125] proposed that prolonged growth in an O_2 limited environment mimics the situation experienced by soft-rotting bacteria prior to encountering the oxidative burst produced by the host. This suggests that *D. dadantii* is able to anticipate the host induced oxidative environment before its onset, in order to increase the chance of survival in the plant environment.

4.5 Adaptation to Reactive Oxygen Species Produced by Plant in Response to Infection

The plant defense responses to *D. dadantii* infection have been analyzed in the *Arabidopsis* model by Fagard et al.[22] Infection was accompanied by a ROS production mainly achieved via the action of the AtrbohD NADPH oxidase before the appearance of symptoms (Fig. 2), followed by an accumulation in the cell wall structure of hydroxyproline-rich glycoproteins (HRGP) and callose, a ß-1,3-glucan that acts as an additional physical barrier.[22] These observations suggest that the host perceived the bacterium as an intruder and initiated immune responses (Fig. 1B). Consequently, activities of the salicylic acid, jasmonic acid, and ethylene signaling pathways involved in the plant immune response network were induced.[22] To evade plant defense reaction, *D. dadantii* produces antioxidants that are essential for its virulence. For example, *D. dadantii* mutants altered in protein oxidative damage repair systems (*mrsA* and *sufABCDSE*), or in the manganese superoxide dismutase MnSOD (*sodA*), are unable to cause systemic invasion in *S. ionantha.*[126–128] Furthermore, transcriptome analysis revealed that in response to H_2O_2 treatment, *Dickeya* induces the production of antioxidant enzymes such as catalases (KatG, KatE) and peroxidase AhpCF responsible for H_2O_2 decomposition into water and molecular oxygen, whereas the *ychF* gene encoding a catalase inhibitor protein is down-regulated.[49a] Resistance to ROS appears to be mediated by OxyR, OhrR, and PecS.[31,110,129] Interestingly, among the two superoxide dismutases SodA and SodC present in *D. dadantii*, only *sodC* is moderately, yet significantly induced under oxidative stress.[49a] This could be related to the lack of *soxRS* orthologs in *D. dadantii*. *D. dadantii* synthesizes the blue pigment indigoidine (*indA*, *indB*, and *indC* genes) proposed to be a radical scavenger conferring increased resistance to oxidative stress[130] (Fig. 1B). Under oxidative stress the *indA*, *indB*, and *indC* genes are induced, as well as the gene of the organic hydroperoxide resistance protein (ABF-0017804). Furthermore, the genes producing glutathione, glutaredoxin, and thioredoxin (*gshB*, *gor*, *grxA*, *trxC*, *trxB*, *gstB*, ABF-0020419) that reduce disulfide bond formation and counteract oxidation of cytoplasmic proteins are upregulated. Glutathione in particular acts as a cofactor for peroxidase by donating electrons necessary for decomposition of H_2O_2. In addition, the Suf system, which is involved in the repair of damaged Fe/S cluster containing proteins, is induced while the *rsxABCDE* genes involved in iron–sulfur cluster reduction are downregulated.[49a] Most genes encoding chemotaxis proteins are also upregulated probably to allow the bacteria to escape hostile environments and promote further invasion. At this point, there are two possible scenarios: either (1) plant defense reactions are strong enough to stop the bacterial progression leading to abortive infection, or (2) bacteria adapt to the oxidative stress, move into nearby tissues and continue to grow until a quorum bacterial density is reached, initiating the synthesis of CWDE.[131] In the latter case, maceration of plant tissues and cell lysis occur and consequently, the release of vacuolar contents in the apoplast creates osmotic shock conditions for the bacteria.

4.6 Adaptation to Osmotic Stress Resulting From Plant Cell Lysis

The osmotic strength of the macerated tissue was estimated at approximately 350 mOsm[132] (Fig. 2). To deal with hyperosmotic upshifts during infection, *D. dadantii* induces two lines of temporally separated responses. The fast response at the level of protein activity involves potassium uptake, whereas the secondary response consists of the uptake of osmoprotectants such as proline, trimethylammonium glycine betaine, choline, tetrahydropyrimidine ectoine, and pipecolate present in the surroundings, and synthesis of high amounts of compatible solutes to avoid deleterious increases in ion concentration and maintain cytoplasm electroneutrality. *Dickeya* is capable of uptake of all the aforementioned osmoprotectants[133] via two transporters: OusA, related to the *E. coli* H^+ symporter ProP and responsible for glycine betaine, proline, ectoine and pipecolic acid uptake,[134] and OusB, related to the *E. coli* ABC transporter ProU and responsible for glycine betaine uptake.[135] In contrast to ProU, OusB was shown to transport both glycine betaine and choline with high affinity.[135] Consequently, *Dickeya* lacks the *betT* gene encoding the choline high-affinity transporter in *E. coli*. The synthesis of glycine betaine from choline is possible in *D. dadantii* since it contains two genes necessary for choline oxidation (*betB* and *betA*), which are encoded in an operon containing also the regulatory gene *betI*. Accumulation of these osmoprotectants reduces the production of pectate lyases[133] indicating that the specific induction of *pelE* gene under high osmolarity is inhibited in presence of osmoprotectants. Accordingly, mutations of *ousA* highly enhance the virulence of *D. dadantii*.[132] To this end it is noteworthy that recent study in the phytopathogen *Pectobacterium wasabiae* revealed the requirement of Trk potassium transporter for activation of virulence.[136] In the absence of osmoprotectants in the environment, *D. dadantii* promotes the accumulation of glutamine, glutamate and α-glucosylglycerate, a ionic osmolyte also present in some halophilic archae.[137] α-Glucosylglycerate synthesis is catalyzed by the glucosyltransferase PggS, which condenses ADP-Glucose with D-3-phosphoglycerate. The resulting phosphorylated intermediate is subsequently converted to glucosylglycerate by the action of the phosphatase PggP. Notably, upon osmotic stress in *D. dadantii*, both the *pggS-pggP* operon involved in α-glucosylglycerate biosynthesis and the *betABI* operon enabling the synthesis of glycine betaine from choline were found to be up-regulated.[49a]

In addition to K^+ influx, high salinity induces the genes encoding the NhaA and the YrbG antiporter systems predominantly mediating sodium ion extrusion. Sodium ion extrusion is crucial to enable normal cell metabolism, as high cytosolic concentrations of Na^+ affect the water economy and are toxic for the activity of numerous enzymes. Since protein stability and folding are profoundly affected by changes in water activity, cells have to sustain correct protein folding under conditions of increased external osmotic pressure. Therefore it is not surprising that the RNA polymerase RpoH and RpoE sigma factors, responsible for heat-shock and extracytoplasmic stress responses, respectively, were found to be upregulated under conditions of osmotic stress.[49a] The increased expression of the antisigma factor *rsd* observed under the same conditions would facilitate the competition of RpoH and RpoE with the major sigma factor RpoD for RNA polymerase holoenzyme. Consequently, osmotic stress induced the genes of heat shock chaperones and refolding proteins as well as a large number of genes encoding proteins related to cell envelope integrity, including the two-component systems *cpxR-cpxA, envZ-ompR*, and *rcsF* belonging to the RcsCDB phosphorelay. The increase in osmolarity is one of the major stimuli sensed by the RcsCDB-RcsF system, which is responsible for exopolysaccharide accumulation to protect cells from desiccation, and inhibition of motility by directly repressing the master regulator of flagellar gene transcription FlhDC.[138,139] This is consistent with the absence of motility requirement at the late stage of infection when bacteria massively degrade pectin. At this stage, as the infecting bacteria achieve their ultimate goal, the cells also cease to produce the Vfm quorum-sensing signal (Section 5.1).

5. REGULATORY MECHANISMS COORDINATING THE VIRULENCE PROGRAM

In unicellular organisms, intricate communication and signaling networks evolved to enable the sensing of variations in both the environment and internal conditions, and the adjusting of the cellular physiology accordingly. In bacteria, regulation of gene expression is exerted primarily at the level of transcription initiation and involves numerous transcription factors (TFs) whose concentrations and activities depend on the external and internal conditions.[140] The distribution of the transcriptional apparatus along the genome is in turn modulated by the physical state of the bacterial DNA polymer (superhelical level or topology), which is also highly responsive to environmental conditions.[141] On this basic substrate, a relatively small number of DNA architectural factors act as global regulators by influencing a large number of genes. In particular, the abundant *n*ucleoid-*a*ssociated-*p*roteins (NAPs) influence both the chromatin structure and transcription initiation. TFs with much more local effects on transcription then act upon the regulatory background imposed by NAPs.[141] The bacterial genetic response to a challenge thus depends on how this network is poised.

Since pathogens have to cope with numerous specific challenges including various types of stress, host defense reactions, access to nutrients and specific oligoelements, and triggering of a timely attack on the host, the production of virulence factors must be highly coordinated to optimize energy consumption and disease development. As have most pathogenic bacteria, *D. dadantii* has evolved a sophisticated system to detect distinct hostile environments and trigger compensatory gene expression to survive within the host. Sensing of anaerobiosis, iron concentration, acidic, oxidative, and osmotic stresses have been described in the previous sections and how *Dickeya* senses plant signal molecules such as pectin and couples CWDE expression to central metabolism is described by Hugouvieux-Cotte-Pattat in Chapter 8. Therefore, this section will focus on the regulatory network implicated in the control of multiple *D. dadantii* virulence factors with emphasis on the impacts of the NAPs and changing chromosomal dynamics in coordination of the virulent growth.

5.1 Regulatory Network Coordinating Multiple Virulence Factor Expression

In *D. dadantii*, the production of virulence factors is fine-tuned by intricate and interconnected regulatory circuits. Several global regulators are involved in coordinating the expression of multiple virulence factors. Among these, PecS, which belongs to the MarR family, directly represses the production of CWDEs and flagellar proteins, biosurfactant synthesis, the systems of resistance to oxidative stress, and T3SS[31,130,142,143] (Fig. 3). PecS is supposed to respond to an element of the plant innate immune system such as salicylate, or to xanthine and urate, produced as by-products of the reactive oxygen

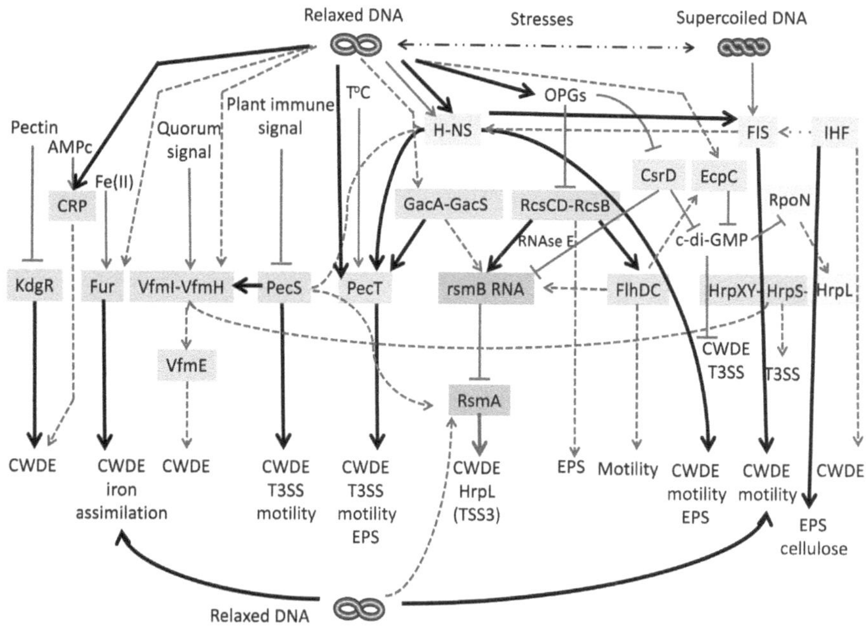

Figure 3 Model of the regulatory network of *D. dadantii* virulence genes. The different actors of the regulatory network including the NAPs, the regulators, the two-component systems, the posttranscriptional RsmA-rsmB system, the c-di-GMP phosphodiesterases and the sigma factors HrpL and RpoN are indicated. *Thick and dashed arrows* indicate repression and activation of transcription, respectively. RsmA binding to CWDE mRNAs as well to *hrpL* mRNAs promotes their degradation. *Grey arrows and grey barred lines* indicate a positive and negative modulation of regulator activity respectively. For example, the plant immune signal is supposed to inhibit PecS DNA-binding activity and relieve PecS repression, whereas Fe (II) acts as a cofactor for the Fur repressor. Similarly, DNA relaxation and high temperatures promote PecT binding to its targets. DNA relaxation modulates the H-NS repression, whereas DNA supercoiling promotes FIS binding. DNA relaxation also induces the components of the CWDE, motility, and iron assimilation systems. High c-di-GMP concentration negatively regulates CWDE and T3SS expression. Degradation of c-di-GMP by EcpC contributes to T3SS activation whereas degradation of c-di-GMP by CsrD promotes both CWDE and T3SS expression. IHF is a strong activator of *fis (dashed arrow)* in *E. coli* but this effect has not yet been explored in *D. dadantii.*

burst during plant defense. Indeed, in *Agrobacteriumtumefasciens*, DNA-binding activity of a PecS homolog is modulated by these coumpounds.[144] In *Dickeya*, the *pecS* mutant produces symptoms more rapidly and efficiently than the wild-type strain, indicating that PecS prevents the premature expression of virulence genes and plays a critical role in the switch from the asymptomatic to symptomatic phase.[31,145] PecS binds DNA via a degenerate consensus motif, which consists of a 23-base pair palindrome-like sequence. This variable consensus is in agreement with the observation that the members of the MarR family are able to bind various DNA targets as dimers by means of a winged helix DNA-binding motif.[146] PecS also appears to be important for insect infection by *D. dadantii*[147] as it activates the insect virulence functions.[148]

The second global transcriptional regulator PecT belongs to the LysR family and regulates CWDE, exopolysaccharide production, motility, and T3SS[97,143,149,150] (Fig. 3). PecT modulates *pel* gene expression in a temperature-dependent manner[151] such that the binding of PecT to *pel* promoters and repression increases with increasing temperature. This effect is linked to the temperature-dependent variation in the topological state of the *pel* promoter, as high temperatures relax the DNA and promote PecT binding. Indeed, by experimentally inducing relaxation of DNA, it is possible to increase PecT binding and repress *pel* expression even at low temperatures, whereas deletion of *pecT* was found to increase the capacity of *D. dadantii* to initiate soft-rot symptoms at high temperatures. These data suggest that the dependence of PecT binding on DNA topology fine-tunes the *D. dadantii* virulence in response to changing temperature. However, during plant infection, bacterial DNA topology strongly fluctuates in response to distinct environmental conditions. For example, acidic and oxidative conditions encountered during early stages of infection induce DNA relaxation and probably also promote repression of PecT targets. Understanding of this interplay between DNA topology and a conventional transcriptional regulator is crucial for the conceptual integration of various mechanisms of virulence regulation in response to multiple signals.[151] Interestingly, PecT was reported to be essential for the virulence of recently emerged potato pathogen in Europe, *D. solani*.[152] Bacteria of this species are characterized by a wide temperature optimum for growth and disease development and can initiate the development of the disease from low inoculum levels. Whether the observed increased role of PecT in *D. solani* virulence is associated with the emergence of this pathogen in Europe remains an open question.

Expression of the *pecT* regulatory gene is itself under the control of the two-component system GacA-GacS[131,145] (Fig. 3). This two-component system has a dual role in the coordinated regulation of virulence gene expression by repressing *pecT* and upregulating the global posttranscriptional regulation system *rsm*[153] (Fig. 3). The *rsm* system is involved in the regulation of various virulence functions including CWDEs and T3SS and is described in detail by Hugouvieux-Cotte-Pattat in Chapter 8. Overall, GacA positively controls CWDE[131] and the T3SS[153] probably in response to carboxylic acids accumulating during infection (Fig. 3). GacA, in conjunction with other regulators, is required for the activation of virulence genes at the onset of symptom occurrence.

The RcsCD-RcsB regulatory system is another two-component system coordinating the regulation of CWDEs, exopolysaccharide, motility, and T3SS[154] (Fig. 3). Current knowledge of the conditions that lead to Rcs activation is based on exposure to antimicrobial peptides or high osmolarity, and growth at low temperature or on solid surface[139,155]. RcsC and RcsD can act as histidine kinases, which after autophosphorylation, are able to independently transfer phosphate to RcsB. These phosphate flow pathways can be activated by different environmental signals, leading to different levels of phosphorylated RcsB and accordingly differential gene modulation.[155] Importantly, the osmoregulated periplasmic glucans (OPGs) are directly involved in perception of the environment through the Rcs phosphorelay. The concentration of OPGs increases during the course of infection and inhibits the Rcs phosphorelay[138,139] (Fig. 3). Inhibition of RcsB phosphorylation allows the expression of the motility, the CWDE and the T3SS genes.[154] Indeed, RcsB~P represses the expression of the *rsmB* gene and the *flhDC* master operon, which controls the structural genes of the flagellum (Fig. 3). The T3SS is regulated by a dedicated signal transduction chain, that includes a two-component system (HrpXY), a sigma 54 enhancer binding protein (HrpS), and an alternative sigma factor, HrpL[81,156] (Fig. 3). HrpXY senses natural phenolic compounds in the plant, which either induce or inhibit the expression of T3SS genes.[157,158] In absence of RcsB~P, the high level of *rsmB* sRNA titrates RsmA away from the CWDE and HrpL mRNAs allowing their translation (Fig. 3). Turnover of *rsmB* sRNAs is tightly regulated by a GGDEF-EAL domain protein, CsrD, which targets them for cleavage by RNase E[154] (Fig. 3). Thus, the RcsCD-RcsB regulatory system coordinates the sensing of OPGs with the motility and pectinolytic activity of *Dickeya*.

Furthermore, a recent study showed that c-di-GMP negatively regulates CWDE and T3SS expression in *D. dadantii*.[159] CsrD via its c-di-GMP phosphodiesterase activity is responsible for the up-regulation of CWDE and T3SS. Activity of CsrD seems to be modulated by OPGs[154] (Fig. 3). Interestingly, a recent study in *E. coli* demonstrated that the unphosphorylated form of $EIIA^{Glc}$ of the glucose-specific PTS system binds to the EAL domain of CsrD and activates CsrB/C turnover in vivo.[160] Whether the membrane-bound proteins OpgG or OpgH responsible for OPG production can bind to the EAL domain of CsrD to modulate its activity is an interesting question for the future.

FlhDC, the master regulator of flagellar genes, also controls the expression of CWDE and T3SS by activating the *rsmB* regulatory RNA and the *ecpC* gene product,a c-di-GMP degrading enzyme[161] (Fig. 3). EcpC positively regulates the alternative sigma factor RpoN, which is required to activate the transcription of *hrpL* in *D. dadantii* 3937[159] (Fig. 3). Expression of *ecpC* gene *in trans* in a *ΔflhDC* mutant restores both the T3SS and CWDE production. In contrast, expression of *rsmB* in Δ*flhDC* restores CWDE production, but not T3SS gene expression, whereas neither *ecpC* nor *rsmB* expression restore the motility in Δ*flhDC*.[161] Therefore, FlhDC controls CWDE via the FlhDC-*rsmB*-RsmA pathway and T3SS both by the FlhDC-*rsmB*-RsmA-HrpL and the FlhDC-EcpC-RpoN-HrpL pathways. However, flagellar gene expression by FlhDC is independent of either EcpC or RsmB.

The last regulatory system coordinating several virulence factors corresponds to the Vfm cell–cell signaling system, which allows bacteria to massively produce CWDE at high population density[57] (Fig. 3). The *vfm* system is specific to the *Dickeya* genus and directs the biosynthesis of an extracellular signaling molecule of unknown chemical structure. This signal is perceived by the VfmI-VfmH TCS, which activates the expression of the *vfmE* gene encoding a regulator of the AraC family. VfmE then activates the expression of both the CWDE genes and the *vfm* operons[57] (Fig. 3). Expression of the *vfm* genes is repressed by PecS[31] (Fig. 3). Thus, overproduction of the Vfm signal in *pecS* mutant may explain the advanced induction of virulence factors observed in this mutant and the early development of disease symptoms.[145] Some genes in the *vfm* cluster (*vfmH* and *vfmQ*) are also activated by the HrpY or HrpL regulators of the T3SS,[162,163] suggesting additional cross-talk between the regulatory networks controlling CWDE production and T3SS expression (Fig. 3). Furthermore, a recent transcriptomic analysis revealed that *vfm* genes are induced by antimicrobial peptides produced by (some) plants in response to bacterial infection,[122] whereas this induction of *vfm* genes allows the subsequent expression of virulence genes. Taken together, these data demonstrate that *D. dadantii* can intercept compounds released during host stress and integrate them into quorum sensing circuitry leading to enhanced virulence. *D. dadantii* also possesses a paradigmatic ExpI-ExpR quorum-sensing system, which is analogous to the conserved LuxI/R system of *Photobacterium fischeri* but has no significant impact on the virulence of *D. dadantii*.[145,164,165] However, recent data revealed that this system modulates the virulence of *D. solani*,[152] illustrating again variability in the function of some regulatory systems in the genus *Dickeya*.

Communications between the virulence regulators and global DNA architectural factors such as the NAPs appear to be crucial for coordination of the virulence program. For instance, transcriptomic analysis showed that PecS directly or indirectly activates RsmA and H-NS production[31] (Fig. 3). This result, surprising at that time, has since been supported by data obtained in *Salmonella enterica* and showing that OmpR, a master regulator of stress response, also modulates *hns* expression.[166] As stated previously, NAPs modulate chromosome structural dynamics according to environmental conditions, and although the significance of the control of NAPs by specific regulators is not yet fully understood, it is reasonable to assume that this cross-talk could allow the bacterium to optimize the implementation of various regulatory mechanisms needed to efficiently adapt its physiology to changing conditions.[167] In the case of *D. dadantii*, possible neutralization of PecS activity by an innate immune system compound would abolish the H-NS activation by PecS during the intermediate stage of plant infection. The resulting decrease in H-NS concentration would lead to de-repression of the adaptive systems, in addition to those controlled by PecS, to allow *D. dadantii* to resist more effectively the stress conditions encountered in the host. Furthermore, by modulating both the *vfm* and *hns* genes, PecS might establish a link between two important barometers of virulence gene regulation: H-NS for environmental conditions and Vfm for bacterial cell density. Such a "back and forth" interactive process between global chromosome-shaping factors and specialized regulators would allow bacteria to rapidly adjust their genetic programs to changes in environmental conditions.

5.2 NAPs and Chromosome Dynamics

D. dadantii virulence program is regulated by compositional changes of the NAPs in response to environmental stimuli and physiological state transitions such as growth phase, which has an impact on the synthesis of most of the virulence factors including CWDEs.[124,168] Three abundant NAPs, FIS (factor for inversion stimulation), H-NS (heat-stable nucleoid-structuring protein) and IHF (integration host factor), affect the expression of various virulence genes[16,169–171] (Hérault et al., unpublished data) (Fig. 3).

FIS plays a pivotal role in the temporal expression of virulence genes by activating the factors required at the early stages (bactericidal compound detoxification factors) and repressing those factors needed at the advanced stages (Pels and cellulase) of infection[171] (Fig. 3). Furthermore, FIS modulates plant surface colonization by *D. dadantii* via regulation of biofilm formation[96] and motility[172] (Fig. 3). In contrast to FIS, IHF activates the synthesis of CWDE (Pels and cellulase), whereas it represses the production of extracellular compounds (cellulose and exopolysaccharides) involved in formation of biofilms (Hérault et al., unpublished data) (Fig. 3). Also H-NS modulates motility and the production of Pels and exopolysaccharide[169,170] (Fig. 3). Accordingly, the virulence of *D. dadantii ihf*, *fis*, and *hns* mutants is strongly attenuated[169,171] (Jiang, unpublished results).

H-NS exerts an unusual activator phenotype related to the production of Pels. Inactivation of H-NS leads to a drastic reduction of Pel enzyme synthesis mainly because of a negative control exerted by H-NS on the transcription of the strong repressor of *pel* genes, *pecT* (Fig. 3). Thus inactivation of H-NS results in overproduction of PecT, which in turn represses the transcription of *pel* genes (Fig. 3). However, in addition to its action on Pel production via PecT, H-NS also directly represses *pel* gene expression by binding to extended DNA segments in the regulatory regions of these genes[170] (Fig. 3). Thus, the double mutant *hns-pecT* produces more Pels than the *pecT* mutant.

Recent studies uncovered that cross-talk between the DNA topology and the two abundant NAPs, FIS and H-NS, is involved in optimizing the virulence program of *D. dadantii*.[169,173] It was found that expression of the *pel* genes strongly decreased under conditions of chromosomal DNA relaxation (Fig. 3) induced either by treatment with DNA gyrase inhibitors or by applying environmental (oxidative and acidic) stresses, whereas FIS and H-NS modulated the response of *pel* promoters [to supercoiling both in vivo and in vitro (Fig. 3)] in distinctly different manners. FIS was found to act as a repressor under physiological conditions but buffered the negative impact of DNA relaxation on *pel* gene transcription, whereas H-NS fine-tuned the response of *pel* genes, precluding their expression under suboptimal conditions of supercoiling. These observations constitute the first evidence of virulence gene regulation by DNA topology in phytopathogenic bacteria;

an important aspect of regulation revealed by these studies is the discovery of the dependence of H-NS binding on the superhelical state of the DNA.[173] Whereas previous studies concerning different modes of H-NS binding placed emphasis primarily on changing the oligomerization propensity of the protein depending on the temperature and/or ionic conditions,[174,175] the studies performed in *D. dadantii* showed for the first time that under otherwise identical conditions a change in the superhelical density of DNA dramatically altered the effect of H-NS binding on the transcription of a virulence gene. How would sensing DNA topology by *D. dadantii* underpin pathogenesis? The relaxation of DNA induced by both the acidic and oxidative stresses experienced by *D. dadantii* early during infection leads to a significant reduction in Pel synthesis, preventing a premature detection of the pectin breakdown products that signal the presence of the pathogen to the host. At a later stage during infection, alkalinization of the apoplast would result in an increased DNA superhelical density and, consequently, a decrease in H-NS mediated repression of the *pel* genes. This would stimulate Pel production and consequently, result in a switch from biotrophic to necrogenic lifestyle.[173]

Subsequent studies suggested that the regulation of bacterial virulence involves gross alterations of the bacterial chromosome structure, revealing that the local influence of DNA topology on *pel* gene expression is just one detail of a global effect of DNA superhelicity and NAPs operating at the level of the entire chromosome. Explorations in *Dickeya* provided the first evidence for the determinative role of the physicochemical properties of the primary DNA sequence, DNA topology, and NAPs in organizing transient structural functional entities in the chromosome, dubbed chromosomal stress-response domains (CSRDs). The CSRDs react distinctly to adverse conditions encountered by the infecting bacteria, ultimately determining their phenotypic behaviour.[105] The identified domains highlight distinct characteristic combinations of DNA sequence parameters (such as thermodynamic stability and base composition) and harbor different virulence determinants. Furthermore, it was found that in *D. dadantii* the boundaries of CSRDs often coincide with strong transcription units and that selective induction of these domains involves a cross-talk between DNA supercoiling and the abundant NAPs, FIS, and H-NS. The CSRDs thus appear as modular structural–functional entities coordinating the genetic expression of the chromosome and sustaining bacterial pathogenicity. Overall, the organizational properties of at least some CSRDs closely resemble those of the horizontally acquired genomic pathogenicity islands, comprising large genomic regions (up to 200 kb in size) present in the genomes of pathogenic strains but absent from the genomes of nonpathogenic members of the same or related species.[176] Indeed, some of the CSRDs observed in *D. dadantii* are likely to have acquired AT-rich DNA sequences by horizontal gene transfer and strongly resemble pathogenicity islands,[39] suggesting variations on a common theme (pathogenicity). However, the emerging picture is that the entire bacterial chromosome may have a modular structure, displaying transient structural functional domains as basic organizational units that are responsive to and coordinate gene expression during bacterial adaptation.[105,177]

6. CONCLUDING REMARKS

In this review we have described, albeit briefly, the intricate communication networks of the bacterial pathogen *D. dadantii* involved in sensing and responding to changing environmental conditions. As in other enterobacteria, the control of gene expression in *D. dadantii* is exerted primarily at the level of transcription initiation and involves numerous transcription factors, which regulate the distribution of the transcriptional apparatus along the chromosome. The activation timing and effects of these TFs is linked tightly to particular environmental cues that emerge during the plant–pathogen interaction, including various types of stress. *Dickeya* evolved an intricate strategy to protect itself from adverse conditions encountered during infection and override the defense reactions of the host. In addition to numerous specific protection systems involved in the bacterial stress response, a small number of highly abundant NAPs influence both the chromatin structure and transcription initiation. These global regulators facilitate the adaptation of the pathogen to distinct kinds of stress induced by host defense reactions and coordinate the temporal production of virulence factors involved in disease development.

Recent studies strongly suggest that the virulence program is underpinned by dynamic organization of so-called chromosomal stress response domains—CSRDs—stabilized by cooperative effects of the NAPs and DNA supercoiling. In addition to the need for further study of the role of CSRDs in bacterial pathogenic behavior, the discovery of CSRDs fundamentally changes our concepts of bacterial pathogenesis, previously thought to be based solely on regulatory interactions between particular genes. It appears that the coordination of the virulence program is primarily dictated by chromosomal dynamics involving stabilization of transient CSRDs in response to specific environmental challenges. These domains (and various combinations thereof) act as discrete entities coordinating the spatiotemporal expression of the adaptation/virulence determinants that they harbor, and by doing so, manage the entire virulence program. The dynamic alterations of DNA topology and composition of NAPs involved in organization of CSRDs are thus central to bacterial virulence, as evident also from the dramatic attenuation of virulence in mutants lacking the NAPs.

REFERENCES

1. Chakraborty SK, Newton AC. Climate change, plant diseases and food security: an overview. *Plant Pathol.* 2011;60:2–14.
2. Oerke EC. Crop losses to pests. *J Agric Sci.* 2006;144:31–43.
3. Engering A, Hogerwerf L, Slingenbergh J. Pathogen–host-environment interplay and disease emergence. *Emerg Microbes Infect.* 2013;2(2):e5.
4. Chakraborty S. Migrate or evolve: options for plant pathogens under climate change. *Global Change Biol.* 2013;19(7):1985–2000.
5. Mansfield J, Genin S, Magori S, et al. Top 10 plant pathogenic bacteria in molecular plant pathology. *Mol Plant Pathol.* 2012;13(6):614–629.
6. Ma B, Hibbing ME, Kim HS, et al. Host range and molecular phylogenies of the soft rot enterobacterial genera pectobacterium and dickeya. *Phytopathology.* 2007;97(9): 1150–1163.
7. Samson R, Legendre JB, Christen R, Fischer-Le Saux M, Achouak W, Gardan L. Transfer of *Pectobacterium chrysanthemi* (Burkholder et al., 1953) Brenner et al., 1973 and *Brenneria paradisiaca* to the genus Dickeya gen. nov. as *Dickeya chrysanthemi* comb. nov. and *Dickeya paradisiaca* comb. nov. and delineation of four novel species, *Dickeya dadantii* sp. nov., *Dickeya dianthicola* sp. nov., *Dickeya dieffenbachiae* sp. nov. and *Dickeya zeae* sp. nov. *Int J Syst Evol Microbiol.* 2005;55(Pt 4):1415–1427.
8. Brady CL, Cleenwerck I, Denman S, et al. Proposal to reclassify *Brenneria quercina* (Hildebrand and Schroth 1967) Hauben et al., 1999 into a new genus, Lonsdalea gen. nov., as *Lonsdalea quercina* comb. nov., descriptions of *Lonsdalea quercina* subsp. quercina comb. nov., *Lonsdalea quercina* subsp. iberica subsp. nov. and *Lonsdalea quercina* subsp. britannica subsp. nov., emendation of the description of the genus Brenneria, reclassification of *Dickeya dieffenbachiae* as *Dickeya dadantii* subsp. dieffenbachiae comb. nov., and emendation of the description of *Dickeya dadantii*. *Int J Syst Evol Microbiol.* 2012;62(Pt 7):1592–1602.
9. Janse JD, Ruissen MA. Characterization and classification of *Erwinia chrysanthemi* strains from several hosts in The Netherlands. *Phytopathology.* 1988;78:800–808.
10. Tsror L, Erlich O, Lebiush S, et al. Assessment of recent outbreaks of Dickeya sp. (syn. *Erwinia chrysanthemi*) slow wilt in potato crops in Israel. *Eur J Plant Pathol.* 2009;123: 311–320.
11. Sławiak M, Beckhoven J, Speksnijder A, Czajkowski R, Grabe G, van der Wolf JM. Biochemical and genetical analysis reveal a new clade of biovar 3 Dickeya spp. strains isolated from potato in Europe. *Eur J Plant Pathol.* 2009;125:245–261.
12. Czajkowski R, Pérombelon MCM, van Veen JA, van der Wolf JM. Control of blackleg and tuber soft rot of potato caused by *Pectobacterium* and *Dickeya* species: a review. *Plant Pathol.* 2011;60:999–1013.

13. van der Wolf JM, Nijhuis EH, Kowalewska MJ, et al. Dickeya solani sp. nov., a pectinolytic plant-pathogenic bacterium isolated from potato (*Solanum tuberosum*). *Int J Syst Evol Microbiol.* 2014;64:768–774.
14. Parkinson N, DeVos P, Pirhonen M, Elphinstone J. Dickeya aquatica sp. nov., isolated from waterways. *Int J Syst Evol Microbiol.* 2014;64(Pt 7):2264–2266.
15. Marrero G, Schneider KL, Jenkins DM, Alvarez AM. Phylogeny and classification of Dickeya based on multilocus sequence analysis. *Int J Syst Evol Microbiol.* 2013;63(Pt 9): 3524–3539.
16. Reverchon S, Nasser W. Dickeya ecology, environment sensing and regulation of virulence programme. *Environ Microbiol Rep.* 2013;5(5):622–636.
17. Czajkowski R, de Boer WJ, Velvis H, van der Wolf JM. Systemic colonization of potato plants by a soilborne, green fluorescent protein-tagged strain of Dickeya sp. biovar 3. *Phytopathology.* 2010;100(2):134–142.
18. Tsror L, Lebiush S, Erlich O, Ben-Daniel B, van der Wolf J. First report of latent infection of *Cyperus rotundus* caused by a biovar 3 Dickeya sp. (syn. *Erwinia chrysanthemi*) in Israel. *New Dis Rep.* 2012;22:Article 14.
19. Toth I, van der Wolf JM, Saddler G, et al. *Dickeya* species: an emerging problem for potato production in Europe. *Plant Pathol.* 2011;60:385–399.
20. Effantin G, Rivasseau C, Gromova M, Bligny R, Hugouvieux-Cotte-Pattat N. Massive production of butanediol during plant infection by phytopathogenic bacteria of the genera *Dickeya* and *Pectobacterium*. *Mol Microbiol.* 2011;82(4):988–997.
21. Grenier AM, Duport G, Pages S, Condemine G, Rahbe Y. The phytopathogen *Dickeya dadantii* (*Erwinia chrysanthemi* 3937) is a pathogen of the pea aphid. *Appl Environ Microbiol.* 2006;72(3):1956–1965.
22. Fagard M, Dellagi A, Roux C, et al. *Arabidopsis thaliana* expresses multiple lines of defense to counterattack *Erwinia chrysanthemi*. *Mol Plant Microbe Interact.* 2007;20(7): 794–805.
23. Hugouvieux-Cotte-Pattat N, Condemine G, Shevchik VE. Bacterial pectate lyases, structural and functional diversity. *Environ Microbiol Rep.* 2014;6(5):427–440.
24. Keen NT, Boyd C, Henrissat B. Cloning and characterization of a xylanase gene from corn strains of *Erwinia chrysanthemi*. *Mol Plant Microbe Interact.* 1996;9(7):651–657.
25. Delangle A, Prouvost AF, Cogez V, Bohin JP, Lacroix JM, Cotte-Pattat NH. Characterization of the *Erwinia chrysanthemi gan* locus, involved in galactan catabolism. *J Bacteriol.* 2007;189(19):7053–7061.
26. Py B, Bortoli-German I, Haiech J, Chippaux M, Barras F, Cellulase EGZ. of *Erwinia chrysanthemi:* structural organization and importance of His98 and Glu133 residues for catalysis. *Protein Eng.* 1991;4(3):325–333.
27. Wandersman C, Delepelaire P, Letoffe S, Schwartz M. Characterization of *Erwinia chrysanthemi* extracellular proteases: cloning and expression of the protease genes in *Escherichia coli*. *J Bacteriol.* 1987;169(11):5046–5053.
28. Pineau C, Guschinskaya N, Robert X, Gouet P, Ballut L, Shevchik VE. Substrate recognition by the bacterial type II secretion system: more than a simple interaction. *Mol Microbiol.* 2014;94(1):126–140.
29. Delepelaire P. Erwinia metalloprotease permease: aspects of secretion pathway and secretion functions. *Methods Enzymol.* 1998;292:67–81.
30. Andro T, Chambost JP, Kotoujansky A, et al. Mutants of *Erwinia chrysanthemi* defective in secretion of pectinase and cellulase. *J Bacteriol.* 1984;160(3):1199–1203.
31. Hommais F, Oger-Desfeux C, Van Gijsegem F, et al. PecS is a global regulator of the symptomatic phase in the phytopathogenic bacterium *Erwinia chrysanthemi* 3937. *J Bacteriol.* 2008;190(22):7508–7522.
32. Barras F, Van Gijsegem F, Chatterjee AK. Extracellular enzymes and pathogenesis of soft-rot Erwinia. *Ann Rev Phytopathol.* 1994;32:201–234.

33. Caffall KH, Mohnen D. The structure, function, and biosynthesis of plant cell wall pectic polysaccharides. *Carbohydr Res.* 2009;344(14):1879–1900.
34. O'Neill MA, Ishii T, Albersheim P, Darvill AG. Rhamnogalacturonan II: structure and function of a borate cross-linked cell wall pectic polysaccharide. *Ann Rev Plant Biol.* 2004;55:109–139.
35. Ishii T. Structure and functions of feruloylated polysaccharides. *Plant Sci.* 1997;127: 111–127.
36. Marschner H, Romheld V, Kissel M. Different strategies in higher-plants in mobilization and uptake of iron. *J Plant Nutr.* 1986;9:695–713.
37. Laatu M, Condemine G. Rhamnogalacturonate lyase RhiE is secreted by the out system in *Erwinia chrysanthemi*. *J Bacteriol.* 2003;185(5):1642–1649.
38. Hassan S, Hugouvieux-Cotte-Pattat N. Identification of two feruloyl esterases in *Dickeya dadantii* 3937 and induction of the major feruloyl esterase and of pectate lyases by ferulic acid. *J Bacteriol.* 2011;193(4):963–970.
39. Ferrandez Y, Condemine G. Novel mechanism of outer membrane targeting of proteins in Gram-negative bacteria. *Mol Microbiol.* 2008;69(6):1349–1357.
40. Ferrari S, Savatin DV, Sicilia F, Gramegna G, Cervone F, Lorenzo GD. Oligogalacturonides: plant damage-associated molecular patterns and regulators of growth and development. *Front Plant Sci.* 2013;4:49.
41. Côté F, Hahn MG. Oligosaccharins: structures and signal transduction. *Plant Mol Biol.* 1994;26:1379–1411.
42. Braccini I, Perez S. Molecular basis of Ca(2+)-induced gelation in alginates and pectins: the egg-box model revisited. *Biomacromolecules.* 2001;2:1089–1096.
43. Cabrera JC, Boland A, Messiaen J, Cambier P, Van Cutsem P. Egg box conformation of oligogalacturonides: the time-dependent stabilization of the elicitor-active conformation increases its biological activity. *Glycobiology.* 2008;18:473–482.
44. Roy C, Kester H, Visser J, et al. Modes of action of five different endopectate lyases from *Erwinia chrysanthemi* 3937. *J Bacteriol.* 1999;181(12):3705–3709.
45. Decreux A, Messiaen J. Wall-associated kinase WAK1interacts with cell wall pectins in a calcium-induced conformation. *Plant Cell Physiol.* 2005;46:268–278.
46. Brutus A, Sicilia F, Macone A, Cervone F, De Lorenzo G. A domain swap approach reveals a role of the plant wall-associated kinase 1 (WAK1) as a receptor of oligogalacturonides. *Proc Natl Acad Sci USA.* 2010;107(20):9452–9457.
47. Davidsson PR, Kariola T, Niemi O, Palva ET. Pathogenicity of and plant immunity to soft rot pectobacteria. *Front Plant Sci.* 2013;4:191.
48. Boccara M, Diolez A, Rouve M, Kotoujansky A. The role of the individual pectate lyases of *Erwinia chrysanthemi* strain 3937 in pathogenicity on saintpaulia plants. *Physiol Mol Plant P.* 1988;33:95–104.
49. Beaulieu C, Boccara M, Vangijsegem F. Pathogenic behavior of pectinase-defective *Erwinia-chrysanthemi* mutants on different plants. *Mol Plant Microbe Interact.* 1993;6(2): 197–202.
49a. Jiang X, Zghidi-Abouzid O, Oger-Desfeux C, Hommais F, Greliche N, Muskhelishvili G, et al. Global transcriptional response of Dickeya dadantii to environmental stimuli relevant to the plant infection. *Environ Microbiol.* 2016. [Epub ahead of print].
50. Nachin L, Barras F. External pH: an environmental signal that helps to rationalize *pel* gene duplication in *Erwinia chrysanthemi*. *Mol Plant Microbe Interact.* 2000;13(8):882–886.
51. Kester HC, Magaud D, Roy C, et al. Performance of selected microbial pectinases on synthetic monomethyl-esterified di- and trigalacturonates. *J Biol Chem.* 1999;274(52): 37053–37059.
52. Nasser W, Shevchik VE, Hugouvieux-Cotte-Pattat N. Analysis of three clustered polygalacturonase genes in *Erwiniachrysanthemi* 3937 revealed an anti-repressor function for the PecS regulator. *Mol Microbiol.* 1999;34(4):641–650.

53. Flego D, Marits R, Eriksson AR, et al. A two-component regulatory system, pehR-pehS, controls endopolygalacturonase production and virulence in the plant pathogen Erwinia carotovora subsp. carotovora. *Mol Plant Microbe Interact.* 2000;13(4):447–455.
54. Haque MM, Yamazaki A, Tsuyumu S. Virulence, accumulation of acetyl-coenzyme A, and pectate lyase synthesis are controlled by PhoP-PhoQ two-component regulatory system responding to organic acids of *Erwinia chrysanthemi* 3937. *J Gen Plant Pathol.* 2005;71:133–138.
55. Chapelle E, Alunni B, Malfatti P, et al. A straightforward and reliable method for bacterial *in planta* transcriptomics: application to the *Dickeya dadantii/Arabidopsis thaliana* pathosystem. *PlantJ.* 2015;82(2):352–362.
56. D'Ovidio R, Mattei B, Roberti S, Bellincampi D. Polygalacturonases, polygalacturonase-inhibiting proteins and pectic oligomers in plant-pathogen interactions. *Biochim Biophys Acta.* 2004;1696(2):237–244.
57. Nasser W, Dorel C, Wawrzyniak J, et al. Vfm a new quorum sensing system controls the virulence of *Dickeya dadantii*. *Environ Microbiol.* 2013;15(3):865–880.
58. Kazemi-Pour N, Condemine G, Hugouvieux-Cotte-Pattat N. The secretome of the plant pathogenic bacterium *Erwinia chrysanthemi*. *Proteomics.* 2004;4(10):3177–3186.
59. Toth IK, Birch PRJ. Rotting softly and stealthily. *Curr Opin Plant Biol.* 2005;8:424–429.
60. Xin X, He S. *Pseudomonas syringae* pv. tomato DC3000: a model pathogen for probing disease susceptibility and hormone signaling in plants. *Annu Rev Phytopathol.* 2013;51: 473–498.
61. Bell KS, Sebaihia M, Pritchard L, et al. Genome sequence of the enterobacterial phytopathogen *Erwinia carotovora* subsp. atroseptica and characterization of virulence factors. *Proc Natl Acad Sci USA.* 2004;101(30):11105–11110.
62. Glasner JD, Yang CH, Reverchon S, et al. Genome sequence of the plant-pathogenic bacterium *Dickeya dadantii* 3937. *J Bacteriol.* 2011;193(8):2076–2077.
63. Dodds PN, Rathjen JP. Plant immunity: towards an integrated view of plant–pathogen interactions. *Nat Rev Genet.* 2010;11(8):539–548.
64. Dubiella U, Seybold H, Durian G, et al. Calcium-dependent protein kinase/NADPH oxidase activation circuit is required for rapid defense signal propagation. *Proc Natl Acad Sci USA.* 2013;110(21):8744–8749.
65. Zipfel C. Plant pattern-recognition receptors. *Trends Immunol.* 2014;35(7):345–351.
66. Jones JD, Dangl JL. The plant immune system. *Nature.* 2006;444(7117):323–329.
67. Cui H, Tsuda K, Parker JE. Effector-triggered immunity: from pathogen perception to robust defense. *Annu Rev Plant Biol.* 2015;66:487–511.
68. Hirakawa Y, Nomura T, Hasezawa S, Higaki T. Simplification of vacuole structure during plant cell death triggered by culture filtrates of *Erwinia carotovora*. *J Integr Plant Biol.* 2015;57(1):127–135.
69. Choi MS, Kim W, Lee C, Oh CS. Harpins, multifunctional proteins secreted by gram-negative plant-pathogenic bacteria. *Mol Plant Microbe Interact.* 2013;26(10): 1115–1122.
70. Oh J, Kim JG, Jeon E, et al. Amyloidogenesis of type III-dependent harpins from plant pathogenic bacteria. *J Biol Chem.* 2007;282(18):13601–13609.
71. Yang CH, Gavilanes-Ruiz M, Okinaka Y, et al. *hrp* genes of *Erwinia chrysanthemi* 3937 are important virulence factors. *Mol Plant Microbe Interact.* 2002;15(5):472–480.
72. Bauer DW, Wei ZM, Beer SV, Collmer A. *Erwinia chrysanthemi* harpinEch: an elicitor of the hypersensitive response that contributes to soft-rot pathogenesis. *Mol Plant Microbe Interact.* 1995;8(4):484–491.
73. Yap MN, Rojas CM, Yang CH, Charkowski AO. Harpin mediates cell aggregation in *Erwinia chrysanthemi* 3937. *J Bacteriol.* 2006;188(6):2280–2284.
74. Toth I, Humphris S, Campbell E, Pritchard L. Why genomics research on Pectobacterium and Dickeya makes a difference. *Am J Potato Res.* 2015;92(2):218–222.

75. Degrave A, Siamer S, Boureau T, Barny MA. The AvrE superfamily: ancestral type III effectors involved in suppression of pathogen-associated molecular pattern-triggered immunity. *Mol Plant Pathol.* 2015;16(8):899–905.
76. Jacobs JM, Milling A, Mitra RM, et al. *Ralstonia solanacearum* requires PopS, an ancient AvrE-family effector, for virulence and to overcome salicylic acid-mediated defenses during tomato pathogenesis. *MBio.* 2013;4(6):e00875–00813.
77. Siamer S, Guillas I, Shimobayashi M, Kunz C, Hall MN, Barny MA. Expression of the bacterial type III effector DspA/E in *Saccharomyces cerevisiae* down-regulates the sphingolipid biosynthetic pathway leading to growth arrest. *J Biol Chem.* 2014;289 (26):18466–18477.
78. Berkey R, Bendigeri D, Xiao S. Sphingolipids and plant defense/disease: the "death" connection and beyond. *Front Plant Sci.* 2012;3:68.
79. Chen M, Han G, Dietrich CR, Dunn TM, Cahoon EB. The essential nature of sphingolipids in plants as revealed by the functional identification and characterization of the *Arabidopsis* LCB1 subunit of serine palmitoyltransferase. *Plant Cell.* 2006;18 (12):3576–3593.
80. Xin X, Nomura K, Ding X, et al. *Pseudomonas syringae* effector AvrE localizes to the host plasma membrane and down regulates the expression of the NDR/HIN1-like 13 gene required for antibacterial immunity in *Arabidopsis. Plant Physiol.* 2015 (Epub ahead of print)
81. Peng Q, Yang S, Charkowski AO, et al. Population behavior analysis of dspE and pelD regulation in *Erwinia chrysanthemi* 3937. *Mol Plant Microbe Interact.* 2006;19(4):451–457.
82. Munzinger M, Budzikiewicz H, Expert D, Enard C, Meyer JM. Achromobactin, a new citrate siderophore of *Erwinia chrysanthemi. Z Naturforsch [C].* 2000;55(5–6):328–332.
83. Persmark M, Expert D, Neilands JB. Isolation, characterization, and synthesis of chrysobactin, a compound with siderophore activity from *Erwinia chrysanthemi. J Biol Chem.* 1989;264(6):3187–3193.
84. Franza T, Mahe B, Expert D. *Erwinia chrysanthemi* requires a second iron transport route dependent of the siderophore achromobactin for extracellular growth and plant infection. *Mol Microbiol.* 2005;55(1):261–275.
85. Enard C, Diolez A, Expert D. Systemic virulence of *Erwinia chrysanthemi* 3937 requires a functional iron assimilation system. *J Bacteriol.* 1988;170(6):2419–2426.
86. Dellagi A, Rigault M, Segond D, et al. Siderophore-mediated upregulation of *Arabidopsis ferritin* expression in response to *Erwinia chrysanthemi* infection. *Plant J.* 2005;43(2):262–272.
87. Dellagi A, Segond D, Rigault M, et al. Microbial siderophores exert a subtle role in *Arabidopsis* during infection by manipulating the immune response and the iron status. *Plant Physiol.* 2009;150(4):1687–1696.
88. Aznar A, Chen NW, Rigault M, et al. Scavenging iron: a novel mechanism of plant immunity activation by microbial siderophores. *Plant Physiol.* 2014;164(4):2167–2183.
89. Aznar A, Dellagi A. New insights into the role of siderophores as triggers of plant immunity: what can we learn from animals? *J Exp Bot.* 2015;66(11):3001–3010.
90. Aznar A, Patrit O, Berger A, Dellagi A. Alterations of iron distribution in *Arabidopsis* tissues infected by *Dickeya dadantii. Mol Plant Pathol.* 2015;16(5):521–528.
91. Boughammoura A, Matzanke BF, Bottger L, et al. Differential role of ferritins in iron metabolism and virulence of the plant-pathogenic bacterium *Erwinia chrysanthemi* 3937. *J Bacteriol.* 2008;190(5):1518–1530.
92. Franza T, Michaud-Soret I, Piquerel P, Expert D. Coupling of iron assimilation and pectinolysis in *Erwinia chrysanthemi* 3937. *Mol Plant Microbe Interact.* 2002;15(11): 1181–1191.
93. Masclaux C, HugouvieuxCottePattat N, Expert D. Iron is a triggering factor for differential expression of *Erwinia chrysanthemi* strain 3937 pectate lyases in pathogenesis of African violets. *Mol Plant Microbe Interact.* 1996;9(3):198–205.

94. Masclaux C, Expert D. Signalling potential of iron in plant—microbe interactions: the pathogenic switch of iron transport in *Erwinia chrysanthemi*. *Plant J.* 1995;7(1):121–128.
95. Jahn CE, Selimi DA, Barak JD, Charkowski AO. The *Dickeya dadantii* biofilm matrix consists of cellulose nanofibers and is an emergent property dependent upon the type III secretion system and the cellulose synthesis operon. *Microbiology.* 2011;157: 2733–2744.
96. Prigent-Combaret C, Zghidi-Abouzid O, Effantin G, Lejeune P, Reverchon S, Nasser W. The nucleoid-associated protein Fis directly modulates the synthesis of cellulose, an essential component of pellicle-biofilms in the phytopathogenic bacterium *Dickeya dadantii*. *Mol Microbiol.* 2012;86(1):172–186.
97. Condemine G, Castillo A, Passeri F, Enard C. The PecT repressor coregulates synthesis of exopolysaccharides and virulence factors in *Erwinia chrysanthemi*. *Mol Plant Microbe Interact.* 1999;12(1):45–52.
98. Rojas CM, Ham JH, Deng WL, Doyle JJ, Collmer A. HecA, a member of a class of adhesins produced by diverse pathogenic bacteria, contributes to the attachment, aggregation, epidermal cell killing, and virulence phenotypes of *Erwinia chrysanthemi* EC16 on *Nicotiana clevelandii* seedlings. *Proc Natl Acad Sci USA.* 2002;99(20): 13142–13147.
99. Perez-Mendoza D, Coulthurst SJ, Humphris S, et al. A multi-repeat adhesin of the phytopathogen, *Pectobacterium atrosepticum*, is secreted by a Type I pathway and is subject to complex regulation involving a non-canonical diguanylate cyclase. *Mol Microbiol.* 2011;82(3):719–733.
100. Nykyri J, Mattinen L, Niemi O, et al. Role and regulation of the Flp/Tad pilus in the virulence of *Pectobacterium atrosepticum* SCRI1043 and *Pectobacterium wasabiae* SCC3193. *PLoS One.* 2013;8(9):e73718.
101. Rossez Y, Wolfson EB, Holmes A, Gally DL, Holden NJ. Bacterial flagella: twist and stick, or dodge across the kingdoms. *PLoS Pathog.* 2015;11(1):e1004483.
102. Antunez-Lamas M, Cabrera-Ordonez E, Lopez-Solanilla E, et al. Role of motility and chemotaxis in the pathogenesis of *Dickeya dadantii* 3937 (ex *Erwinia chrysanthemi* 3937). *Microbiology.* 2009;155(Pt 2):434–442.
103. Antunez-Lamas M, Cabrera E, Lopez-Solanilla E, et al. Bacterial chemoattraction towards jasmonate plays a role in the entry of *Dickeya dadantii* through wounded tissues. *Mol Microbiol.* 2009;74:662–671.
104. Rio-Alvarez I, Munoz-Gomez C, Navas-Vasquez M, et al. Role of *Dickeya dadantii* 3937 chemoreceptors in the entry to *Arabidopsis* leaves through wounds. *Mol Plant Pathol.* 2015;16(7):685–698.
105. Jiang X, Sobetzko P, Nasser W, Reverchon S, Muskhelishvili G. Chromosomal "stress-response" domains govern the spatiotemporal expression of the bacterial virulence program. *Mbio.* 2015;6(3):e00353–00315.
106. Llama-Palacios A, Lopez-Solanilla E, Rodriguez-Palenzuela P. The *ybiT* gene of *Erwinia chrysanthemi* codes for a putative ABC transporter and is involved in competitiveness against endophytic bacteria during infection. *Appl Environ Microbiol.* 2002;68 (4):1624–1630.
107. Barabote RD, Johnson OL, Zetina E, San Francisco SK, Fralick JA, San Francisco MJ. *Erwinia chrysanthemi* tolC is involved in resistance to antimicrobial plant chemicals and is essential for phytopathogenesis. *J Bacteriol.* 2003;185(19):5772–5778.
108. Maggiorani Valecillos A, Rodriguez Palenzuela P, Lopez-Solanilla E. The role of several multidrug resistance systems in *Erwinia chrysanthemi* pathogenesis. *Mol Plant Microbe Interact.* 2006;19(6):607–613.
109. Ravirala RS, Barabote RD, Wheeler DM, et al. Efflux pump gene expression in *Erwinia chrysanthemi* is induced by exposure to phenolic acids. *Mol Plant Microbe Interact.* 2007;20(3):313–320.

110. Reverchon S, Van Gijsegem F, Effantin G, Zghidi-Abouzid O, Nasser W. Systematic targeted mutagenesis of the MarR/SlyA family members of *Dickeya dadantii* 3937 reveals a role for MfbR in the modulation of virulence gene expression in response to acidic pH. *Mol Microbiol.* 2010;78(4):1018–1037.
111. Grignon C, Sentenac H. pH and ionic conditions in the apoplast. *Annu Rev Plant Physiol Plant Mol Biol.* 1991;42:103–125.
112. Duca D, Lorv J, Patten CL, Rose D, Glick BR. Indole-3-acetic acid in plant-microbe interactions. *Antonie Van Leeuwenhoek.* 2014;106(1):85–125.
113. Kazan K, Manners JM. Linking development to defense: auxin in plant–pathogen interactions. *Trends Plant Sci.* 2009;14(7):373–382.
114. Piisila M, Keceli MA, Brader G, et al. The F-box protein MAX2 contributes to resistance to bacterial phytopathogens in *Arabidopsis thaliana. BMC Plant Biol.* 2015;15:53.
115. Llama-Palacios A, Lopez-Solanilla E, Poza-Carrion C, Garcia-Olmedo F, Rodriguez-Palenzuela P. The *Erwinia chrysanthemi phoP-phoQ* operon plays an important role in growth at low pH, virulence and bacterial survival in plant tissue. *Mol Microbiol.* 2003;49(2):347–357.
116. Llama-Palacios A, Lopez-Solanilla E, Rodriguez-Palenzuela P. Role of the PhoP-PhoQ system in the virulence of *Erwinia chrysanthemi* strain 3937: involvement in sensitivity to plant antimicrobial peptides, survival at acid pH, and regulation of pectolytic enzymes. *J Bacteriol.* 2005;187(6):2157–2162.
117. Venkatesh B, Babujee L, Liu H, et al. The *Erwinia chrysanthemi* 3937 PhoQ sensor kinase regulates several virulence determinants. *J Bacteriol.* 2006;188(8):3088–3098.
118. Haque MM, Hirata H, Tsuyumu S. SlyA regulates motA and motB, virulence and stress-related genes under conditions induced by the PhoP-PhoQ system in *Dickeya dadantii* 3937. *Res Microbiol.* 2015;166(6):467–475.
119. Bontemps-Gallo S, Madec E, Lacroix JM. The two-component system CpxAR is essential for virulence in the phytopathogen bacteria *Dickeya dadantii* EC3937. *Environ Microbiol.* 2015;17(11):4415–4428.
120. Hyytiainen H, Sjoblom S, Palomaki T, Tuikkala A, Tapio Palva E. The PmrA-PmrB two-component system responding to acidic pH and iron controls virulence in the plant pathogen *Erwinia carotovora* ssp. carotovora. *Mol Microbiol.* 2003;50(3):795–807.
121. Condemine G, Ghazi A. Differential regulation of two oligogalacturonate outer membrane channels, KdgN and KdgM, of *Dickeya dadantii* (*Erwinia chrysanthemi*). *J Bacteriol.* 2007;189(16):5955–5962.
122. Rio-Alvarez I, Rodriguez-Herva JJ, Cuartas-Lanza R, et al. Genome-wide analysis of the response of *Dickeya dadantii* 3937 to plant antimicrobial peptides. *Mol Plant Microbe Interact.* 2012;25(4):523–533.
123. Lopez-Solanilla E, Garcia-Olmedo F, Rodriguez-Palenzuela P. Inactivation of the sapA to sapF locus of *Erwinia chrysanthemi* reveals common features in plant and animal bacterial pathogenesis. *Plant Cell.* 1998;10(6):917–924.
124. Hugouvieux-Cotte-Pattat N, Dominguez H, Robert-Baudouy J. Environmental conditions affect transcription of the pectinase genes of *Erwinia chrysanthemi* 3937. *J Bacteriol.* 1992;174(23):7807–7818.
125. Babujee L, Apodaca J, Balakrishnan V, et al. Evolution of the metabolic and regulatory networks associated with oxygen availability in two phytopathogenic enterobacteria. *BMC Genomics.* 2012;13:110.
126. El Hassouni ME, Chambost JP, Expert D, Van Gijsegem F, Barras F. The minimal gene set member *msrA*, encoding peptide methionine sulfoxide reductase, is a virulence determinant of the plant pathogen *Erwinia chrysanthemi. Proc Natl Acad Sci USA.* 1999;96(3):887–892.

127. Santos R, Franza T, Laporte ML, Sauvage C, Touati D, Expert D. Essential role of superoxide dismutase on the pathogenicity of *Erwinia chrysanthemi* strain 3937. *Mol Plant Microbe Interact.* 2001;14(6):758–767.
128. Nachin L, Loiseau L, Expert D, Barras F. SufC: an unorthodox cytoplasmic ABC/ATPase required for [Fe-S] biogenesis under oxidative stress. *Embo J.* 2003;22(3):427–437.
129. Miguel E, Poza-Carrion C, Lopez-Solanilla E, et al. Evidence against a direct antimicrobial role of H_2O_2 in the infection of plants by *Erwinia chrysanthemi*. *Mol Plant Microbe Interact.* 2000;13(4):421–429.
130. Reverchon S, Rouanet C, Expert D, Nasser W. Characterization of indigoidine biosynthetic genes in *Erwinia chrysanthemi* and role of this blue pigment in pathogenicity. *J Bacteriol.* 2002;184(3):654–665.
131. Lebeau A, Reverchon S, Gaubert S, et al. The GacA global regulator is required for the appropriate expression of *Erwinia chrysanthemi* 3937 pathogenicity genes during plant infection. *Environ Microbiol.* 2008;10(3):545–559.
132. Gloux K, Touze T, Pagot Y, Jouan B, Blanco C. Mutations of *ousA* alter the virulence of *Erwinia chrysanthemi*. *Mol Plant Microbe Interact.* 2005;18(2):150–157.
133. Gouesbet G, Jebbar M, Bonnassie S, Hugouvieux-Cotte-Pattat N, Himdi-Kabbab S, Blanco C. *Erwinia chrysanthemi* at high osmolarity: influence of osmoprotectants on growth and pectate lyase production. *Microbiology.* 1995;141:1407–1412.
134. Gouesbet G, Trautwetter A, Bonnassie S, Wu LF, Blanco C. Characterization of the *Erwinia chrysanthemi* osmoprotectant transporter gene ousA. *J Bacteriol.* 1996;178 (2):447–455.
135. Choquet G, Jehan N, Pissavin C, Blanco C, Jebbar M. OusB, a broad-specificity ABC-type transporter from Erwinia chrysanthemi, mediates uptake of glycine betaine and choline with a high affinity. *Appl Environ Microbiol.* 2005;71(7):3389–3398.
136. Valente R, Xavier K. The Trk potassium transporter is required for RsmB-mediated activation of virulence in the phytopathogen *Pectobacterium wasabiae*. *J Bacteriol.* 2015;198(2):248–255.
137. Goude R, Renaud S, Bonnassie S, Bernard T, Blanco C. Glutamine, glutamate, and alpha-glucosylglycerate are the major osmotic solutes accumulated by *Erwinia chrysanthemi* strain 3937. *Appl Environ Microbiol.* 2004;70(11):6535–6541.
138. Bontemps-Gallo S, Madec E, Dondeyne J, et al. Concentration of osmoregulated periplasmic glucans (OPGs) modulates the activation level of the RcsCD RcsB phosphorelay in the phytopathogen bacteria *Dickeya dadantii*. *Environ Microbiol.* 2013;15(3): 881–894.
139. Madec E, Bontemps-Gallo S, Lacroix JM. Increased phosphorylation of the RcsB regulator of the RcsCDB phosphorelay in strains of *Dickeya dadantii* devoid of osmoregulated periplasmic glucans revealed by Phos-tag gel analysis. *Microbiology.* 2014;160 (Pt 12):2763–2770.
140. Browning DF, Busby SJ. The regulation of bacterial transcription initiation. *Nat Rev Microbiol.* 2004;2(1):57–65.
141. Dillon SC, Dorman CJ. Bacterial nucleoid-associated proteins, nucleoid structure and gene expression. *Nat Rev Microbiol.* 2010;8(3):185–195.
142. Reverchon S, Nasser W, Robert-Baudouy J. pecS: a locus controlling pectinase, cellulase and blue pigment production in *Erwinia chrysanthemi*. *Mol Microbiol.* 1994;11 (6):1127–1139.
143. Nasser W, Reverchon S, Vedel R, Boccara M. PecS and PecT coregulate the synthesis of HrpN and pectate lyases, two virulence determinants in *Erwinia chrysanthemi* 3937. *Mol Plant Microbe Interact.* 2005;18(11):1205–1214.
144. Perera IC, Grove A. Urate is a ligand for the transcriptional regulator PecS. *J Mol Biol.* 2010;402(3):539–551.

145. Mhedbi-Hajri N, Malfatti P, Pedron J, Gaubert S, Reverchon S, Van Gijsegem F. PecS is an important player in the regulatory network governing the coordinated expression of virulence genes during the interaction between *Dickeya dadantii* 3937 and plants. *Environ Microbiol.* 2011;13(11):2901–2914.
146. Rouanet C, Reverchon S, Rodionov DA, Nasser W. Definition of a consensus DNA-binding site for PecS, a global regulator of virulence gene expression in *Erwinia chrysanthemi* and identification of new members of the PecS regulon. *J Biol Chem.* 2004;279(29):30158–30167.
147. Costechareyre D, Dridi B, Rahbe Y, Condemine G. Cyt toxin expression reveals an inverse regulation of insect and plant virulence factors of *Dickeya dadantii. Environ Microbiol.* 2010;12(12):3290–3301.
148. Costechareyre D, Chich JF, Strub JM, Rahbe Y, Condemine G. Transcriptome of *Dickeyadadantii* infecting *Acyrthosiphonpisum* reveals a strong defense against antimicrobial peptides. *PloS one.* 2013;8(1):e54118.
149. Surgey N, Robert-Baudouy J, Condemine G. The *Erwinia chrysanthemi pecT* gene regulates pectinase gene expression. *J Bacteriol.* 1996;178(6):1593–1599.
150. Castillo A, Nasser W, Condemine G, Reverchon S. The PecT repressor interacts with regulatory regions of pectate lyase genes in *Erwinia chrysanthemi. Biochim Biophys Acta.* 1998;1442(2–3):148–160.
151. Hérault E, Reverchon S, Nasser W. Role of the LysR-type transcriptional regulator PecT and DNA supercoiling in the thermoregulation of pel genes, the major virulence factors in *Dickeya dadantii. Environ Microbiol.* 2014;16(3):734–745.
152. Potrykus M, Golanowska M, Hugouvieux-Cotte-Pattat N, Lojkowska E. Regulators involved in Dickeya solani virulence, genetic conservation, and functional variability. *Mol Plant Microbe Interact.* 2014;27(7):700–711.
153. Yang S, Peng Q, Zhang Q, et al. Dynamic regulation of GacA in type III secretion, pectinase gene expression, pellicle formation, and pathogenicity of *Dickeya dadantii* (*Erwinia chrysanthemi* 3937). *Mol Plant Microbe Interact.* 2008;21(1):133–142.
154. Wu X, Zeng Q, Koestler BJ, et al. Deciphering the components that coordinately regulate virulence factors of the soft rot pathogen *Dickeya dadantii. Mol Plant Microbe Interact.* 2014;27(10):1119–1131.
155. Pescaretti Mde L, Farizano JV, Morero R, Delgado MA. A novel insight on signal transduction mechanism of RcsCDB system in *Salmonellaenterica* serovar typhimurium. *PLoS One.* 2013;8(9):e72527.
156. Yap MN, Yang CH, Barak JD, Jahn CE, Charkowski AO. The *Erwinia chrysanthemi* type III secretion system is required for multicellular behavior. *J Bacteriol.* 2005;187(2):639–648.
157. Li Y, Peng Q, Selimi D, et al. The plant phenolic compound p-coumaric acid represses gene expression in the *Dickeyadadantii* type III secretion system. *Appl Environ Microbiol.* 2009;75(5):1223–1228.
158. Yang S, Peng Q, San Francisco M, Wang Y, Zeng Q, Yang CH. Type III secretion system genes of *Dickeya dadantii* 3937 are induced by plant phenolic acids. *PLoS One.* 2008;3(8):e2973.
159. Yi X, Yamazaki A, Biddle E, Zeng Q, Yang CH. Genetic analysis of two phosphodiesterases reveals cyclic diguanylate regulation of virulence factors in *Dickeyadadantii. Mol Microbiol.* 2010;77(3):787–800.
160. Leng Y, Vakulskas C, Zere T, et al. Regulation of CsrB/C sRNA decay by EIIAGlc of the phosphoenolpyruvate: carbohydrate phosphotransferase system. *Mol Microbiol.* 2015;99(4):627–639.
161. Yuan X, Khokhani D, Wu X, et al. Cross-talk between a regulatory small RNA, cyclic-di-GMP signalling and flagellar regulator FlhDC for virulence and bacterial behaviours. *Environ Microbiol.* 2015;17(11):4745–4763.

162. Yap MN, Yang CH, Charkowski AO. The response regulator HrpY of *Dickeya dadantii* 3937 regulates virulence genes not linked to the hrp cluster. *Mol Plant Microbe Interact.* 2008;21(3):304–314.
163. Yang S, Peng Q, Zhang Q, et al. Genome-wide identification of HrpL-regulated genes in the necrotrophic phytopathogen *Dickeya dadantii* 3937. *PLoS One.* 2010;5 (10):e13472.
164. Nasser W, Bouillant ML, Salmond G, Reverchon S. Characterization of the *Erwinia chrysanthemi* expI-expR locus directing the synthesis of two *N*-acyl-homoserine lactone signal molecules. *Mol Microbiol.* 1998;29(6):1391–1405.
165. Crepin A, Barbey C, Beury-Cirou A, et al. Quorum sensing signaling molecules produced by reference and emerging soft-rot bacteria (Dickeya and Pectobacterium spp.). *PloS One.* 2012;7(4):e35176.
166. Cameron AD, Dorman CJ. A fundamental regulatory mechanism operating through OmpR and DNA topology controls expression of *Salmonella* pathogenicity islands SPI-1 and SPI-2. *PLoS Genet.* 2012;8(3):e1002615.
167. Marr C, Geertz M, Hutt MT, Muskhelishvili G. Dissecting the logical types of network control in gene expression profiles. *BMC Syst Biol.* 2008;2:18.
168. Kepseu WD, Sepulchre JA, Reverchon S, Nasser W. Toward a quantitative modeling of the synthesis of the pectate lyases, essential virulence factors in *Dickeya dadantii*. *J Biol Chem.* 2010;285(37):28565–28576.
169. Nasser W, Faelen M, Hugouvieux-Cotte-Pattat N, Reverchon S. Role of the nucleoid-associated protein H-NS in the synthesis of virulence factors in the phytopathogenic bacterium *Erwinia chrysanthemi*. *Mol Plant Microbe Interact.* 2001;14(1):10–20.
170. Nasser W, Reverchon S. H-NS-dependent activation of pectate lyases synthesis in the phytopathogenic bacterium *Erwinia chrysanthemi* is mediated by the PecT repressor. *Mol Microbiol.* 2002;43(3):733–748.
171. Lautier T, Nasser W. The DNA nucleoid-associated protein Fis co-ordinates the expression of the main virulence genes in the phytopathogenic bacterium *Erwinia chrysanthemi*. *Mol Microbiol.* 2007;66(6):1474–1490.
172. Lautier T, Blot N, Muskhelishvili G, Nasser W. Integration of two essential virulence modulating signals at the *Erwinia chrysanthemi pel* gene promoters: a role for Fis in the growth-phase regulation. *Mol Microbiol.* 2007;66(6):1491–1505.
173. Ouafa ZA, Reverchon S, Lautier T, Muskhelishvili G, Nasser W. The nucleoid-associated proteins H-NS and FIS modulate the DNA supercoiling response of the pel genes, the major virulence factors in the plant pathogen bacterium *Dickeya dadantii*. *Nucleic Acids Res.* 2012;40(10):4306–4319.
174. Giangrossi M, Wintraecken K, Spurio R, de Vries R. Probing the relation between protein–protein interactions and DNA binding for a linker mutant of the bacterial nucleoid protein H-NS. *Biochim Biophys Acta.* 2014;1844(2):339–345.
175. Vreede J, Dame RT. Predicting the effect of ions on the conformation of the H-NS dimerization domain. *Biophys J.* 2012;103(1):89–98.
176. Hacker J, Kaper JB. Pathogenicity islands and the evolution of microbes. *Annu Rev Microbiol.* 2000;54:641–679.
177. Sobetzko P, Glinkowska M, Travers A, Muskhelishvili G. DNA thermodynamic stability and supercoil dynamics determine the gene expression program during the bacterial growth cycle. *Mol Biosyst.* 2013;9(7):1643–1651.

CHAPTER FOUR

Metabolism and Virulence Strategies in *Dickeya*–Host Interactions

N. Hugouvieux-Cotte-Pattat*
Microbiology Adaptation and Pathogenesis, CNRS, University of Lyon, University Claude Bernard Lyon 1, INSA Lyon, Villeurbanne, France
* Corresponding author. E-mail address: nicole.cotte-pattat@insa-lyon.fr

Contents

Abstract

Dickeya, a genus of the *Enterobacteriaceae* family, all cause plant diseases. They are aggressive necrotrophs that have both a wide geographic distribution and a wide host range. As a plant pathogen, *Dickeya* has had to adapt to a vegetarian diet.

Progress in Molecular Biology and Translational Science, Volume 142
ISSN 1877-1173
http://dx.doi.org/10.1016/bs.pmbts.2016.05.006

Plants constitute a large storage of carbohydrates; they contain substantial amounts of soluble sugars and the plant cell wall is composed of long polysaccharides. Metabolic functions used by *Dickeya* in order to multiply during infection are essential aspects of pathogenesis. *Dickeya* is able to catabolize a large range of oligosaccharides and glycosides of plant origin. Glucose, fructose, and sucrose are all efficiently metabolized by the bacteria. To avoid the formation of acidic products, their final catabolism involves the butanediol pathway, a nonacidifying fermentative pathway. The assimilation of plant polysaccharides necessitates their prior cleavage into oligomers. Notably, the *Dickeya* virulence strategy is based on its capacity to dissociate the plant cell wall and, for this, the bacteria secrete an extensive set of polysaccharide degrading enzymes, composed mostly of pectinases. Since pectic polymers have a major role in plant tissue cohesion, pectinase action results in plant rot. The pectate lyases secreted by *Dickeya* play a double role as virulence factors and as nutrient providers. This dual function implies that the *pel* gene expression is regulated by both metabolic and virulence regulators. The control of sugar assimilation by specific or global regulators enables *Dickeya* to link its nutritional status to virulence, a coupling that optimizes the different phases of infection.

1. INTRODUCTION

For microorganisms, the evolution toward pathogenesis was probably driven mostly by the necessity of finding abundant nutrient resources. Given the abundance of sugars and polysaccharides present in plants, it is not surprising that some bacteria adapted a plant-pathogenic lifestyle to take advantage of this carbon-rich material.

Pectinolytic enterobacteria of the genus *Dickeya* and *Pectobacterium* are the causative agents of the economically significant soft-rot disease[1,2] (see also Chapter 3, by Reverchon et al.). They are omnivorous pathogens that cause disease on hundreds of plants belonging to different families. They are well-studied examples of necrotrophic plant pathogens, present in the Top 10 of phytopathogenic bacteria.[3] The taxonomic classification of these phytopathogens has been revised several times. Initially, they were included in the genus *Erwinia* which comprised all the plant pathogenic enterobacteria. Pectinolytic enterobacteria (*E. chrysanthemi* and *E. carotovora*) were then separated to form the genus *Pectobacterium*.[4] The species *P. chrysanthemi* was subsequently suppressed to create the *Dickeya* genus[5] that, after some refining, now includes seven characterized species: *D. aquatica*,[6] *D. chrysanthemi*, *D. dadantii*, *D. dianthicola*, *D. paradisiaca*, *D. solani*,[7] and *D. zeae*. The model strain 3937 of the species *D. dadantii* has been studied extensively.[8–11]

Since *Dickeya* species are able to survive and multiply in a variety of environmental niches (plants, soil, water, and insects), they have evolved regulatory mechanisms to adapt to changes in nutrient availability. During plant infection, the bacteria need to acquire energy and material from the compounds found in the host and the plant tissue provides an excellent source of nutrients. Simple sugars and short oligosaccharides can be readily absorbed whereas polysaccharides from plant cell walls have to be predigested by extracellular enzymes. During the consecutive phases of infection, bacteria have to adapt their metabolism to the successive assimilation of the different carbon sources available in the plant hosts. Moreover, the secreted virulence factors, particularly extracellular enzymes, function both as virulence factors and nutrient providers. Thus, bacterial pathogenesis implies a coordinated evolution of metabolic and virulence factors. Knowledge of bacterial metabolism within the host is essential for understanding plant–pathogen interactions.

This review will focus on the virulence and metabolic strategies that contribute to plant infection. Two fundamental aspects will be detailed: (1) the nutritional strategies used by the plant pathogenic *Dickeya* and (2) the molecular mechanisms dedicated to coordinating metabolic processes with virulence factors.

2. THE *DICKEYA* VIRULENCE STRATEGY

Plant pathogens are categorized as biotrophs or necrotrophs, based on their nutrition strategy.[12] Biotrophs derive nutrients from living cells and, therefore, have established elaborate infection strategies to maintain host viability. Necrotrophs promote the destruction of host cells in order to feed on their content and so have developed efficient tools that rapidly destroy host tissues.

Dickeya species are aggressive necrotrophs that cause major pre- and postharvest decay in numerous agronomic and horticultural crops. They use a "brute-force" strategy to actively destroy plant host tissue; provoking soft-rot disease.[13] Their weapons are a set of secreted endo-pectate lyases, used as primary virulence determinants. Indeed, purified pectate lyases are able to reproduce in vitro the symptom of maceration observed in the soft-rot diseases.[8,14] These enzymes act by destroying the integrity of the middle lamella and the plant cell wall.[14] Since pectin provides cell wall stability, its rapid degradation irremediably damages the plant tissue, leading to cell wall disorganization, cell lysis and liberation of the cell content. The ubiquity of pectin among all plant families gives these bacteria the capacity to affect a broad-host range of plants.

Thus, *Dickeya* species have a destructive pathogenesis strategy resulting in extensive tissue maceration and plant rot. The success of this pathogen is due to the secretion of an extensive array of enzymes, which efficiently degrade pectin, giving access to other plant components. Despite their potential aggressiveness, *Dickeya* species are considered as facultative saprophytes that cycle between saprophytic and virulent lifestyles. An arsenal of regulatory systems enables them to adapt to changes in lifestyle and to the different phases of infection.[8,11]

2.1 Plant Colonization and the Infectious Phases

During their saprophytic life, *Dickeya* species must survive in nutrient-poor and stressful niches such as soil or water. In these conditions, bacterial chemotaxis and motility play an essential role in the retrieval of nutrients and sensing of compounds from plant origin. The initial establishment of the bacteria at the plant surface corresponds to host colonization (Fig. 1). The bacterial population can reside as epiphytes on leaf or root surfaces (the phyllosphere and the rhizosphere). At this stage, bacteria can remain latent until conditions become favorable for the development of disease.

The pathogenic cycle of *D. dadantii* is a dynamic process usually divided into two main phases (Fig. 1) (see also chapter 7, by Reverchon et al.). Bacteria penetrate their host using either wounds or natural openings.[15] During the first few hours, the bacteria begin to slowly multiply but they reside latently in the plant intercellular space without provoking any symptoms.[1,2,10,11,16] During this "asymptomatic" phase, a controlled repression of aggressive factors allows the bacteria to multiply without being recognized by their host. The bacteria repress their main virulence factors but they begin to assimilate the easily metabolizable plant soluble sugars.[17] This asymptomatic phase shows substantial similarities with a short biotrophic lifestyle. It is followed by a transition stage corresponding to a strong metabolic change with an induction of pectate lyase production. The "symptomatic" phase corresponds to the expansion of soft rot due to a burst of extracellular enzymes. Bacteria fully induce their virulence factors and quickly multiply to large population sizes by obtaining more nutrients through plant tissue degradation (Fig. 1). The whole infection process is rapid, giving the plant little time to develop its defense responses. For instance, after experimental

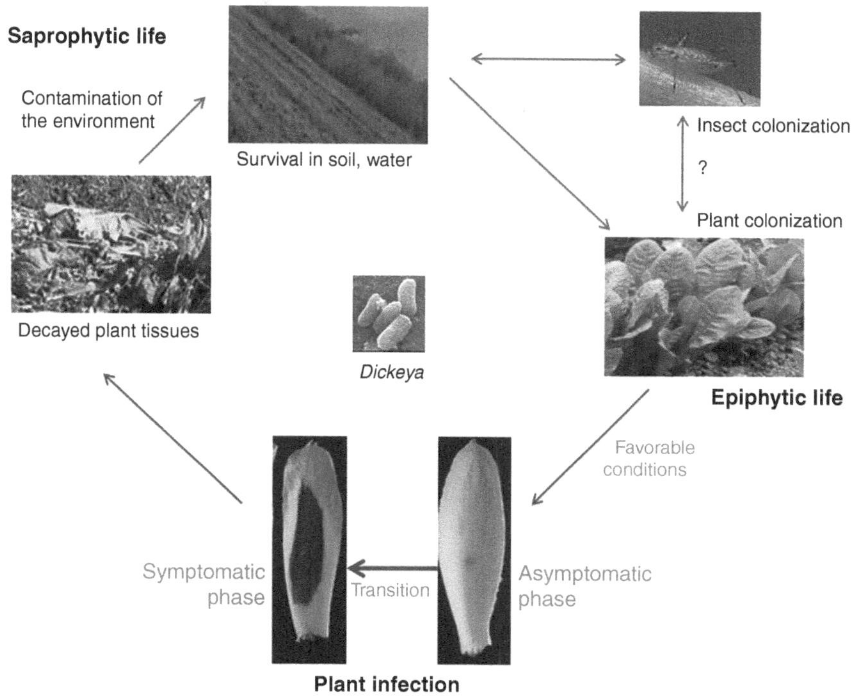

Figure 1 Different niches and lifestyles of *Dickeya*. *Dickeya* species are considered as facultative saprophytes that cycle between saprophytic and virulent lifestyles. They are found in a variety of environmental niches (plants, soil, water, and insects). Plant colonization corresponds to the initial establishment of the bacteria at the surface of leaves, stems, or roots. *D. dadantii* can also infect insects and it is able to kill the pea aphid *Acyrtosiphon pisum*. The role of insects in bacterial transmission is poorly documented. Bacteria reside as epiphytes on plant surfaces where they remain latent until conditions become favorable for the disease development. The phytopathogenic cycle is usually divided into two main phases. During the first few hours, the bacteria begin to slowly multiply without provoking any symptoms. This "asymptomatic" phase is followed by a transition stage corresponding to a strong metabolic change with an induction of pectate lyase production. The "symptomatic" phase corresponds to the expansion of soft rot due to the pectate lyase action. After experimental infection of chicory by *D. dadantii* 3937, the total plant maceration is observed in less than 48 h (photos: 4 h and 24 h postinfection).

infection of chicory by *D. dadantii* 3937, the duration of the asymptomatic and the symptomatic phase were estimated to be 8 h and 12 h, respectively, with total plant maceration observed in less than 48 h.[17]

Although the plant defense mechanisms are not sufficient to avoid disease, some host reactions seem to be involved in the pathogenic cycle of *D. dadantii*. For example, there is evidence of the generation of reactive oxygen species during infection.[18–20] This reaction is the result of bacterial or plant elicitors produced during infection. The primary target of *Dickeya* virulence is pectin, but the oligogalacturonides resulting from pectin cleavage are able to function as defense elicitors.[21] Such elicitation of plant defense systems is highly dependent on the degree of oligogalacturonide polymerization, since the large oligomers generated by partial pectin degradation are more efficientat eliciting plant defenses than the small final products (di- and trigalacturonides). The activity of accessory pectinases can also release signaling molecules. Pectin acetyl- and methylesterases liberate acetate and methanol, respectively, during plant attack. Acetate is also secreted as an end product of bacterial metabolism and it is suspected of being a signal affecting virulence via the GacA/GacS system (see later). Methanol is an effector recognized by plants, after the penetration of a pathogen, and released either by plant or by pathogen enzymes. This volatile signal transmits information to the intact leaves of the infected plant and also to neighboring plants.[22,23]

In addition to these plant damage-associated molecular patterns (DAMPs), *Dickeya* secretes end products of sugar metabolism that are pathogen-associated molecular patterns (PAMPs). Plants can perceive butanediol and acetoin, two compounds that are largely produced during host infection by plant pathogenic enterobacteria,[17] and these volatile molecules are able to induce plant systemic resistance.[24] They play a significant role in plant defense against pathogens by activating the ethylene signaling pathway, which is effective against necrotrophs.[25] Butanediol and acetoin have been shown to be efficient in reducing disease severity caused by *P. carotovorum*.[24,26] While pectate lyase activity is directly responsible for plant maceration, the butanediol pathway is necessary for obtaining the alkaline pH favorable for pectate lyase activity.[17] Thus, bacteria release both oligosaccharides and volatile molecules able to induce plant defenses. However, the production of these elicitors is repressed during the asymptomatic phase of infection, and delayed until the symptomatic phase. Pectinase action and the butanediol pathway are essential for the development of pathogenesis, but the correct timing of oligogalacturonide, methanol, and butanediol production is crucial. The bacteria have to avoid a too rapid and a too strong plant response; fine control of oligogalacturonide and volatile compound production allows the pathogen to limit or adjust the host immune response accordingly.

2.2 Pectate Lyases and Plant Cell Wall Degradation

Dickeya use a very simple and efficient phytopathogenic mechanism that involves attacking the plant cell wall with a battery of secreted enzymes. The soft-rot symptoms are caused by the cumulative action of pectinases (especially pectate lyases) that destroy pectin, the plant cell wall adhesive material.[14] Additional plant cell wall degrading enzymes, such as cellulases, xylanases, and proteases, are also secreted by the bacteria and can complement the pectinase activity for the breakdown of the cell wall components, even though they are not essential for pathogenesis. Since pectate lyases are the main determinant of the symptom of maceration caused by *D. dadantii*, the pectinolytic arsenal of *D. dadantii* is one of the most thoroughly characterized[14] (Table 1).

The plant cell wall is an important barrier against pathogen attack. The composition and structure of the cell wall differ significantly among plant lineages. Nevertheless, they share some basic principles; principally, all cell walls comprise cellulose microfibrils embedded in a matrix of pectin and hemicellulose. Cellulose is composed of long linear glucose chains that are packed together to form microfibrils. These fibrils are noncovalently linked to hemicelluloses, a general term encompassing different neutral branched polysaccharides containing main chains of D-xylose (xylan), D-mannose (mannan), or D-glucose (xyloglucan) and side chains composed of D-galactose, D-xylose, L-arabinose and D-glucuronate.[50] The cellulose/hemicellulose network is embedded in a jelly-like matrix of acidic polysaccharides, commonly referred as "pectins." Pectins are composed of three types of polysaccharide, homogalacturonan (or polygalacturonate, PGA), rhamnogalacturonan I (RGI), and rhamnogalacturonan II (RGII).[51] PGA is the most abundant pectic polymer; it has a linear structure composed of long chains of D-galacturonate residues. The carboxylic groups of D-galacturonate are methyl-esterified to various degrees, up to 70–80%, and these residues are, to a lesser extent, acetylated. In contrast to PGA, RGI, and RGII are highly branched polymers that form "hairy" regions. The RGI backbone consists of alternating L-rhamnose and D-galacturonate residues; it is substituted, to a high degree, with variable side chains of galactan and arabinan. RGII contains a short galacturonan backbone carrying four well-defined side chains composed of a series of conserved glycosyl residues, including very rare monosaccharides.

The pectic polysaccharides represent between 30% and 50% of the cell walls of dicotyledonous plants. They are the major constituents of the middle

Table 1 Polysaccharide Active Enzymes Encoded by the *D. dadantii* 3937 Genome.

Enzyme Family	Name/ID Number	Activity	Cell Compartment	References
Pectinases and Related Proteins				
PSL1	PelA	Endopectate lyase	Extracellular (Out)	[27]
	PelB	Endopectate lyase	Extracellular (Out)	[27]
	PelC	Endopectate lyase	Extracellular (Out)	[27]
	PelD	Endopectate lyase	Extracellular (Out)	[27]
	PelE	Endopectate lyase	Extracellular (Out)	[27]
	PelZ	Endopectate lyase	Extracellular (Out)	[28]
	PnlH	*Pectin lyase*	Outer membrane (Stt)	[102]
	ID19932	*Pectin lyase*		
PSL2	PelW	Exopectate lyase	Cytoplasm	[30]
PSL3	PelI	Endopectate lyase	Extracellular (Out)	[71]
	HrpW	*Harpin, pectate-binding*		[32]
PSL4	RhiE	Rhamnogalacturonate lyase	Extracellular (Out)	[103]
PSL9	PelL	Endopectate lyase	Extracellular (Out)	[34]
	PelN	Endopectate lyase	Extracellular (Out)	[35]
	PelX	Exopectate lyase	Periplasm	[36]
PSL22	Ogl	Oligogalacturonate lyase	Cytoplasm	[30]
CE8	PemA	Pectin methyl esterase	Extracellular (Out)	[37]
	PemB	Pectin methyl esterase	Outer membrane	[38]
CE10	PaeX	Pectin acetyl esterase	Periplasm	[39]
	FaeD	Feruloyl esterase	Extracellular (Out)	[40]
	FaeT	Feruloyl esterase	Cytoplasm	[40]

(Continued)

Table 1 Polysaccharide Active Enzymes Encoded by the *D. dadantii* 3937 Genome.—cont'd

Enzyme Family	Name/ID Number	Activity	Cell Compartment	References
CE12	PaeY	Pectin acetyl esterase	Extracellular (Out)	[41]
GH28	PehV	Exopolygalacturonase	Periplasm	[29]
	PehW	Exopolygalacturonase	Periplasm	[29]
	PehX	Exopolygalacturonase	Periplasm	[29]
	PehN	Endopolygalacturonase	Extracellular (Out)	[32]
	PehK	*Polygalacturonase*		
GH105	RhiN	*Rhamnogalacturonyl hydrolase*	Cytoplasm	[43]
	LfgN			[42]
Other Glycosidases				
GH1	ArbB	Phospho β-glucosidase	Cytoplasm	[44]
	BglA	*Phospho β-glucosidase*	Cytoplasm	
	ID18471	*Phospho β-glucosidase*	Cytoplasm	
	ID18484	*Phospho β-glucosidase*	Cytoplasm	
	ID18485	*Phospho β-glucosidase*	Cytoplasm	
GH2	LacZ	β-Galactosidase	Cytoplasm	[45]
GH3	BgxA	β-Glucosidase	Periplasm	[46]
	BglX	*β-Glucosidase*	*Periplasm*	
	NagZ	*Exo-N-acetyl-β-glucosaminidase*	*Cytoplasm*	
GH5	Cel5Z	Endoglucanase	Extracellular (Out)	[47]
GH8	CelY	Endoglucanase	Periplasm	[47]
GH13	GlgB	*Glycogen branching enzyme*	*Cytoplasm*	
	GlgX	*Glycogen debranching enzyme*	*Cytoplasm*	
GH19	ID00553			
	ID03139			
	ID03232			
GH26	ID17828			
GH30	XynA	Glucuronoxylanase	Extracellular	[48]
GH31	LfaA			[42]
GH32	ScrB	Sucrose hydrolase/ invertase	Cytoplasm	[49]

(Continued)

Table 1 Polysaccharide Active Enzymes Encoded by the *D. dadantii* 3937 Genome.—cont'd

Enzyme Family	Name/ID Number	Activity	Cell Compartment	References
GH33	ID02530 ID02632			
GH36	RafA	α-Galactosidase	Cytoplasm	[49]
GH42	GanB	Exo-galactanase/ β-galactosidase	Cytoplasm	[45]
GH43	ID16411			
GH53	GanA	Endo-galactanase	Periplasm	[45]

PSL, polysaccharide lyase; CE, carbohydrate esterase; GH, glycosyl hydrolase. The enzymes of phage origin (GH23 and GH24) and those involved in peptigoglycane synthesis (GH73, GH77, GH102, GH103, and GH104) are not listed (classification http://www.cazy.org/).
Protein names shown in italics have a potential function, predicted from homology data but not experimentally validated in *D. dadantii*. (Out) indicates that the Out system is involved in the secretion of this protein.

lamella, a gelatinous adhesive layer that is often described as the glue preserving the plant tissue stability. Given its abundance and structural importance in plant integrity, pectin is a main target for many plant pathogens. The complexity of pectin is mirrored by the diversity of enzymes produced by microorganisms. The term "pectinases" covers different enzymes cleaving the glycosidic linkages or the esterification of the pectic polymers. Enzymes that cleave glycosidic bonds are classified as glycoside hydrolases or polysaccharide lyases, depending on their mode of action, hydrolysis or β-elimination, respectively[52] (Table 1). Of the three types of pectic polymers, PGA is the preferred target of the microbial enzymes. Pectate lyases cleave the glycosidic bonds between two D-galacturonate residues of the pectin linear regions, PGA or moderately methylesterified PGA. They release unsaturated oligogalacturonates, presenting an unsaturation in the residue situated at the nonreducing end.

D. dadantii secretes nine endo-pectate lyases cleaving randomly the PGA main chains. They belong to different protein families (Table 1) and present some variations in their optimal activities.[14] The reason for this multiplicity is a debatable issue. The predominant hypothesis is that multiplicity contributes to more efficient substrate degradation as a result of synergistic activities and/or contrasting affinities to slightly different substrate forms.[14] *D. dadantii* exports three complementary depolymerases: a secreted rhamnogalacturonate lyase cleaving the RGI backbone,[33] a periplasmic exo-pectate lyase cleaving from the reducing end of the substrate, and three periplasmic exo-polygalacturonases that generate (saturated) digalacturonates, by hydrolytic cleavage from the nonreducing end.[36] The complete degradation of pectin requires accessory esterases that remove the substitutions. *D. dadantii* produces three types of esterases: two methyl-esterases,[37,38] two acetyl-esterases[39,41] and two feruloyl-esterases.[40] By removing the pectin esterifications, these esterases render the substrate more suitable for the action of pectate lyases.

A protein secretion system is required to export the extracellular enzymes. In *D. dadantii*, the endo-pectate lyases are secreted in the extracellular medium by a type II secretion system (T2SS) named Out. T2SSs are large transmembrane protein complexes associated with both the inner and the outer membrane.[53,54] The Out system mediates the secretion of at least 15 proteins: 9 endo-pectate lyases (PelA, PelB, PelC, PelD, PelE, PelI, PelL, PelN, and PelZ), 3 esterases (PemA, PaeY, and FaeD), a rhamnogalacturonate lyase (RhiE), a cellulase (CelZ), and a protein of unknown function (AvrL).[55] These proteins are translocated to the periplasm by the classical Sec pathway. After their final folding in the periplasm, the Out system enables the enzymes to traverse the outer membrane and to reach the external medium.

Despite frequent statements that *Dickeya* virulence is due to the secretion of plant cell wall degrading enzymes, enzymes other than the pectinases have, apparently, no direct role in pathogenesis. The degradation of cellulose, the major plant cell wall polysaccharide, is not necessary for *D. dadantii* virulence.[47] Similarly, the glucuronoxylanase of *D. dadantii* strain D1 and the galactanase of *D. dadantii* 3937 are not involved in virulence.[45,48] Thus, pectic degrading enzymes, essentially pectate lyases, are the sole cell wall degrading enzymes indispensable for the *D. dadantii* virulence and they are sufficient to provoke tissue maceration. The degradation products of cellulose, xylan, and galactan may provide additional nutrients at the end of the infection process. The capacity to degrade and to assimilate a wide range of plant polysaccharides may also favor the saprophytic life of the bacteria, during which they can assimilate any plant remains present in the soil.

3. THE *DICKEYA* CATABOLIC CAPACITIES

The identification of the bacterial metabolic pathways active during infection is important in order to understand bacterial pathogenesis. Functional studies have given only partial information on the metabolic pathways, which *Dickeya* uses in order to grow in plant tissues. The identification of these pathways by classic mutagenesis approaches is complicated by the variety of carbon sources available in the plant host, at very diverse concentrations. The contribution of any one carbon source can be relatively small, and inactivation of only one of these pathways may not clearly affect the *in planta* bacterial growth.

Genomic and postgenomic approaches were expected to give more accurate data to identify the catabolic pathways expressed during plant infection. However, transcriptomic, proteomic, and metabolomics studies are sometimes difficult to conduct during plant infection, especially considering the particular state of macerated tissues. Moreover, only a few of the catabolic functions identified in such studies have been investigated further.

3.1 What Information From the Genome?

For efficient host colonization, pathogens have to adapt their metabolism to the host nutritional composition. The completion of the *D. dadantii* genome provided a genome-scale view into its potential metabolic capacities.[9] A substantial part of this genome is predicted to encode transporters and enzymes involved in carbohydrate catabolism. This is consistent with the diversity of carbon sources utilized by these bacteria (Table 2). Only part of the potential catabolic pathways probably contributes to the pathogenic lifestyle. However, metabolic diversity could provide a competitive growth advantage within the large soil microbiome.

Outer-membrane porins serve to facilitate diffusion of large molecules such as oligosacharides into the periplasm. Then, different types of inner-membrane transporters allow bacteria to effectively accumulate nutrients into their cytoplasm. Three mechanisms (ABC, PTS, and MSF) are very common in sugar uptake. The ATP-binding cassette (ABC) transporters involve protein transmembrane complexes. The sugar is captured by a periplasmic binding protein, which passes it to the cognate membrane spanning protein. The energy for the sugar uptake is released by ATP hydrolysis catalyzed by the associated ATPase. The phosphoenolpyruvate (PEP)-dependent phosphotransferase system (PTS) transporters couples sugar phosphorylation with the concomitant sugar entry. They consist of protein complexes including the general energy-coupling proteins (Enzyme I and HPr) and a sugar-specific permease (Enzymes IIA, IIB, and IIC). Energy from PEP is transferred via a phosphorylation cascade by which the sugar is ultimately phosphorylated as it enters the cytoplasm. The secondary

Table 2 Genes Involved in Sugar Catabolic Pathways in *D. dadantii* 3937.

Carbon Source	Genes Involved in the Pathway: Transport(s)	Enzyme(s)	Regulator(s)	Differences in *E. coli*
Monosaccharides				
Arabinose	*araFGH*	*araBA araD*	*araC*	+ *araE* transport
Fructose	*fruA fruB*	*fruK*	*fruR*	None
Galactose	*mglBAC*	*galETKM*	*galR galS*	+ *galP* transport
Galactarate/glucarate	*gudP*	*garD garLRK*		+ *garP* transport
		gurD	*cdaR*	Genes in two loci
Galacturonate	*exuT*	*uxaCBA*	*exuR*	Genes in two loci
Gluconate	*gntT*	*gntK*	*gntR*	+ *gntP, gntU* and *idnT* transports
Glucose	*ptsG crr*	(Glycolysis)		None
Glucose-6-phosphate	*uhpT*		*uhpABC*	None
KDG	*kdgT*	*kdgK kdgA*	*kdgR*	None
Mannitol	*mtlA*	*mtlD*	*mtlR*	None
Mannose	*manXYZ*	*manA*	—	None
N-acetylglucosamine	*nagE*	*nagA nagB*	*nagC*	None
Rhamnose	—	*rhaBAD fucO*	*rhaSR*	+ *rhaT* transport
Ribose	*rbsDACB*	*rbsK*	*rbsR*	None
Xylose	*xylFGH*	*xylAB*	*xylR*	+ *xylE* transport
Glycosides and Oligosaccharides				
Arbutin/salicin[44]	*arbF*	*arbB*	*arbG*	Cryptic operon
Galactans[45]	*ganL ganEFGK*	*ganA ganB*	*ganR*	Absent
Galacturonides[14,64–67]				
	kdgM kdgN	*ogl kduI kduD*	*kdgR*	Absent
	togT togMNAB			

(*Continued*)

Table 2 Genes Involved in Sugar Catabolic Pathways in *D. dadantii* 3937.—cont'd

Carbon Source	Genes Involved in the Pathway: Transport(s)	Enzyme(s)	Regulator(s)	Differences in *E. coli*
Rhamnogalacturonides[33,43]				
	rhiT	*rhiE rhiN*	*rhaSR kdgR*	Absent
Melibiose[49]	*rafB*	*rafA*	*rafR*	Absent
Raffinose[49]	*scrY rafB*	*rafA scrB scrK*	*rafR scrR*	Absent
Sucrose[49]	*scrY scrA*	*scrB scrK*	*scrR*	Absent

transporters of the major facilitator superfamily (MFS) usually consist of single polypeptide chains. The driving force is often the proton motive force or, alternatively, the gradient of a co- or antitransported ion.

Genome analysis confirmed that *D. dadantii* is able to transport glucose using a PTS transporter and to catabolize this sugar using the Embden–Meyerhof pathway (glycolysis) or the pentose phosphate pathway, with a complete tricarboxylic acid cycle. However, in contrast to most enterobacteria,[56] *D. dadantii* is unable to metabolize glucose via the Entner–Doudoroff pathway, due to the absence of 6-phosphogluconate dehydratase (Edd).

D. dadantii catabolizes several monosaccharides using pathways very similar to those described in *Escherichiacoli* (Table 2). However, *D. dadantii* clearly possesses fewer monosaccharide transport systems than *E. coli*. For the uptake of galactose, xylose and arabinose, *E. coli* uses both a low affinity MSF and a high affinity ABC transport system, whereas *D. dadantii* has only the ABC transporter. For gluconate, *D. dadantii* has a single MSF transporter, while four transporters have been identified in *E. coli*. For rhamnose, *D. dadantii* has no specific transporter, indicating that it assimilates rhamnose present in oligosaccharides rather than the monomeric sugar.[43] The genes involved in a catabolic pathway tend to be more clustered in *D. dadantii* than in *E. coli*. For instance, the genes involved in galacturonate catabolism are gathered together in one locus (*exuT–uxaCBA*) in *D. dadantii* while they are divided into two loci in *E. coli* (*exuT–uxaCA, uxaB*), a trend also observed for the galactarate and glucarate catabolic genes (Table 2).

Despite its β-galactosidase activity, *D. dadantii* is not able to grow with lactose (a disaccharide of animal origin) as the sole carbon source. The truncated lactose operon found in the *D. dadantii* genome (*lacI–lacZ*) has minor functional significance. The *lacZ* gene is weakly transcribed and the β-galactosidase activity is mostly due to GanB, an enzyme involved in galactan degradation.[45] Thus, in this enterobacterium adapted to a plant host, lactose cleavage results from the secondary activity of a protein involved in the catabolism of plant oligosaccharides.

The potential of *D. dadantii* to degrade β-glucosides appears to be well developed, with 16 genes that may be involved in β-glucoside assimilation. Glucosides are common in plant tissues and are also generated during cellulose cleavage by endoglucanases. In addition to the disaccharide cellobiose resulting from cellulose degradation, *D. dadantii* is able to metabolize different β-glucosides such as arbutin, salicin, and esculin. Genetic analysis has revealed the existence of at least two independent assimilatory pathways. The *arb* pathway allows growth with arbutin and salicin only, while the second hypothetical pathway, called *cbr*, enables the bacteria to use cellobiose, arbutin, and salicin.[44] Two genomic regions encode both transporters of the PTS family, phospho-β-glucosidases and transcriptional antiterminators homologous to *E. coli* BglF, BglB, and BglG, respectively. The previously characterized operon *arbGFB* encodes only one protein for each function. The second region, which may correspond to the *cbr* pathway, contains the genes $bglGFB_1B_2$, *ascG-celC-celB-celH-celA* encoding an antiterminator (BglG), a PTS transporter (BglF), three phospho-β-glucosidases ($BglB_1$, $BglB_2$, and CelH), a regulator of the LacI family (AscG), and the three domains of another PTS transporter (CelABC). In *E. coli*, the locus containing *celA, celB*, and *celC* homologs is able to mediate the transport of cellobiose, arbutin, and salicin but its primary function is chitobiose utilization.[57] Another isolated gene, *bglA*, is predicted to encode a fifth phospho-β-glucosidase.

Thus, there is evidence of functional redundancy in *D. dadantii* β-glucoside assimilation, but the specificity of the three transporters, the five phospho-β-glucosidases and the three regulators remains to be clarified. In addition to these whole β-glucoside assimilation systems, involving cytoplasmic phospho-β-glucosidases, *D. dadantii* produces two periplasmic β-glucosidases, BglX and BgxA. BgxA is primarily involved in the cleavage of aromatic β-glucosides.[46] Its location in the periplasm may allow it to avoid the release of the toxic phenolic aglycone into the cytoplasm, while still allowing for assimilation of the glucose moiety via the classical glucose uptake system. Thus, the BgxA subcellular compartmentation could be an important element in its specialization for aromatic glycosides originating from the host or the environment.

The *D. dadantii* genome appears to encode some catabolic pathways previously described in plant-associated bacteria for the utilization of opines, inositol and many organic acids such as citrate, oxalate, and malonate. These compounds are specific to plants but they are not present in large concentrations in plant tissues. Their assimilation might favor the growth and survival of the bacteria during the saprophytic life of the bacteria.

3.2 The Transcriptomic Approaches

Transcriptomic approaches could be fruitful in identifying the catabolic pathways expressed during plant infection. However, transcriptional data derived from different studies are often inconsistent. The upregulated genes depend on the plant model, the type of infection and/or the timing after inoculation. With *D. dadantii* 3937, a few transcriptome analyses have been performed during plant infection to identify genes required for pathogenesis. These studies have revealed several metabolic genes whose expression is induced *in planta*.

The first analysis performed at the genomic level consisted of microarrays constructed with small genomic regions (3-kb DNA fragments).[18] After a 14 h infection of African violet leaves (the 3937 original host), 74 upregulated regions were recovered. Some of them contain genes involved in pectin catabolism (*pelE–pelD*, *togB–kdgM–paeX*), glucose transport (*ptsI–crr–ptsG*), xylose catabolism (*xylB*), butanediol production (*budR–budA–budB*), or the tricarboxylic acid cycle (*sucA*).

Another approach developed to identify *D. dadantii* genes upregulated in plants, was a green fluorescent protein (GFP)-based in vivo expression technology (IVET) array.[58] After an 18 h infection of spinach leaves, 61 regions (1.5-kb DNA fragments) containing plant upregulated promoters were selected. They included some genes involved in pectin catabolism (*ogl* and *rhiT*), in the tricarboxylic acid cycle (*ppc*) and global or specific regulators of sugar catabolism (*crp* and *xylR*).

The data obtained from these two studies almost certainly do not cover all the potential targets and, as the experimental conditions differed, comparison is difficult. It is noticeable, however, that both the glucose transport and some pectin catabolic steps were induced together, in both studies. New analyses suggest that the assimilation of glucose and pectin are expressed at two different phases of plant infection, the asymptomatic and the symptomatic phase, respectively (see later). For a reliable transcriptomic analysis, it is essential to have precise knowledge of the infection kinetics and to strictly control the infection conditions.

A more precise transcriptomic analysis was recently performed to analyze gene expression at the onset of infection in the model *Arabidopsis thaliana*.[16] Bacteria recovered in the asymptomatic phase (8 h after inoculation) were compared to "epiphytic" bacteria recovered from the leaf surface 4 h after inoculation. The infection was followed by microscopic observations and quantification of the bacterial population. After 8 h, no symptoms were visible and no massive bacterial multiplication had occurred, with only 10^5–10^6 bacteria present per leaf. Gene transcription was measured using NimbleGen microarrays containing five probes per *D. dadantii* coding sequence. Among approximately 100 differentially expressed genes, only a few were involved in carbohydrate assimilation. Some genes downregulated in the asymptomatic phase encode proteins involved in the assimilation of arabinose (*araF*), oligogalactans (*ganC*), and oligogalacturonides (*kdgN, togM, togN*, and *kduD*). Only one pectate lyase gene, *pelC*, was upregulated, indicating that pectin degradation was weakly activated at this early colonization step.[16] A strong catabolite repression is probably exerted at this step in which bacteria assimilate their preferred carbon sources—glucose, fructose, and sucrose (see later).

D. dadantii is not only a plant pathogen but can also infect insects and is able to kill the pea aphid *Acyrtosiphonpisum*.[59] The transcriptomes of bacteria isolated from infected aphids were analyzed to identify factors required for pathogenesis in the insect.[60] Concerning sugar metabolism, this study revealed that several genes involved in the catabolism of plant sugars (PGA, galactan, galacturonate, galactose, and sucrose) were downregulated but no information about the carbon sources used in the insect could be deduced from the list of upregulated genes.

3.3 Contribution of Metabolomics

Metabolomics can give a direct image of the metabolic activity of the pathogen during the plant–bacterium interaction. Thus, this approach could be attractive, although it is limited by the difficulty of differentiating metabolites from plant and bacterial origin.

Metabolomic profiling, based on ^{13}C-NMR spectroscopy, was used to monitor the evolution of the main soluble carbon metabolites during plant infection by *D. dadantii*.[17] In healthy chicory leaves, the major soluble carbon compounds are glucose, fructose, and sucrose, accompanied by malate and several amino acids, mainly glutamine, arginine, and asparagine. In the profiles of infected tissues, these components showed a drastic reduction but new peaks appeared. A kinetic analysis, at 4 h intervals, was performed to follow the infection. In these experimental conditions, the asymptomatic and the symptomatic phases of the infection process corresponded to 1–8 h and 12–24 h, respectively. The bacterial population began to increase during the asymptomatic phase but bacterial growth was particularly intense during the transition from the asymptomatic to the symptomatic phase. Throughout the asymptomatic phase, only a slight modification of the plant metabolite profile was observed. During the transition and symptomatic phases, a regular decrease in the plant metabolites glucose, fructose, sucrose, and malate was observed, accompanied by a decline in the concentration of free amino acids. This decline was concomitant with the appearance of new compounds corresponding to known end-products of fermentative bacterial metabolism, such as the organic acids succinate and acetate. However, the predominant metabolite at the end of infection was identified as 2,3-butanediol. It accumulated throughout the symptomatic phase of the disease. Inactivation of the *bud* genes, involved in butanediol production, revealed that the virulence of the mutants is clearly reduced. These genes are highly expressed in the symptomatic phase. With the wild-type strain, a pH increase was observed in the macerated tissue, whereas the *bud* mutants were deficient for this pH change during infection. Since the *D. dadantii* pectate lyases need an alkaline pH for their activity,[27] the butanediol pathway participates in the pH adjustment necessary for maceration. These results highlighted the role of the butanediol pathway in limiting acidification of the plant tissue and reconfiguring pH during the development of the soft-rot disease.[17,61]

In this first metabolomics study, infected plant extracts were prepared by lyophilization, allowing some volatile compounds to evaporate. Acetate and butanediol were only partially volatile in these conditions. However, further analysis of exometabolites secreted by *D. dadantii* demonstrated that other volatiles, such as acetoin (the butanediol precursor) and ethanol, are excreted during fermentative metabolism.[17]

Plant soluble sugars (glucose, fructose, and sucrose) provide an easily metabolizable source of carbohydrates but their catabolism generates organic acids as end products. The butanediol pathway is an alternative fermentative pathway that generates only neutral products (Fig. 2). In other enterobacteria, it is induced by glucose abundance, acidic pH, high cell population, and/or a low oxygen level.[62,63] These conditions are compatible with those encountered by bacteria at the beginning of infection, when the plant tissue

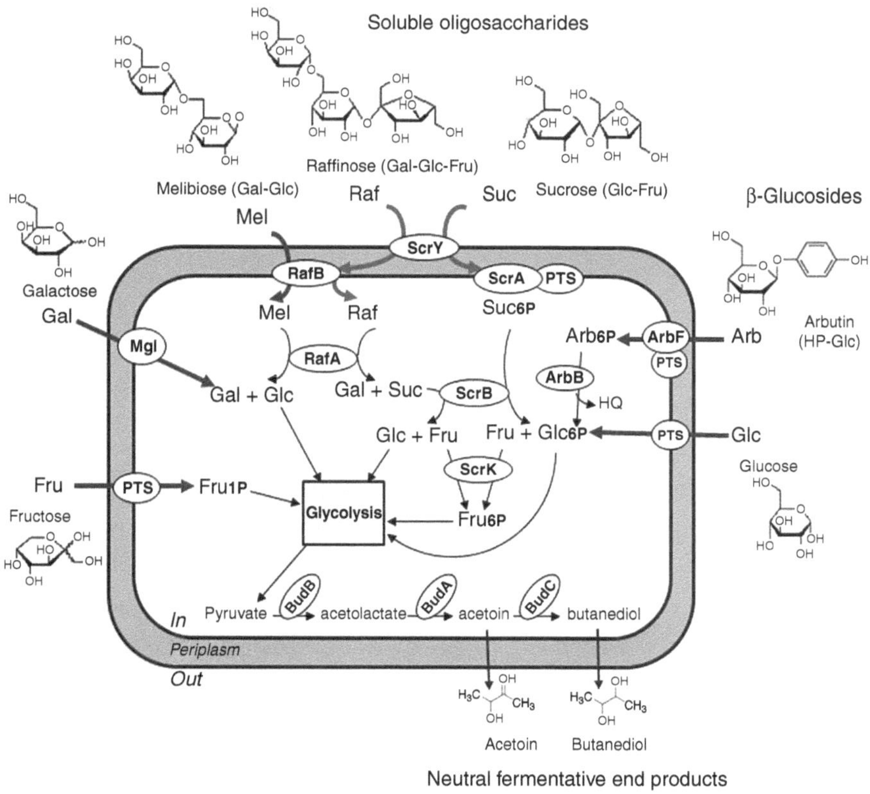

Figure 2 Assimilation of plant soluble sugars in *Dickeya dadantii*. The entry of glucose *(Glc)* and fructose *(Fru)* involves transporters of the PTS family *(PTS)* that couple sugar entry with sugar phosphorylation. The addition of "1P" or "6P" to a sugar abbreviation indicates that it is phosphorylated. The galactose *(Gal)* uptake is mediated by the Mgl permease. The catabolism of sucrose *(Suc)* successively involves the outer membrane porin ScrY, the PTS transporter ScrA, the sucrose hydrolase ScrB, and the fructose kinase ScrK. The assimilation of melibiose *(Mel)* involves the permease RafB and the α-galactosidase RafA that hydrolyses melibiose into galactose and glucose. The uptake of raffinose *(Raf)* is performed by the porin ScrY and the permease RafB. The enzyme RafA hydrolyses raffinose into galactose and sucrose, which is cleaved by the hydrolase ScrB. The *arb* pathway allows growth with arbutin *(Arb)* and salicin. It involves the PTS transporter ArbF and the phospho-β-glucosidase ArbB (*HQ*, hydroquinone). The butanediol pathway *(BudB, BudA,* and *BudC)* generates neutral products from pyruvate.

is acidic and has high sugar concentrations. The use of different carbon sources shows that *D. dadantii* produces butanediol when glucose, fructose, or sucrose is used as the sole carbon source but not with any other tested sugars.[17] Thus, the catabolism of these three major soluble sugars promotes butanediol production during plant infection.

4. THE MAJOR CARBON SOURCES ASSIMILATED BY BACTERIA IN PLANT TISSUES

Plant tissue maceration caused by *D. dadantii* is concomitant with a significant increase in bacterial population. Plant tissue provides a large source of nutrients for these bacteria. Simple sugars, such as glucose and fructose, or small oligosaccharides, such as sucrose and raffinose, are abundant and could serve as primary carbon sources for bacterial multiplication (Fig. 2). Polysaccharides from plant cell walls must be predigested by extracellular enzymes before bacteria can assimilate the resulting oligosaccharides to obtain carbon and energy.

4.1 Utilization of the Plant Soluble Sugars

In macerated tissues, the *D. dadantii* population can exceed 4 10^9 cells per gram. To reach such levels, bacterial metabolism must find suitable nutritional resources. Glucose, fructose and sucrose are the best carbon sources for in vitro growth of *D. dadantii*, with generation times of only 100 minutes.[49] *D. dadantii* also assimilates raffinose and melibiose, two common plant soluble oligosaccharides (Table 2). The *scr* gene cluster is involved in the assimilation of the disaccharide sucrose. The *raf* cluster is required for the assimilation of the trisaccharide raffinose and of the related disaccharide, melibiose (Fig. 2).

The catabolism of sucrose successively involves the outer membrane porin ScrY, the inner membrane PTS transporter ScrA, the sucrose hydrolase ScrB, and the fructose kinase ScrK.[49] The assimilation of melibiose involves the inner membrane transporter RafB and the α-galactosidase RafA that hydrolyses melibiose into galactose and glucose. RafA and RafB are sufficient to enable *D. dadantii* to use melibiose as a sole carbon source for growth and this is probably their primary function (the denominations MelA and MelB would be more appropriate).

RafA and RafB are also necessary, but not sufficient alone, for the assimilation of raffinose (Fig. 2). The permease RafB allows uptake of raffinose and the enzyme RafA hydrolyses raffinose into galactose and sucrose. The two additional steps of raffinose catabolism are supplied by the sucrose pathway, with the porin ScrY and the sucrose hydrolase ScrB.[49] Inactivation of sucrose, raffinose, or melibiose assimilation did not affect *D. dadantii* virulence but the wild-type strain showed a selective advantage over the mutants during plant infection.

4.2 Utilization of Sugars Derived From Pectic Polysaccharides

Pectate lyases are employed by pathogenic bacteria to degrade host tissue and to provide nutrients for bacterial growth. Pectinolytic enterobacteria produce a series of enzymes and transporters to use PGA or pectin as sole carbon and energy sources. These bacteria produce a series of enzymes and transporters for this catabolism (Fig. 3).[8] Given the length of PGA chains, complete degradation involves sequential cleavages which take place in different cell compartments. Pectate lyases, secreted by the Out system to reach their polymeric substrate, catalyze the first cleavages, giving rise to an oligomer mixture. A second round of cleavages occurs in the periplasm, generating small oligomers. The final cleavages, giving rise to monomers, take place in the cytoplasm. Since pectin catabolism is achieved in successive cell compartments, as just described, it is strongly dependent on efficient enzyme export and metabolite import across the bacterial outer and inner membranes (Fig. 3).

Unsaturated oligogalacturonides, extracellularly generated by Out-secreted pectate lyases, are transported in two steps, first into the periplasm and then into the cytoplasm. They enter the periplasm using the two specific outer-membrane porins, KdgM and KdgN.[64,65] The short oligomers resulting from digestion by periplasmic enzymes, mostly di- and trimers, enter the cytoplasm using an ABC and a MFS specific transporter, TogMNAB and TogT, respectively.[66,67] They are finally cleaved to give monomers by the action of two cytoplasmic lyases, PelW and Ogl.[30] Since pectin depolymerization is mostly dependent on lyase activities, the major monomeric end product is an unsaturated sugar, 4-deoxy-L-*threo*-5-hexosulose uronic acid (or 5-keto-4-deoxyuronate, DKI). Only a small quantity of galacturonate is formed, resulting from the action of periplasmic polygalacturonases. The two pathways involved in the catabolism of DKI and galacturonate converge to produce a common intermediate, 2-keto-3-deoxygluconate (KDG). KDG is then converted to pyruvate and 3-phosphoglyceraldehyde, which enter the general sugar metabolism. While only a few monomers are formed extracellularly, specific transporters could enable them to enter the cells, namely KdgT for DKI, KDG, and DKII (3-deoxy-D-glycero-2,5-hexodiulosonate or 2,5-diketo-3-deoxygluconate)[68] and ExuT for galacturonate.[69]

The RGI backbone of pectin is also catabolized by a pathway involving different cell compartments (Fig. 3). Unsaturated rhamnogalacturonides are generated extracellularly by the Out-secreted enzyme RhiE.[33] After their diffusion into the periplasm, probably by the porins KdgM or KdgN, they use the inner-membrane transporter RhiT to enter the cytoplasm where

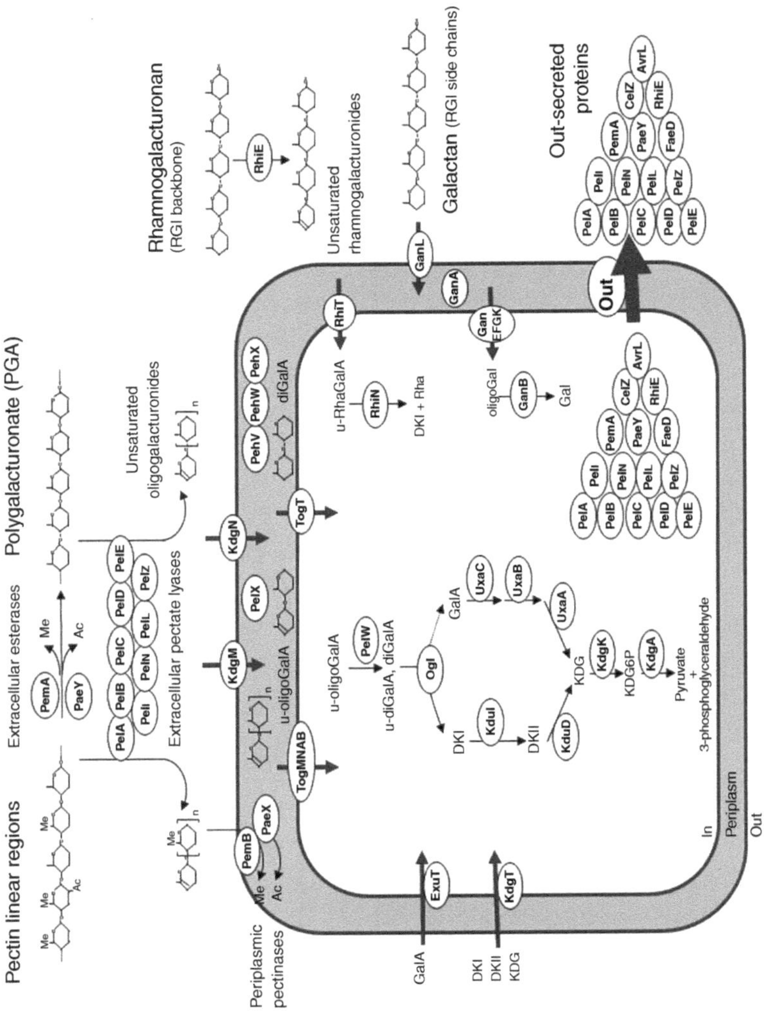

Pectin linear regions
Polygalacturonate (PGA)
Rhamnogalacturonan
(RGI backbone)
Galactan (RGI side chains)
Extracellular esterases
PemA
PaeY
Me
Ac
Extracellular pectate lyases
PelA
PelB
PelC
PelD
PelE
PelI
PelN
PelL
PelZ
Unsaturated
oligogalacturonides
Unsaturated
rhamnogalacturonides
RhiE
GanL
GanA
Gan
EFGK
RhiT
Out
Out-secreted
proteins
CelZ
AvrL
FaeD
Periplasmic
pectinases
PemB
PaeX
TogMNAB
KdgM
KdgN
TogT
PelX
PehV
PehW
PehX
u-oligoGalA
diGalA
u-RhaGalA
RhiN
DKI + Rha
oligoGal
GanB
Gal
u-oligoGalA
PelW
u-diGalA, diGalA
Ogl
GalA
UxaC
UxaB
UxaA
DKI
KduI
DKII
KduD
KDG
KdgK
KDG6P
KdgA
Pyruvate
+
3-phosphoglyceraldehyde
ExuT
KdgT
GalA
DKI
DKII
KDG
In
Periplasm
Out

they are digested by the enzyme RhiN.[43] The final monomeric products are DKI and L-rhamnose, which are further catabolized by their specific pathways. The assimilation of the galactan side chains of RGI depends on periplasmic and cytoplasmic enzymes, together with specific transporters (Fig. 3).[45] The digestion of the pectin backbone (PGA and RGI) by the set of Out-secreted pectinases provokes liberation of the pectic side chains. The galactan chains diffuse across the outer membrane through the porin GanL.[45] They are digested into the periplasm by the endo-galactanase GanA, which cleaves randomly inside the oligosaccharide chain. The resulting small oligogalactans are transported across the inner membrane by a specific ABC transporter, GanEFGK. The cytoplasmic exo-galactanase (or β-galactosidase) GanB

◀**Figure 3** Degradation and assimilation of pectic polymers in *Dickeya dadantii.* The Out system mediates the secretion of at least 15 proteins: 9 endo-pectate lyases *(PelA, PelB, PelC, PelD, PelE, PelI, PelL, PelN,* and *PelZ),* 3 esterases *(PemA, PaeY,* and *FaeD),* a rhamnogalacturonate lyase *(RhiE),* a cellulase *(CelZ),* and a protein of unknown function *(AvrL).* Pectin acetyl- and methyl-esterases remove acetate *(Ac)* and methanol *(Me),* respectively, from the pectin backbone. They liberate moderately esterified polygalacturonate *(PGA),* a good substrate for pectate lyases. Unsaturated oligogalacturonides *(u-oligoGalA)* are extracellularly generated by the Out-secreted pectate lyases. They enter the periplasm using the two specific outer-membrane porins, KdgM and KdgN. Their digestion by periplasmic enzymes generates short oligomers up to dimers *(diGalA,* digalacturonide; *u-diGalA,* unsaturated digalacturonide). The short oligomers enter the cytoplasm using two specific transporters, TogMNAB and TogT. They are cleaved to give monomers by the action of two cytoplasmic lyases, PelW and Ogl. Since pectin depolymerization is mostly dependent on lyase activities, the major monomeric end product is an unsaturated sugar, 5-keto-4-deoxyuronate *(DKI).* Only a small quantity of galacturonate *(GalA)* is formed, mostly resulting from the action of periplasmic polygalacturonases *(PehV, W,* and *X).* The two pathways involved in the catabolism of DKI and galacturonate converge to produce a common intermediate, 2-keto-3-deoxygluconate *(KDG)* which is converted to pyruvate and 3-phosphoglyceraldehyde. While only a few monomers are formed extracellularly, specific transporters could enable them to enter the cells, namely KdgT for DKI, KDG, and 2,5-diketo-3-deoxygluconate *(DKII)* and ExuT for galacturonate. The RGI backbone of pectin is cleaved extracellularly by the Out-secreted enzyme RhiE. Unsaturated rhamnogalacturonides *(u-RhaGal)* diffuse into the periplasm, probably by the porins KdgM or KdgN. They use the inner-membrane transporter RhiT to enter the cytoplasm where they are digested by the enzyme RhiN. The final monomeric products are DKI and L-rhamnose *(Rha).* The galactan side chains, liberated after digestion of the RGI backbone, diffuse across the outer membrane through the porin GanL. They are cleaved into the periplasm by the endo-galactanase GanA. The resulting small oligogalactans *(oligoGal)* are transported across the inner membrane by a specific transporter, GanEFGK. The cytoplasmic exo-galactanase GanB liberates the monomeric product, D-galactose *(Gal).*

cleaves from the oligomer reducing end to liberate the monomeric product, D-galactose, which is catabolized by its specific pathway.

It is notable that metabolomics analyses did not reveal the presence of soluble pectic oligomers in plant tissues during *D. dadantii* infection[17] (and unpublished data), indicating that these oligomers may not accumulate in macerated tissue. Either they are produced only in very small amounts, in comparison to the other identified carbon compounds, or they are very quickly assimilated by the bacteria.

5. THE REGULATORS COORDINATING VIRULENCE AND METABOLISM

The secreted pectate lyases provoke the general disorganization of the plant cell wall and also release oligosaccharides, used as carbon sources by *D. dadantii*. Their dual function in virulence and nutrition implies that expression of the pectate lyase genes (*pel*) may be regulated by both metabolic and virulence regulators. Specific and global regulators serve to adapt gene expression to available substrates. The specific repressor KdgR ensures that the whole pectin catabolic pathway is repressed in the absence of the appropriate substrate(s). The global activator Crp ensures the preferential utilization of glucose or other easily metabolizable carbohydrates, when bacteria are in the presence of a mixture of carbon sources. These mechanisms are involved in different phases of the infection process. Indeed, the successive use of carbon sources may serve as a signal for the proper timing of virulence gene expression.

pel gene expression is influenced by several physiological conditions; in particular, metabolic status, growth phase and population density. In cultures in vitro, pectate lyases are synthesized during the transition from exponential to stationary growth phase.[70] This timing results primarily from two complementary mechanisms involving different regulators (see also chapter 7, by Reverchon et al.). During the lag and log growth phases, the *pel* genes are repressed by Fis and KdgR. At the end of the exponential growth phase, the *pel* genes are activated by the Crp–AMPc complex and KdgR is fully inactivated in the presence of pectin. Similar regulations take place during *inplanta* growth, with an increase in *pel* gene expression starting during the transition from asymptomatic to symptomatic phases. Indeed, a parallel can be drawn between in vitro culture and *in planta* growth. The lag, exponential and stationary growth phases approximately correspond to the asymptomatic, transition, and symptomatic phases of infection, respectively. Thus, in vitro cultures in a medium supplemented with sucrose and PGA are an approximate model of the conditions encountered in plants.

A number of observations support the existence of strong coupling between central metabolism and *pel* regulation. Three main mechanisms are involved: (1) CRP, which activates virulence genes when efficient carbon sources are exhausted; (2) KdgR, which inhibits pectinase production in the absence of pectin breakdown products and participates in the coordination of central metabolism; and (3) the GacA/GacS two-component system, critical for *pel* gene expression and closely linked to Rsm posttranscriptional regulation.

5.1 The Pectin Sensor KdgR

KdgR is the major transcriptional repressor of the pectin degradation pathway. It allows *Dickeya* to induce pectinase production only in the presence of pectin or its metabolic derivatives.[8,11] In the absence of pectin, there is only a low basal level of pectate lyase production but when the bacteria encounter pectin, this level is sufficient to initiate PGA cleavage and to ensure the formation of intracellular metabolites. DKI, DKII, and KDG are the inducers that, by binding to KdgR, provoke the dissociation of KdgR from its DNA binding sites.[31] As soon as the inducer level reaches a sufficient intracellular concentration, KdgR repression is suppressed, leading to the expression of the KdgR-repressed genes. Thus, KdgR acts as a pectin sensor in response to the formation of the PGA metabolites, KDG being the most efficient inducer.

The "noninduced" level of pectate lyases is important for initiation of the induction cycle. It mainly results from a slight transcriptional leakage of some *pel* genes. The transcription of *pelL*, a gene which is not regulated by KdgR,[34] could also greatly influence this basal level. In fact, the regulation of the different *pel* genes is similar but not identical; each of them showing some particularity.[28,34,35,70,71]

Due to the role of pectate lyases in inducer formation, KdgR control resembles an autoinduction mechanism. The inducers, formed by PGA catabolism, auto-induce their own production and the cells respond to an inducer threshold. When the inducer threshold is reached, *pel* gene expression increases and pectate lyases enhance the inducer formation. This autoinduction is responsible for a rapid amplification of pectate lyase production, leading to a strong burst of pectate lyases. In addition, KdgR regulation is linked to cell density. The intracellular inducer concentration is dependent on the concentration of extracellular pectate lyases, which is itself dependent on bacterial density. Indeed, during in vitro growth, the KdgR derepression only begins at the end of the exponential growth phase. Most pathogenic bacteria use quorum-sensing mechanisms to induce virulence factors when the population density has reached a level high enough to initiate the pathogenic process. While pectate lyase production is dependent on cell density, the quorum sensing system involving *N*-acyl homoserine lactones has a minor role in *pel* expression in *D. dadantii*.[72] This could be explained by the cell density effect of KdgR regulation, as well as by a second type of quorum sensing molecule produced, and sensed, by the specific *Dickeya* Vfm system.[73]

A modeling approach of pectate lyase production predicted that the minimal, but essential, subnetwork underlying their synthesis is the feedback loop involving the repressor KdgR and the metabolic steps leading to KDG formation.[74] A dynamic mathematical model was then developed and validated, simultaneously taking into account the cleavage of PGA by the extracellular enzymes and the regulation of the *pel* genes by the intracellular metabolite KDG. This work demonstrated that, during in vitro growth, PGA is almost totally consumed before the strong burst of pectate lyase production.[75] This paradoxical behavior suggests that, during infection, pectate lyases function as a virulence weapon rather than a nutrient provider. Pectate lyases diffuse more rapidly than do bacteria in plant tissues surrounding the infection site; this allows for the maceration of as-yet-uninfected tissues adjacent to the infection site. Such an adaptation strategy enhances the spread of the disease.

Comparative genomics has been used to determine the whole *D. dadantii* KdgR regulon (the set of KdgR-regulated genes) by searching for KdgR-binding sites in the genomes of various γ-proteobacteria.[76] The *D. dadantii* KdgR regulon is large, with at least 32 target operons, most of them related to pectin degradation. The KdgR-regulated genes encode the pectate lyases and accessory pectinases, the different transporters and intracellular enzymes involved in the catabolism of oligogalacturonides and rhamnogalacturonides, the Out secretion system, chemotaxis receptors and sugar efflux transporters. Surprisingly, KdgR also has an effect on some steps of central carbohydrate catabolism.[76] KdgR directly controls the genes *pykF*, encoding pyruvate kinase I, an enzyme involved in glycolysis, and *ppsA*, encoding phosphoenolpyruvate synthase, a key gluconeogenic enzyme.[77] Interestingly, PpsA and PykF catalyze reverse reactions and they are regulated by KdgR in opposite ways; that is, KdgR acts as a repressor for *pykF* but as an activator for *ppsA*. The effect of KdgR would be to favor carbon flow through glycolysis in the presence of pectin, but through the gluconeogenic pathway when pectin is not metabolized. Thus, the role of KdgR is not restricted to the negative control of pectin catabolism but is extended to the coordination of central sugar metabolism by directing the intracellular carbon flux toward glycolysis or gluconeogenesis.

Another gene, *pecT*, was found to be positively regulated by KdgR. PecT is a negative regulator of the LysR family involved in the control of pectate lyase production.[78] Variations in PecT concentration have drastic effects on the target genes. The regulation of PecT by KdgR provides a further example of interactions between the different regulators involved in the network controlling pectate lyase expression in *D. dadantii*.

Interestingly, a dual-regulatory mechanism controls the two steps involved in the transport and catabolism of rhamnogalacturonides, carried out by RhiT and RhiN. The *rhiT-rhiN* genes are under the control of both KdgR and RhaS, the transcriptional activator of rhamnose catabolism, and their expression is therefore dependent on the intracellular formation of the two inducers, DKI and rhamnose.[43] This double regulation allows for coupling between the catabolism of the polymer and that of the corresponding monomers. In contrast, the *rhiE* gene, encoding the secreted rhamnogalacturonate lyase, is controlled only by RhaS.[43] The regulation of this gene differs from that of other pectinases in that it is strongly induced in the presence of rhamnose but not in the presence of PGA metabolites.[33]

The regulators KdgR and RhaS are conserved in several enterobacteria. In *E. coli*, they control several genes involved in the catabolism of KDG and L-rhamnose, respectively (Table 2). In *D. dadantii*, they exhibit an extended function whereby RhaS also controls the steps involved in the assimilation of the rhamnose-rich polysaccharides of the plant cell wall. KdgR is a major regulator that controls several genes involved in pectin degradation and associated functions (secretion, chemotaxis, sugar efflux, etc). Predicted members of the KdgR regulon have been experimentally validated as KdgR targets in *D. dadantii*. Notably, most of them are also predicted to be regulated by the Crp activator.[76]

5.2 The cAMP Receptor Protein (Crp) for Preferential Sugar Utilization

The plant host presents a very abundant and varied source of sugars for the pathogen. The major role of Crp is to direct the right choice of nutrients by favoring the utilization of the most efficiently metabolizable carbohydrates. Crp is a highly conserved transcription factor that coordinates the assimilation of alternative carbon sources in a process termed carbon catabolite repression.[79] It is a crucial global regulator that controls the transcription of multiple genes in enterobacteria.[80] Crp is activated by binding the signal metabolite cyclic adenosine mono-phosphate (cAMP). The assimilation of glucose, or other efficiently metabolizable sugars, leads to a low intracellular cAMP level, which prevents bacteria from catabolizing alternative sugars.

A *crp* mutation has serious consequences for the virulence of *D. dadantii* in that it strongly decreases pectate lyase production.[81] The *D. dadantii crp* mutant retains the ability to grow only with glucose, fructose, or sucrose, three efficiently metabolized carbon sources that are highly abundant in plant tissues.[49]

Crp has a pivotal role in coordinating the expression of virulence factors during the infection process. The high concentration of easily metabolizable sugars available during the asymptomatic phase prevents Crp activation. After consumption of these sugars, cAMP levels increase and the Crp–AMPc complex activates the expression of the *pel* genes and other virulence genes involved in the symptomatic phase. This system insures that the virulence factors are not expressed inappropriately during the asymptomatic phase. Thus, Crp plays a crucial role in *D. dadantii* pathogenesis by tuning virulence gene expression according to the nutrient conditions encountered during the two phases of infection.

5.3 The GacA/GacS Two-Component System and Rsm Posttranscriptional Regulation

The GacA/GacS two-component system (also referred as BarA/UvrY) is highly conserved in the gamma-proteobacteria. The GacA regulator is essential for *D. dadantii* virulence.[82,83] It was recently suggested that the GacA/GacS system serves as a barometer for the metabolic state of the cell.[84] However, the environmental signal to which it responds is still a subject of discussion. In *E. coli*, it has been reported that metabolic end products, such as formate and acetate, are able to stimulate the BarA sensor kinase.[85] In *Pseudomonas fluorescens*, it has been proposed that intermediates of the Krebs cycle or short organic acids are involved in GacS activation.[86] Levels of carboxylic acids fluctuate throughout infection[17]; these changes could influence GacA/GacS activity in *D. dadantii*. This hypothesis implies that metabolic end products may be important factors linking metabolic status to the expression of virulence factors.

GacA regulation is associated with carbon catabolism as it is part of a regulatory network involving both Crp and the Rsm system (see later). Crp represses the response regulator GacA, which is itself required for RsmB activation.

RsmB is a small RNA involved in posttranscriptional control together with the RNA-binding protein RsmA (repressor of secondary metabolites). The RsmA homolog, CsrA (carbon storage regulator), was originally described in *E. coli* as a regulator of glycogen biosynthesis whose expression is maximal in the stationary growth phase.[87] Its inactivation had pleiotropic effects on genes involved in carbon flux pathways. CsrA homologs are conserved in most enteric bacteria and deletion of the corresponding gene often impairs virulence. The Rsm/Csr system links carbon metabolism and virulence in a variety of animal and plant pathogens, such as *Salmonella, Yersinia, Pseudomonas, Xanthomonas*, and *Pectobacterium*.[83,88] The RsmA/RsmB system is conserved in *Dickeya* but the lack of information about its role in this genus stems from an inability to inactivate these genes. However, several observations suggest that this system may exert its posttranscriptional control in a manner similar to that described in *P. carotovorum*, where the orthologous system influences pectate lyase production by interacting with GacA/GacS.[89–93]

The RNA-binding protein, RsmA, affects the production of pectate lyases by binding directly to *pel* mRNAs. The small RNA RsmB carries multiple RsmA binding sites and, therefore, titrates RsmA away from its mRNA targets. In the absence of antagonist small RNAs, the binding of RsmA to its targets results in rapid mRNA degradation.[94]

The model predicts that the catabolism of efficiently assimilated sugars, observed during the early phases of plant infection, generates the formation of organic acids leading, successively, to GacA phosphorylation, RsmB activation, RsmA inactivation and, finally, pectate lyase production. The two regulatory systems, GacA/S and RsmA/B, contribute, in combination with Crp and KdgR, to the nutritional adjustment and the control of virulence factor expression. Regulation of virulence factors in response to nutrient availability is necessary for bacteria to express early or late virulence factors at precisely the appropriate moment for successful infection. Pectinolytic enterobacteria use these systems to coordinate the expression of virulence factors according to successive phases of infection, corresponding to changes in the host microenvironment.

5.4 Other Regulatory Systems

In *D. dadantii*, pectate lyase production is dependent on a large set of environmental conditions, such as the presence of pectin, catabolite repression, growth phase, temperature, pH, osmolarity, oxygen, and iron availability.[70] In addition to the catabolic regulators KdgR and Crp, a complex network of interconnected regulators modulate *pel* gene expression. This network includes the transcriptional regulators PecS, PecT, MfbR, Fur, H-NS, and Fis, as has been recently detailed in two reviews.[10,11] The role of some major regulators has also been analyzed in *D. solani*, a species mostly associated with potato, suggesting that the two species (*dadantii* and *solani*) have adopted closely related mechanisms, with some variations, for adjusting the expression of virulence genes during infection.[95]

Additional regulatory interactions may participate in the coordination between the virulence regulators and metabolism. For instance, transcriptomic analysis has shown that PecS directly or indirectly plays a role in activation of RsmA.[96] Transcription of the *pel* genes was observed to increase under iron depletion, a transcriptional switch which is directly mediated by the ferric iron uptake regulator Fur.[97] In the presence of iron, binding of the metallorepressor Fur may inhibit *pel* gene activation by CRP via an anti-activation mechanism. In contrast, the intracellular accumulation of KDG induces transcription of the genes implicated in iron assimilation.[97] The various functions involved in pathogenicity are frequently coregulated by responding to common signals encountered *in planta*.

Furthermore, *pel* gene expression is affected by the topology of the bacterial chromosome, such as its supercoiling state. The supercoiling level is dependent on different nucleoid-associated proteins, such as Fis and H-NS, which are known to affect the expression of the *D. dadantii* virulence factors.[98,99] The metabolic state of the cell affects the composition of nucleoid-associated proteins such as H-NS and Fis.[100] Thus, a higher level of coordination could be mediated by the dynamics of genome topology which is itself modulated by the DNA architectural proteins.[101]

6. CONCLUDING REMARKS

Bacterial pathogenesis is multifaceted and virulence factors alone are not sufficient to ensure bacterial multiplication during infection. Metabolism is an essential aspect of the life of the pathogen, in or outside of the host. Therefore, the regulators responsible for coordinating virulence and metabolism are important elements for the ensuring proper progression through the successive phases that are necessary for successful infection. In the case of plant pathogens, pathways specific to the metabolism of plant compounds and the systems responsible for transcriptional and posttranscriptional regulation of those pathways may be viable targets for pathogen inhibition. Improving our knowledge of the nutritional strategies of phytopathogenic bacteria may yield innovative strategies for limiting plant diseases.

ACKNOWLEDGMENTS

I would like to thank Guy Condemine and Vladimir Shevchik for their critical reading of the manuscript and Valerie James for the English corrections.

REFERENCES

1. Czajkowski R, Pérombelon MCM, van Veen JA, van der Wolf JM. Control of blackleg and tuber soft rot of potato caused by *Pectobacterium* and *Dickeya* species: a review. *Plant Pathol.* 2011;60:999–1013.
2. Toth I, van der Wolf JM, Saddler G, Lojkowska E, Hélias V, Pirhonen M, Tsror (Lakhim) L, Elphinstone JG. *Dickeya* species: an emerging problem for potato production in Europe. *Plant Pathol.* 2011;60:385–399.
3. Mansfield J, Genin S, Magori S, Citovsky V, Sriariyanum M, Ronald P, Dow M, Verdier V, Beer SV, Machado MA, Toth I, Salmond G, Foster GD. Top 10 plant pathogenic bacteria in molecular plant pathology. *Molec Plant Pathol.* 2012;13:614–629.
4. Hauben L, Moore ERB, Vauterin L, Steenackers M, Mergaert J, et al. Phylogenetic position of phytopathogens within the *Enterobacteriaceae*. *Syst Appl Microbiol.* 1998;21:384–397.
5. Samson R, Legendre JB, Christen R, Saux MFL, Achouak W, Gardan L. Transfer of *Pectobacterium chrysanthemi* (Burkholder et al. 1953) Brenner et al. 1973 and *Brenneria paradisiaca* to the genus *Dickeya* gen. nov. as *Dickeya chrysanthemi* comb. nov. and *Dickeya paradisiaca* comb. nov. and delineation of four novel species, *Dickeya dadantii* sp. nov., *Dickeya dianthicola* sp. nov., *Dickeya dieffenbachiae* sp. nov. and *Dickeya zeae* sp. *Int J System Evol Microbiol.* 2005;55:1415–1427.
6. Parkinson N, DeVos P, Pirhonen M, Elphinstone J. *Dickeya aquatica* sp. nov., isolated from waterways. *Int J System Evol Microbiol.* 2014;64:2264–2266.
7. van der Wolf JM, Nijhuis EH, Kowalewska MJ, Saddler GS, Parkinson N, Elphinstone JG, Pritchard L, Toth IK, Lojkowska E, Potrykus M, Waleron M, de Vos P, Cleenwerck I, Pirhonen M, Garlant L, Hélias V, Pothier JF, Pflüger V, Duffy B, Tsror L, Manulis S. *Dickeya solani* sp. nov., a pectinolytic plant pathogenic bacterium isolated from potato (*Solanum tuberosum*). *Int J System Evol Microbiol.* 2014;64:768–774.
8. Hugouvieux-Cotte-Pattat N, Condemine G, Nasser W, Reverchon S. Regulation of pectinolysis in *Erwinia chrysanthemi*. *Annu Rev Microbiol.* 1996;50:213–257.
9. Glasner JD, Yang CH, Reverchon S, Hugouvieux-Cotte-Pattat N, Condemine G, Bohin JP, Van Gijsegem F, Yang S, Franza T, Expert D, Plunkett 3rd G, San Francisco MJ, Charkowski AO, Py B, Bell K, Rauscher L, Rodriguez-Palenzuela P, Toussaint A, Holeva MC, He SY, Douet V, Boccara M, Blanco C, Toth I, Anderson BD, Biehl BS, Mau B, Flynn SM, Barras F, Lindeberg M, Birch PR, Tsuyumu S, Shi X, Hibbing M, Yap MN, Carpentier M, Dassa E, Umehara M, Kim JF, Rusch M, Soni P, Mayhew GF, Fouts DE, Gill SR, Blattner FR, Keen NT, Perna NT. Genome sequence of the plant-pathogenic bacterium *Dickeya dadantii* 3937. *J Bacteriol.* 2011;193:2076–2077.

10. Charkowski A, Blanco C, Condemine G, Expert D, Franza T, Hayes C, Hugouvieux-Cotte-Pattat N, Lopez Solanilla E, Low D, Moleleki L, Pirhonen M, Pitman A, Perna N, Reverchon S, Rodriguez Palenzuela P, San Francisco M, Toth I, Tsuyumu S, van der Walls J, van der Wolf J, Van Gijsegem F, Yang CH, Yedidia I. The role of secretion systems and small molecules in soft-rot enterobacteriaceae pathogenicity. *Annu Rev Phytopathol.* 2012;50:425–449.
11. Reverchon S, Nasser W. *Dickeya* ecology, environment sensing and regulation of virulence programme. *Environ Microbiol Rep.* 2013;5:622–636.
12. Glazebrook J. Contrasting mechanisms of defense against biotrophic and necrotrophic pathogens. *Annu Rev Phytopathol.* 2005;43:205–227.
13. Alfano JR, Collmer A. Bacterial pathogens in plants: life up against the wall. *Plant Cell.* 1996;8:1683–1698.
14. Hugouvieux-Cotte-Pattat N, Condemine G, Shevchik VE. Bacterial pectate lyases, structural and functional diversity. *Environ Microbiol Rep.* 2014;6:427–440.
15. Río-Álvarez I, Muñoz-Gómez C, Navas-Vásquez M, Martínez-García PM, Antúnez-Lamas M, Rodríguez-Palenzuela P, López-Solanilla E. Role of *Dickeya dadantii* 3937 chemoreceptors in the entry to *Arabidopsis* leaves through wounds. *Mol Plant Pathol.* 2015;16:685–698.
16. Chapelle E, Alunni B, Malfatti P, Solier L, Pédron J, Kraepiel Y, Van Gijsegem F. A straightforward and reliable method for bacterial in planta transcriptomics: application to the *Dickeya dadantii/Arabidopsis thaliana* pathosystem. *Plant J.* 2015;82: 352–362.
17. Effantin G, Rivasseau C, Gromova M, Bligny R, Hugouvieux-Cotte-Pattat N. Massive production of butanediol during plant infection by phytopathogenic bacteria of the genera *Dickeya* and *Pectobacterium*. *Mol Microbiol.* 2011;82:988–997.
18. Okinaka Y, Yang CH, Perna NT, Keen NT. Microarray profiling of *Erwiniachrysanthemi* 3937 genes that are regulated during plant infection. *Mol Plant Microbe Interact.* 2002;15:619–629.
19. Fagard M, Dellagi A, Roux C, Périno C, Rigault M, Boucher V, Shevchik VE, Expert D. *Arabidopsis thaliana* expresses multiple lines of defense to counterattack *Erwinia chrysanthemi*. *Mol Plant Microbe Interact.* 2007;20:794–805.
20. Fu SF, Tsai TM, Chen YR, Liu CP, Haiso LJ, Syue LH, Yeh HH, Huang HJ. Characterization of the early response of the orchid, *Phalaenopsis amabilis*, to *Erwinia chrysanthemi* infection using expression profiling. *Physiol Plant.* 2012;145:406–425.
21. De Lorenzo G, Brutus A, Savatin DV, Sicilia F, Cervone F. Enginering plant resistance by constructing chimeric receptors that recognize damage-associated molecular patterns (DAMPs). *FEBS Lett.* 2011;585:1521–1528.
22. Dorokhov YL, Komarova TV, Petrunia IV, Frolova OY, Pozdyshev DV, Gleba YY. Airborne signals from a wounded leaf facilitate viral spreading and induce antibacterial resistance in neighboring plants. *PLoS Pathog.* 2012;8(4):e1002640.
23. Dorokhov YL, Komarova TV, Petrunia IV, Kosorukov VS, Zinovkin RA, Shindyapina AV, Frolova OY, Gleba YY. Methanol may function as a cross-kingdom signal. *PLoS One.* 2012;7(4):e36122.
24. Ryu CM, Farag MA, Hu CH, Reddy MS, Kloepper JW, Paré PW. Bacterial volatiles induce systemic resistance in *Arabidopsis*. *Plant Physiol.* 2004;134:1017–1026.
25. Kim YC, Leveau J, McSpadden Gardener BB, Pierson EA, Pierson LS, Ryu CM. The multifactorial basis for plant health promotion by plant-associated bacteria. *Appl Environ Microbiol.* 2011;77:1548–1555.
26. Han SH, Lee SJ, Moon JH, Park KH, Yang KY, Cho BH, et al. GacS-dependent production of 2R, 3R-butanediol by *Pseudomonas chlororaphis* O6 is a major determinant for eliciting systemic resistance against *Erwinia carotovora* but not against *Pseudomonas syringae* pv. *tabaci* in tobacco. *Mol Plant Microbe Interact.* 2006;19:924–930.

27. Tardy F, Nasser W, Robert-Baudouy J, Hugouvieux-Cotte-Pattat N. Comparative analysis of the five major *Erwinia chrysanthemi* pectate lyases: enzyme characteristics and potential inhibitors. *J Bacteriol.* 1997;179:2503–2511.
28. Pissavin C, Robert-Baudouy J, Hugouvieux-Cotte-Pattat N. Biochemical characterization of the pectate lyase PelZ of *Erwinia chrysanthemi* 3937. *Biochim Biophys Acta.* 1998;1383:188–196.
29. Nasser W, Shevchik VE, Hugouvieux-Cotte-Pattat N. Analysis of three clustered polygalacturonase genes in *Erwiniachrysanthemi* 3937 revealed an anti-repressor function for the PecS regulator. *Mol Microbiol.* 1999;34:641–650.
30. Shevchik VE, Condemine G, Robert-Baudouy J, Hugouvieux-Cotte-Pattat N. The exopolygalacturonate lyase PelW and the oligogalacturonate lyase Ogl, two cytoplasmic enzymes of pectin catabolism in *Erwinia chrysanthemi* 3937. *J Bacteriol.* 1999;181: 3912–3919.
31. Nasser W, Condemine G, Plantier R, Anker D, Robert-Baudouy J. Inducing properties of analogs of 2-keto-3-deoxygluconate on the expression of pectinase genes of *Erwinia chrysanthemi*. *FEMS Microbiol Lett.* 1991;65:73–78.
32. Hugouvieux-Cotte-Pattat N, Shevchik VE, Nasser W. PehN, a polygalacturonase homologue with a low hydrolase activity, is coregulated with the other *Erwinia chrysanthemi* polygalacturonases. *J Bacteriol.* 2002;184:2664–2673.
33. Laatu M, Condemine G. Rhamnogalacturonate lyase RhiE is secreted by the Out system in *Erwinia chrysanthemi*. *J Bacteriol.* 2003;185:1642–1649.
34. Lojkowska E, Masclaux C, Boccara M, Robert-Baudouy J, Hugouvieux-Cotte-Pattat N. Characterization of the *pelL* gene encoding a novel pectate lyase of *Erwinia chrysanthemi* 3937. *Mol Microbiol.* 1995;16:1183–1195.
35. Hassan S, Shevchik VE, Robert X, Hugouvieux-Cotte-Pattat N. PelN is a new pectate lyase of *Dickeya dadantii* with unusual characteristics. *J Bacteriol.* 2013;195: 2197–2206.
36. Shevchik VE, Kester HC, Benen JA, Visser J, Robert-Baudouy J, Hugouvieux-Cotte-Pattat N. Characterization of the exopolygalacturonate lyase PelX of *Erwinia chrysanthemi* 3937. *J Bacteriol.* 1999;181:1652–1663.
37. Fries M, Ihrig J, Brocklehurst K, Shevchik VE, Pickersgill RW. Molecular basis of the activity of the phytopathogen pectin methylesterase. *EMBO J.* 2007;26: 3879–3887.
38. Shevchik VE, Condemine G, Hugouvieux-Cotte-Pattat N, Robert-Baudouy J. Characterization of pectin methylesterase B, an outer membrane lipoprotein of *Erwinia chrysanthemi* 3937. *Mol Microbiol.* 1996;19:455–466.
39. Shevchik VE, Hugouvieux-Cotte-Pattat N. PaeX, a second pectin acetylesterase of *Erwinia chrysanthemi* 3937. *J Bacteriol.* 2003;185:3091–3100.
40. Hassan S, Hugouvieux-Cotte-Pattat N. Identification of two feruloyl esterases in *Dickeya dadantii* 3937 and induction of the major feruloyl esterase and of pectate lyases by ferulic acid. *J Bacteriol.* 2011;193:963–970.
41. Shevchik VE, Hugouvieux-Cotte-Pattat N. Identification of a bacterial pectin acetyl esterase in *Erwinia chrysanthemi* 3937. *Mol Microbiol.* 1997;24:1285–1301.
42. Van Gijsegem F, Wlodarczyk A, Cornu A, Reverchon S, Hugouvieux-Cotte-Pattat N. Analysis of the LacI family regulators of *Erwinia chrysanthemi* 3937, involvement in the bacterial phytopathogenicity. *Mol Plant Microbe Interact.* 2008;21:1471–1481.
43. Hugouvieux-Cotte-Pattat N. The RhaS activator controls the *Erwinia chrysanthemi* 3937 genes *rhiN*, *rhiT* and *rhiE* involved in rhamnogalacturonan catabolism. *Mol Microbiol.* 2004;51:1361–1374.
44. el Hassouni M, Chippaux M, Barras F. Analysis of the *Erwinia chrysanthemi arb* genes, which mediate metabolism of aromatic beta-glucosides. *J Bacteriol.* 1990;172: 6261–6267.

45. Delangle A, Prouvost AF, Cogez V, Bohin JP, Lacroix JM, Cotte-Pattat NH. Characterization of the *Erwinia chrysanthemi gan* locus, involved in galactan catabolism. *J Bacteriol.* 2007;189:7053–7061.
46. Charaoui-Boukerzaza S, Hugouvieux-Cotte-Pattat N. A family 3 glycosyl hydrolase of *Dickeya dadantii* 3937 is involved in the cleavage of aromatic glucosides. *Microbiology.* 2013;159:2395–2404.
47. Boccara M, Aymeric JL, Camus C. Role of endoglucanases in *Erwinia chrysanthemi* 3937 virulence on *Saintpaulia ionantha. J Bacteriol.* 1994;176:1524–1526.
48. Keen NT, Boyd C, Henrissat B. Cloning and characterization of a xylanase gene from corn strains of *Erwinia chrysanthemi. Mol Plant Microbe Interact.* 1996;9:651–657.
49. Hugouvieux-Cotte-Pattat N, Charaoui-Boukerzaza S. Catabolism of raffinose, sucrose, and melibiose in *Erwinia chrysanthemi* 3937. *J Bacteriol.* 2009;191:6960–6967.
50. Scheller and Uluskov, 2010 Scheller HV, Ulvskov P. Hemicelluloses. *Annu Rev Plant Biol.* 2010; 61:263–289
51. Caffall KH, Mohnen D. The structure, function, and biosynthesis of plant cell wall pectic polysaccharides. *Carbohydr Res.* 2009;344:1879–1900.
52. Cantarel BL, Coutinho PM, Rancurel C, Bernard T, Lombard V, Henrissat B. The Carbohydrate-Active EnZymes database (CAZy): an expert resource for glycogenomics. *Nucleic Acids Res.* 2009;37:D233–D238.
53. Korotkov KV, Sandkvist M, Hol WG. The type II secretion system: biogenesis, molecular architecture and mechanism. *Nat Rev Microbiol.* 2012;10:336–351.
54. Nivaskumar M, Francetic O. Type II secretion system: a magic beanstalk or a protein escalator. *Biochim Biophys Acta.* 2014;1843:1568–1577.
55. Kazemi-Pour N, Condemine G, Hugouvieux-Cotte-Pattat N. The secretome of the plant pathogenic bacterium *Erwinia chrysanthemi. Proteomics.* 2004;4:3177–3186.
56. Peekhaus N, Conway T. What's for dinner?: Entner–Doudoroff metabolism in *Escherichia coli. J Bacteriol.* 1998;180:3495–3502.
57. Keyhani NO, Roseman S. Wild-type *Escherichia coli* grows on the chitin disaccharide, *N,N′*-diacetylchitobiose, by expressing the *cel* operon. *Proc Natl Acad Sci USA.* 1997;94:14367–14371.
58. Yang S, Perna NT, Cooksey DA, Okinaka Y, Lindow SE, Ibekwe AM, Keen NT, Yang CH. Genome-wide identification of plant-upregulated genes of *Erwinia chrysanthemi* 3937 using a GFP-based IVET leaf array. *Mol Plant Microbe Interact.* 2004;17: 999–1008.
59. Grenier AM, Duport G, Pages S, Condemine G, Rahbe Y. The phytopathogen *Dickeya dadantii* (*Erwinia chrysanthemi* 3937) is a pathogen of the pea aphid. *Appl Environ Microbiol.* 2006;72:1956–1965.
60. Costechareyre D, Chich JF, Strub JM, Rahbé Y, Condemine G. Transcriptome of *Dickeya dadantii* infecting *Acyrthosiphon pisum* reveals a strong defense against antimicrobial peptides. *PLoS One.* 2013;8(1):e54118.
61. Marquez-Villavicencio Mdel P, Weber B, Witherell RA, Willis DK, Charkowski AO. The 3-hydroxy-2-butanone pathway is required for *Pectobacterium carotovorum* pathogenesis. *PLoS One.* 2011;6(8):e22974.
62. Johansen L, Bryn K, Stormer FC. Physiological and biochemical role of the butanediol pathway in *Aerobacter* (*Enterobacter*) *aerogenes. J Bacteriol.* 1975;123:1124–1130.
63. Moons P, Van Houdt R, Vivijs B, Michiels CW, Aertsen A. Integrated regulation of acetoin fermentation by quorum sensing and pH in *Serratia plymuthica* RVH1. *Appl Environ Microbiol.* 2011;77:3422–3427.
64. Blot N, Berrier C, Hugouvieux-Cotte-Pattat N, Ghazi A, Condemine G. The oligogalacturonate-specific porin KdgM of *Erwinia chrysanthemi* belongs to a new porin family. *J Biol Chem.* 2002;277:7936–7944.

65. Condemine G, Ghazi A. Differential regulation of two oligogalacturonate outer membrane channels, KdgN and KdgM, of *Dickeya dadantii* (*Erwinia chrysanthemi*). *J Bacteriol.* 2007;189:5955–5962.
66. Hugouvieux-Cotte-Pattat N, Blot N, Reverchon S. Identification of TogMNAB, an ABC transporter which mediates the uptake of pectic oligomers in *Erwinia chrysanthemi* 3937. *Mol Microbiol.* 2001;41:1113–1123.
67. Hugouvieux-Cotte-Pattat N, Reverchon S. Two transporters, TogT and TogMNAB, are responsible for oligogalacturonide uptake in *Erwinia chrysanthemi* 3937. *Mol Microbiol.* 2001;41:1125–1132.
68. Condemine G, Robert-Baudouy J. 2-keto-3-deoxygluconate transport system in *Erwinia chrysanthemi*. *J Bacteriol.* 1987;169:1972–1978.
69. San Francisco MJ, Keenan RW. Uptake of galacturonic acid in *Erwinia chrysanthemi* EC16. *J Bacteriol.* 1993;175:4263–4265.
70. Hugouvieux-Cotte-Pattat N, Dominguez H, Robert-Baudouy J. Environmental conditions affect transcription of the pectinase genes of *Erwinia chrysanthemi* 3937. *J Bacteriol.* 1992;174:7807–7818.
71. Shevchik VE, Robert-Baudouy J, Hugouvieux-Cotte-Pattat N. Pectate lyase PelI of *Erwinia chrysanthemi* 3937 belongs to a new family. *J Bacteriol.* 1997;179:7321–7330.
72. Nasser W, Bouillant ML, Salmond G, Reverchon S. Characterization of the *Erwinia chrysanthemi expI-expR* locus directing the synthesis of two N-acyl-homoserine lactone signal molecules. *Mol Microbiol.* 1998;29:1391–1405.
73. Nasser W, Dorel C, Wawrzyniak J, Van Gijsegem F, Groleau MC, Déziel E, Reverchon S. Vfm a new quorum sensing system controls the virulence of *Dickeya dadantii*. *Environ Microbiol.* 2013;15:865–880.
74. Sepulchre JA, Reverchon S, Nasser W. Modeling the onset of virulence in a pectinolytic bacterium. *J Theor Biol.* 2007;244:239–257.
75. Kepseu WD, Sepulchre JA, Reverchon S, Nasser W. Toward a quantitative modeling of the synthesis of the pectate lyases, essential virulence factors in *Dickeya dadantii*. *J Biol Chem.* 2010;285:28565–28576.
76. Rodionov DA, Gelfand MS, Hugouvieux-Cotte-Pattat N. Comparative genomics of the KdgR regulon in *Erwinia chrysanthemi* 3937 and other gammaproteobacteria. *Microbiology.* 2004;150:3571–3590.
77. Oh MK, Rohlin L, Kao KC, Liao JC. Global expression profiling of acetate-grown *Escherichia coli*. *J Biol Chem.* 2002;277:13175–13183.
78. Surgey N, Robert-Baudouy J, Condemine G. The *Erwinia chrysanthemi pecT* gene regulates pectinase gene expression. *J Bacteriol.* 1996;178:1593–1599.
79. Görke B, Stülke J. Carbon catabolite repression in bacteria: many ways to make the most out of nutrients. *Nat Rev Microbiol.* 2008;6:613–624.
80. Saier Jr MH. Multiple mechanisms controlling carbon metabolism in bacteria. *Biotechnol Bioeng.* 1998;58:170–174.
81. Reverchon S, Expert D, Robert-Baudouy J, Nasser W. The cyclic AMP receptor protein is the main activator of pectinolysis genes in *Erwinia chrysanthemi*. *J Bacteriol.* 1997;179:3500–3508.
82. Lebeau A, Reverchon S, Gaubert S, Kraepiel Y, Simond-Cote E, Nasser W, Van Gijsegem F. The GacA global regulator is required for the appropriate expression of *Erwinia chrysanthemi* 3937 pathogenicity genes during plant infection. *Environ Microbiol.* 2008;10:545–559.
83. Yang S, Peng Q, Zhang Q, Yi X, Choi CJ, Reedy RM, Charkowski AO, Yang CH. Dynamic regulation of GacA in type III secretion, pectinase gene expression, pellicle formation, and pathogenicity of *Dickeya dadantii* (*Erwinia chrysanthemi* 3937). *Mol Plant Microbe Interact.* 2008;21:133–142.

84. Vakulskas CA, Potts AH, Babitzke P, Ahmer BM, Romeo T. Regulation of bacterial virulence by Csr (Rsm) systems. *Microbiol Mol Biol Rev.* 2015;79:193–224.
85. Chavez RG, Alvarez AF, Romeo T, Georgellis D. The physiological stimulus for the BarA sensor kinase. *J Bacteriol.* 2010;192:2009–2012.
86. Takeuchi K, Kiefer P, Reimmann C, Keel C, Dubuis C, Rolli J, Vorholt JA, Haas D. Small RNA-dependent expression of secondary metabolism is controlled by Krebs cycle function in *Pseudomonas fluorescens*. *J Biol Chem.* 2009;284:34976–34985.
87. Romeo T, Gong M, Liu MY, Brun-Zinkernagel AM. Identification and molecular characterization of *csrA*, a pleiotropic gene from *Escherichia coli* that affects glycogen biosynthesis, gluconeogenesis, cell size, and surface properties. *J Bacteriol.* 1993;175: 4744–4755.
88. Papenfort K, Vogel J. Small RNA functions in carbon metabolism and virulence of enteric pathogens. *Front Cell Infect Microbiol.* 2014;4:91.
89. Cui Y, Chatterjee A, Chatterjee AK. Effects of the two-component system comprising GacA and GacS of *Erwinia carotovora* subsp. *carotovora* on the production of global regulatory *rsmB* RNA, extracellular enzymes, and harpinEcc. *Mol Plant Microbe Interact.* 2001;14:516–526.
90. Mukherjee A, Cui Y, Liu Y, Dumenyo CK, Chatterjee AK. Global regulation in *Erwinia* species by *Erwinia carotovora rsmA*, a homologue of *Escherichia coli csrA*: repression of secondary metabolites, pathogenicity and hypersensitive reaction. *Microbiology.* 1996;142:427–434.
91. Liu Y, Cui Y, Mukherjee A, Chatterjee AK. Characterization of a novel RNA regulator of *Erwinia carotovora* ssp. *carotovora* that controls production of extracellular enzymes and secondary metabolites. *Mol Microbiol.* 1998;29:219–234.
92. Hyytiäinen H, Montesano M, Palva ET. Global regulators ExpA (GacA) and KdgR modulate extracellular enzyme gene expression through the RsmA-rsmB system in *Erwinia carotovora* subsp. *carotovora*. *Mol Plant Microbe Interact.* 2001;14: 931–938.
93. Cui Y, Chatterjee A, Liu Y, Dumenyo CK, Chatterjee AK. Identification of a global repressor gene, *rsmA*, of *Erwinia carotovora* subsp. *carotovora* that controls extracellular enzymes, N-(3-oxohexanoyl)-L-homoserine lactone, and pathogenicity in soft-rotting *Erwinia* spp. *J Bacteriol.* 1995;177:5108–5115.
94. Adamson DN, Lim HN. Rapid and robust signaling in the CsrA cascade via RNA-protein interactions and feedback regulation. *Proc Natl Acad Sci USA.* 2013;110: 13120–13125.
95. Potrykus M, Golanowska M, Hugouvieux-Cotte-Pattat N, Lojkowska E. Regulators involved in *Dickeya* solani virulence, genetic conservation, and functional variability. *Mol Plant Microbe Interact.* 2014;27:700–711.
96. Hommais F, Oger-Desfeux C, Van Gijsegem F, Castang S, Ligori S, Expert D, et al. PecS is a global regulator of the symptomatic phase in the phytopathogenic bacterium *Erwinia chrysanthemi* 3937. *J Bacteriol.* 2008;190:7508–7522.
97. Franza T, Michaud-Soret I, Piquerel P, Expert D. Coupling of iron assimilation and pectinolysis in *Erwinia chrysanthemi* 3937. *Mol Plant Microbe Interact.* 2002;15: 1181–1191.
98. Ouafa ZA, Reverchon S, Lautier T, Muskhelishvili G, Nasser W. The nucleoid-associated proteins H-NS and FIS modulate the DNA supercoiling response of the *pel* genes, the major virulence factors in the plant pathogen bacterium *Dickeya dadantii*. *Nucleic Acids Res.* 2012;40:4306–4319.
99. Duprey A, Reverchon S, Nasser W. Bacterial virulence and Fis: adapting regulatory networks to the host environment. *Trends Microbiol.* 2014;22:92–99.
100. Travers A, Muskhelishvili G. DNA supercoiling—a global transcriptional regulator for enterobacterial growth? *Nat Rev Microbiol.* 2005;3:157–169.

101. Jiang X, Sobetzko P, Nasser W, Reverchon S, Muskhelishvili G. Chromosomal "stress-response" domains govern the spatiotemporal expression of the bacterial virulence program. *mBio.* 2015;6:e00353–e00415.
102. Rondelet A, Condemine G. SurA is involved in the targeting to the outer membrane of a Tat signal sequence-anchored protein. *J Bacteriol.* 2012;194:6131–6142.
103. Charkowski AO, Alfano JR, Preston G, Yuan J, He SY, Collmer A. The *Pseudomonas syringae* pv. *tomato* HrpW protein has domains similar to harpins and pectate lyases and can elicit the plant hypersensitive response and bind to pectate. *J Bacteriol.* 1998;180:5211–5217.

CHAPTER FIVE

Multidrug Efflux Pumps in the Genus *Erwinia*: Physiology and Regulation of Efflux Pump Gene Expression

J. Thekkiniath*[,1], R. Ravirala[†], M. San Francisco[‡]

[*] Department of Medicine, University of Massachusetts Medical School, Worcester, MA, United States
[†] Roche Molecular System, Pleasanton, CA, United States
[‡] Department of Biological Sciences, Texas Tech University, Lubbock, TX, United States
[1] Corresponding author. E-mail address: Jose.Thekkiniath@umassmed.edu

Contents

Abstract

Plant pathogens belonging to the genus *Erwinia* cause diseases in several economically important plants. Plants respond to bacterial infection with a powerful chemical arsenal and signaling molecules to rid themselves of the microbes. Although our understanding of how *Erwinia* initiate infections in plants has become clear, a comprehensive understanding of how these bacteria rid themselves of noxious

Progress in Molecular Biology and Translational Science, Volume 142
ISSN 1877-1173
http://dx.doi.org/10.1016/bs.pmbts.2016.05.011

antimicrobial agents during the infection is important. Multidrug efflux pumps are key factors in bacterial resistance toward antibiotics by reducing the level of antimicrobial compounds in the bacterial cell. *Erwinia* induce the expression of efflux pump genes in response to plant-derived antimicrobials. The capability of *Erwinia* to co-opt plant defense signaling molecules such as salicylic acid to trigger multidrug efflux pumps might have developed to ensure bacterial survival in susceptible host plants. In this review, we discuss the developments in *Erwinia* efflux pumps, focusing in particular on efflux pump function and the regulation of efflux pump gene expression.

1. INTRODUCTION

1.1 Plant–Pathogen Interaction: An Overview

Plants live in multifaceted environments and closely associate with microbial pathogens with various lifestyles and infection strategies. The evolutionary arms battle between plants and microbial pathogens has endowed plants with elaborate defense systems. In response to the invasion by bacteria, plants initiate defense mechanisms leading to the production of secondary metabolites including phytoalexins, peptides, and alkaloids that play a role in protection.[1–3] Multidrug resistance (MDR) transporters disseminate an array of antimicrobial compounds and are critical for bacterial survival in hostile host environments.[4,5] To successfully colonize a host, bacteria must be able to surmount defensive barriers. MDR transporters are implicated in the expulsion of a broad variety of antimicrobial chemicals and are very important for bacterial survival in hostile host environment.[6]

2. BACTERIAL EFFLUX PUMPS

2.1 Types

Bacterial efflux pumps are classified into five families including ATP-binding cassette (ABC),[7] major facilitator superfamily (MFS),[8] resistance/nodulation/cell division (RND),[9] small multidrug resistance (SMR),[10] and multidrug and toxic compound extrusion (MATE).[11] On the basis of energy sources, these efflux pumps are grouped into transporters that employ ATP hydrolysis such as ABC transporters and those that employ the proton motive force or Na^+/H^+ for expelling drugs (SMR, MATE, MFS, and RND pumps).[12]

2.2 Occurrence

Of the MDR systems, efflux pumps belonging to RND- and MFS-type transporters are widespread in Gram-negative bacteria. In general, efflux pumps in Gram-negative bacteria consist of cytoplasmic, periplasmic, and outer-membrane proteins that assemble to form multicomponent transporters.[12] Each of the several efflux pumps encoded in a given bacterial genome may have a different function. Several successful pathogens grouped in the family *Enterobactericeae* employ multidrug efflux pumps that utilize an outer membrane protein TolC to form a continuous channel across the cytoplasmic membrane, the periplasm and the outer membrane. The relevance of MDR pumps in bacterial and fungal pathogens of animals have been recently reviewed.[13,14] For example, AcrAB-TolC in *Escherichia coli*[15] and MexABOprM in *Pseudomonas aeruginosa*[16] are well known tripartite efflux pumps belonging to the RND family and are involved in resistance to antibiotics.

3. EFFLUX PUMPS IN *ERWINIA*

In phytopathogens and plant-associated bacteria only very few MDR systems have been studied in detail so far. In plant-associated bacteria, multidrug efflux pumps belonging to the MFS and RND superfamily have been investigated. These bacteria include the symbiotic nitrogen-fixer *Rhizobium etli*,[17] *Bradyrhizobiumjaponicum*,[18] and *Agrobacteriumtumefaciens* strain C58 (*Agrobacterium fabrum*).[19,20] A previous study has reviewed the functional role ofbacterial efflux pumps in natural ecosystems including plant–bacteria interactions.[21] Here we focus on the efflux pumps in one of the major plant pathogens, the genus *Erwinia*. This genus consists of (1) "soft-rot group," which comprises *Erwinia carotovora* and *Erwiniachrysanthemi* that are responsible for tissue-macerating plant diseases and storage rots. The nomenclature for *E. chrysanthemi* has been changed to *Dickeyadadantii*. (2) The "amylovora group," which includes species that are able to produce neither pectolytic enzymes nor yellow pigments, but cause necrotic and/or wilt diseases of plants. *E. amylovora* is the causal organism of the fire blight disease of apple, pear, and other members of the Rosaceae family.

3.1 RND Efflux Pumps

3.1.1 Acr (Acriflavine) Efflux System

A functional RND-type efflux pump constitutes a tripartite complex including the RND-type transporter protein located in the inner membrane, a periplasmic membrane protein, and an outer membrane channel.[22] In the family *Enterobactericeae*, the acrAB operon encodes a membrane fusion protein, AcrA, located in the periplasm and a membrane protein AcrB.[23] These proteins together with the outer membrane protein TolC expel a broad range of antimicrobial chemicals.[24–28]

Among the efflux pumps, AcrAB is known to be the major one in *E. coli* and is highly conserved in Gram-negative bacteria.[12,29] It has been demonstrated that all components of AcrAB efflux system are crucial for effective transport, because mutation in any of these genes leads to hypersusceptibility of *E. coli* to many substrate compounds.[30]

Genomic analyses of phytopathogens such as *D. dadantii* and *E. amylovora* revealed the presence of homologs of AcrAB from *E. coli* and other Gram-negative bacteria.[31–33] Interestingly, *D. dadantii* encodes two AcrAB homologs including Acr1AB and Acr2AB efflux pumps. The significance of the AcrAB efflux system in these bacteria has been demonstrated by mutational analyses and pathogenesis assays.[32–35] In *D. dadantii*, both AcrAB homologs have been shown to be essential for full virulence on chicory leaves. However, only the Acr1AB system contributed to virulence in Saintpaulia leaves.[33] Additionally, this study suggested the mutations in *acr1AB* and *acr2AB* systems cause susceptibility to a wide range of plant antimicrobials. This observation suggests that individual efflux systems may play unrelated roles in diverse hosts. During interaction with plants, *E. amylovora*, is exposed to a wide range of plant-derived antimicrobial chemicals, including isoflavonoids.[34] It has been shown that bacterial efflux pumps implicated in resistance toward structurally dissimilar compounds might provide tolerance to the phytoalexins.[34] AcrAB of *E. amylovora* has been known to exhibit a related substrate range similar to that of *E. coli*. Additionally, in *E. amylovora*, it was demonstrated that a mutant lacking *acrB* was compromised in virulence on apple rootstock MM 106. This mutant was susceptible to phytoalexins including phloretin, naringenin, quercetin, and (+)-catechin,[34] suggesting the significance of AcrAB efflux pump in the virulence of *E. amylovora* in resistance toward apple phytoalexins and for colonization of host plants.[34] Additionally, disruption of *E. amylovora* acrB led to a dramatic decrease in its capacity to cause fire blight symptoms.

Environmental signals have been involved in triggering efflux pump gene expression in bacteria. In certain Gram-negative bacteria, plant compounds including salicylic acid, naphthoquinones, hydrogen peroxide and paraquat have been implicated in inducing efflux pump gene expression. The majority of these chemicals are produced as plant defense mechanisms in response to attack by pathogens. Salicylic acid is known to be a key signaling molecule produced by plants that elicits resistance to plant pathogens.[36] The study by Ravirala et al. demonstrated that efflux pumps are important in facilitating plant-bacteria interplay including the response to noxious compounds.[37] This study showed that in *D. dadantii*, *acrAB* and *emrAB* efflux pump genes were found to be expressed in plant leaves immediately following exposure to bacteria, indicating plant environment may be eliciting efflux pump gene expression. Interestingly, salicylic acid together with its precursors were shown to be inducers for *acrAB* and *emrAB* efflux pump gene expression in this bacterium.[37] The capability of *D. dadantii* to co-opt plant defense signaling molecules such as salicylic acids to trigger multidrug efflux pumps might have developed to ensure bacterial survival in susceptible host plants. Salicylic acid has also been implicated in eliciting efflux pump gene expression in *Burkholderia cepacia*, an important human pathogen which lives in the water and soil.[38] These studies suggest that plant and animal pathogens may employ certain common strategies for virulence.

The outer membrane TolC has been involved in MDR in human pathogens including[39–41] *Salmonella enterica* [42,43] and *Borrelia bugdorfei*.[44] TolC has been characterized in plant pathogens including *D. dadantii* and *E. amylovora*. A study by Barabote et al showed that *D. dadantii* carries a functional homologue of the *E. coli* TolC protein.[32] This study demonstrated that a *D. dadantii* TolC mutant was severely compromised in plant tissue maceration and in causing soft-rot disease. Interestingly, the defect in virulence of the *D. dadantii* TolC mutant was used to elucidate certain survival and virulence tactics used by animal and plant pathogens. Additionally, the TolC mutant showed hypersensitivity to plant-based antimicrobial compounds including berberine, rhein, plumbagin, pyrithione, genistein, p-coumaric acid and t-cinnamic acid (phenolic acids), and esculetin indicating that TolC-mediated resistance to antimicrobial compounds may be essential in phytopathogenesis. Using a plant-derived chemical, berberine, it was confirmed that sensitivity to plant chemicals may result from lack of efflux of the antimicrobial compounds. This suggests that efflux pumps may play a key role in bacterial survival in the hostile plant milieu.[32]

In *E. amylovora*, it was demonstrated that pathogenicity and the fitness of the bacterium to colonize plant tissue were compromised in *tolC* and *acrB/tolC* mutants.[45] Thus in *E. amylovora*, TolC functions both as a virulence and fitness factor by facilitating resistance against phytoalexins via its interplay with AcrAB efflux pump.[45]

AcrD, another efflux pump belonging to Acriflavine system functions similar to AcrAB and has been implicated in the transport of highly hydrophilic aminoglycosides and amphiphilic compounds in *E. coli*. Although AcrAB functions as a tripartite efflux pump, AcrD acts as a single protein and apparently operates with AcrA that has been known to be co-expressed with AcrB.[46] The characterization of AcrD in *E. amylovora* revealed the substrate specificity of this efflux pump. Additionally, the role of this protein in virulence of apple and pear trees has been established. Although AcrD facilitated resistance to a number of amphiphilic compounds including clotrimazole and luteolin, this protein was incapable of pumping out aminoglycosides. The *acrD* mutant showed complete virulence on apple rootstock and immature pear fruits. Additionally, gene expression analysis uncovered an induction of *acrD* expression in infected apple tissue, while there was no such effect on infected pear fruits. It was found that the expression of *acrD* was induced by exposure to the substrates including deoxycholate, naringenin, tetracycline and zinc.

3.1.2 MdtABCD and MdtUVW Efflux Pumps

The MdtABCD efflux pump of *E. coli* consists of RND transporters including MdtB and MdtC and membrane fusion protein, MdtA. Like AcrAB, MdtABC also functions with an outer membrane channel TolC for expelling toxic compounds. The MdtABC efflux pump in *E. coli* has been implicated in conferring resistance to a wide range of hydrophobic antimicrobial agents including novobiocin and deoxycholate.[47,48] Additionally, MdtABC has been involved in detoxification of certain heavy metals such as zinc, copper and tungstate.[49–51]

Analysis of the *E. amylovora* genome revealed the presence of RND efflux pumps such as MdtABC and MdtUVW which are homologues of *E. coli* MdtABC. A recent study in *E. amylovora* characterized the MdtABC and MdtUVW efflux pumps and demonstrated that these efflux pumps play an important role in survival and proliferation of *E. amylovora* in apple rootstocks.[52] Furthermore, this study showed that overexpression of the MdtABC efflux pump in *E. amylovora* resulted in enhanced resistance toward flavonoids including apigenin, daidzein, and kaempferol and other chemicals such as tannin, fusidic acid, josamycin, novobiocin, bile salts, and silver nitrate. Additionally, it was demonstrated that overexpression of MdtUVW efflux pump caused an augmented resistance toward many flavonoids (apigenin, daidzein, genistein, kaempferol, luteolin, naringenin, orobol) and other compounds including fusidic acid, novobiocin and clotrimazole. Further analysis showed that there exists a common substrate specificity between the MdtABC and MdtUVW pumps from *E. amylovora* as well as MdtABC pumps from *E. coli* and *S. enterica.*[47,53,54] The overlapping substrate specificities of these efflux pumps for novobiocin, bile salts and flavonoids suggest that these compounds may be similar to the natural substrates of the pumps.

The study by Pletzer and Weingart further showed that the MdtABCD and MdtUVW efflux pumps are important for cell envelope stress response of the bacterium.[52] Gene expression analysis showed that genes encoding both efflux pumps were upregulated in vivo. Interestingly, this study also showed that the plant polyphenol tannin induces *mdtABC* gene expression suggesting the importance of MdtABC and MdtUVW efflux pumps in resistance toward antimicrobial plant compounds including flavonoids. Analysis of *D. dadantii* genome revealed the presence of genes encoding homologues of *E. coli* MdtABC. A recent study showed that the MdtABC efflux pump confers resistance toward nalidixic acid, indicating substrate specificity of MdtABC efflux pump in *D. dadantii* (Thekkiniath et al., unpublished data). Additionally, this study identified indole acetic acid may serve as an inducer for the *D. dadantii* MdtABC efflux pump.

3.2 MFS Efflux Pumps

Of the secondary transporters, the MFS, is the largest group and has been implicated in the transport of diverse substrates including antibiotics, sugars, phosphate esters, and oligosaccharides.[55] Here we describe important MFS efflux pumps in the members of the *Erwinia* genus.

3.2.1 EmrAB Efflux Pump

Analysis of the *D. dadantii* genome revealed the presence of EmrAB homolog from *E. coli* and other Gram-negative bacteria.[32,33,56–58] The EmrAB system includes a periplasmic-membrane fusion protein, EmrA, and a membrane protein, EmrB, which contains 12-14 transmembrane domains.[23] Like the AcrAB pump, the EmrAB system also utilizes the TolC outer membrane protein. Interestingly, the EmrAB efflux pump has a narrower substrate range as compared to AcrAB. For example, while the AcrAB system is capable of pumping out bile salts, dyes, detergents, organic solvents, and antibiotics, the EmrAB expels only ionophores and antibiotics.[28] In *E. coli*, the Emr efflux pump confers resistance to distinct groups of toxic chemicals. The involvement of members of the Emr family in drug resistance and phytopathogenicity in *D. dadantii* has been demonstrated.[33] For example, the inability of an *emr1AB* mutant to grow in the presence of potato tuber extract suggests reduced virulence in the plant tissue.

3.2.2 *YceE Efflux Pump*

The YceE protein, another MFS family efflux pump in *D. dadantii* functions as a single membrane channel and has been shown to be upregulated following infection in planta.[59] In *E. coli*, the YceE efflux pump has been shown to confer twofold increase in resistance to deoxycholate and fourfold increase in resistance to fosfomycin when exposed to inducer isopropyl β-D-1-thiogalactopyranoside (IPTG).[60]

3.2.3 *Sugar Efflux Pumps*

In addition to pumping out toxic chemicals produced by host cells, efflux pumps have been shown to be involved in transporting noxious sugars out of the bacterial cell as by-products of metabolism. Moreover, these transporters aid the cells in regulation of osmotic pressure within the cell through efflux of solutes out of the cytoplasm.[61,62] In *D. dadantti*, two different families of the MFS such as SotA and SotB have been identified. Of these, SotA belongs to sugar efflux transporter family, while SotB is a member of multidrug efflux family. These efflux pumps may be involved in reducing the intracellular concentration of noxious sugars or sugar metabolites.[61] Additionally, ExuT, a MFS pump has been characterized and its role in pathogenicity in *D. dadantii* EC16 has been studied.[63] Joko et al. demonstrated that the sugar transporter (MfsX) of the MFS is necessary for flagella-facilitated pathogenesis in *D. dadantii* 3937.[62]

3.3 ABC Transporters

ABC transporters are important membrane proteins that have been implicated in uptake and expulsion of a variety of substrates, including ions and small molecules such as sugars, amino acids, xenobiotics, and vitamins as well as large molecules including peptides, proteins, and polysaccharides. Various studies have demonstrated the relevance of ABC transporters in a number of biological processes including uptake of nutrients, protection of xenobiotics, and removal of cellular waste products, microbial virulence, stress response, lipid transport and large molecule distribution during biogenesis, differentiation, and pathogenesis.[64–66] ABC transporters in bacteria have been involved in antibiotic resistance by pumping the antibiotics to outside the cell[67] suggesting the importance of ABC pumps in phytopathogenic bacteria in virulence and survival in planta.

3.3.1 *YbiT Efflux Pump*

To be a successful phytopathogen, the microbe must be capable of acquiring nutrients in the plant environment, surviving or evading plant defenses, and competing well with other microbial epiphytes for food and space. Since many epiphytes are capable of producing antimicrobial compounds, efflux pumps may play a role in intermicrobial competition. Analysis of the *D. dadantii* genome revealed the presence of a putative ABC transporter, YbiT, which may be important in intermicrobial competition.

The protein YbiT showed identity with those of other Gram-negative bacteria including *E. coli* (87%), *Vibrio cholerae* N16961 (74%), *P. aeruginosa* PA01 (72%), and *Neisseria meningitidis* MD58 (63%). Additionally, YbiT showed a structural similarity with an efflux system in *Streptomyces spp.* that confers resistance to antibiotic groups such as macrolides.[31]

It has been shown that a *ybiT* mutant was able to retain virulence in potato tubers and chicory leaves. However, in coculture experiments with saprophytic bacteria such as *Pseudomonas putida* or *Pseudomonas fluorescens*, this mutant showed lower virulence as compared to the wild type strain.[31] This study suggested that YbiT may be important in pumping out noxious compounds produced by saprophytes and when the transporter is lacking, *D. dadantii* may out-competed indicating its relevance in planta fitness of the bacteria.

3.4 MATE Efflux Pump

The MATE efflux pump family consists of proteins with 12 putative transmembrane segments, but they do not contain typical "signature sequences" as seen in other efflux pumps. Of the MATE efflux systems, two family members such as NorM from *Vibrio parahaemolyticus* and its homolog YdeH from *E. coli* have been well characterized. In *E. coli*, the expression of NorM has been important for conferring resistance to several antibiotics and antimicrobial compounds. Although there are several MATE efflux pumps present in bacteria, their functions are poorly understood.

3.4.1 NorM Efflux Pump

Among phytopathogens, the first MATE efflux transporter, NorM, a single-component efflux system, has been characterized in *E. amylovora.* This protein is highly homologous to efflux pumps belonging to MATE family[68] such as NorM of *E. coli* (73% identity)[69] NorM of *V. parahaemolyticus* (55% identity),[69] and VcmA of *V. cholerae* (50% identity).[70] NorM confers tolerance to the noxious compounds synthesized by epiphytic bacteria co-colonizing plant blossoms. Burse and coworkers demonstrated that a *norM* mutant was completely virulent on apple rootstock, but was sensitive to antimicrobial chemicals produced by *Pantoeaagglomerans*, which is an important epiphyte of apple and quince blossoms. This study indicates that in *E. amylovora*, NorM is important for effective infections in planta. Antibiotic resistance has been identified as an important colonization factor in clinical milieu exposed to antibiotics.[71] Consistently, resistance to antimicrobial chemicals made by epiphytes may be an important during colonization of plant hosts by phytopathogens.

4. MECHANISM OF REGULATION OF EFFLUX PUMP GENE EXPRESSION

It is widely accepted that efflux pumps are abundant in bacterial genomes. However, the expression of the majority of efflux pumps is under tight regulation by many transcriptional factors including local transcriptional regulators, activators or repressors and global regulators.[7] In addition to transcriptional regulators, two-component regulatory systems play an important role in the regulation of efflux pumps. Since efflux pumps are able to expel a variety of structurally different compounds, the improper activation of efflux systems may result in removal of metabolites or other signaling molecules which may further lead to harmful effects on bacterial physiology. Thus expression of MDR transporters is generally well regulated and is simply expressed at a low, basal level under normal laboratory environments.[72] While the structure and function of efflux pumps are conserved in many bacterial species, the mechanisms of gene expression of these efflux pumps differ greatly. In this section, we provide an outline for typical regulation of efflux pump gene expression, focusing on efflux pump gene expression in *Erwinia*.

4.1 Regulation by Local Regulators

Typically, the regulators of the efflux pumps include the members of TetR, MarR or MerR family which often serve as transcriptional repressors. Local repressor genes are often seen adjacent to the structural genes of RND type efflux pumps, and control the expression of efflux systems based on the availability of substrate in the medium. Mutations in the local repressor gene are common features in clinical isolates that exhibit MDR phenotypes, indicating that the local repressor may aid in the prevention of excessive production of efflux pumps. A well-studied family is MarR, which includes diverse regulators that control the expression of genes involved in conferring resistance to multiple antibiotics, organic solvents, and oxidative stress that are collectively termed the multiple antibiotic resistance (Mar) phenotype.[73] The *marR* gene was initially identified as a component of the negative regulator encoded by the *marRAB* locus in *E. coli*.[74,75] Examples include PecS and SlyA in *D. dadantii* 3937 that play an important role in pathogenesis.[1,76–81] SlyA was initially characterized in *Salmonella*, where it was shown to regulate hemolysin and flagella production, both of which are critical for *S. typhimurium* infection.[82–84] In *E. coli*, MarR represses the transcription from *marAB* operon that includes a MarA transcriptional regulator, belonging to the AraC family that regulates a global network of over 60 genes.[73,85]

4.2 Regulation by Global Response Regulators

A well-characterized mechanism is the regulation of AcrAB efflux pump in *E. coli* by three XylS/AraC family regulators, MarA, Rob, and SoxS. The three regulators bind to the marbox on the promoter of *acrAB* in response to different environmental signals and further activate *acrAB* expression.[86] Analysis of *D. dadantii* 3937 genome revealed the presence of a putative *marA* gene. Although the *D. dadantii* 3937 MarA is similar to the *E. coli* counterpart at the N-terminus, it is approximately 150 amino acids longer compared to that of *E. coli*. It is 55% identical and 76% similar to the *E. coli* MarA (Fig. 1). A comparison of the predicted three-dimensional structures of the *E. coli* MarA and *D. dadantii* MarA show that they have similar configurations, suggesting that they may play a similar role in modulating drug resistance (Fig. 2).

Functional characterization of *D. dadantii 3937* MarA revealed that it plays a role in regulating expression of AcrAB and MdtABCD efflux pump genes in presence of phenolic acids similar to *E. coli*. When over-expressed from a multicopy plasmid, MarA was able to confer a Mar phenotype in both the *D. dadantii* as well as *E. coli* backgrounds.

```
E. coli   ---------  MTMSRRNTDAITIHSILDWIEDNLESPLSLEKVSERSGYSKWHLQRMFK
3937      VFLVVSKEDFAMMLSTMPSMVRFTADLVGWIENHLESPLMINDVTAKSGYSKWHLQRIFK
                     * :*   : .    .::.***::***** ::.*: :**********:**

E. coli   KETGHSLGQYIRSRKMTEIAQKLKESNEPILYLAERYGFESQQTLTRTFKNYFDVPPHKY
3937      KETGVSLGSYIRSRRLSKAAIELKLTSQTIQDVALRYCFDSQQSFTRTFKKQFGMSPGHY
          **** ***.*****:::: * :** :.:.*  :* ** *:***::*****: *.:.* :*
```

Figure 1 Amino acid alignment of *E. coli* MarA and amino terminus of the putative *D. dadantii* 3937 MarA. (*) represents identical residues; (:) represents the conserved substitutions; (.) represents the semi-conserved substitutions.

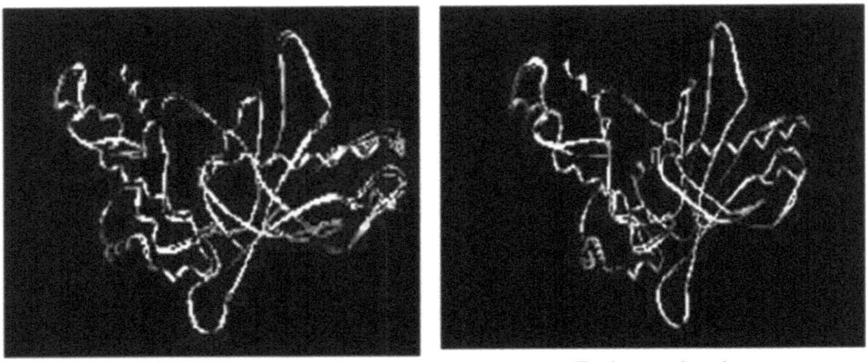

Figure 2 A comparative three-dimensional structure predicted using Swiss-Pdb program for both *E. coli* MarA and *D. dadantii* 3937 MarA (Peitsch MC. Protein modeling by E-mailBio/Technology. 1995;13:658–660; Guex N, Peitsch MC. SWISS-MODEL and the Swiss-PdbViewer: an environment for comparative protein modelling. Electrophoresis. 1997;18:2714–2723; Schwede T, Kopp J, Guex N, Peitsch MC. SWISS-MODEL: an automated protein homology-modeling server. Nucleic Acids Res. 2003;31:3381–3385).

The *D. dadantii 3937 marA* mutant was also defective in pathogenesis, so it is possible that the bacterium is sensitive to the toxic antimicrobial chemicals specifically present in the plant environment. This suggests that the survival of the bacterium in the host is dependent on the functionality of the *mar* regulon (Ravirala et al., unpublished data).

In enteric bacteria, one of the mechanisms of increased resistance to antimicrobial agents is a decrease in expression of the OmpF porin, reducing the permeability of the outer membrane. The expression of the *marA* gene causes decreased expression of OmpF porin in *mar* mutants of *E. coli* by positively activating *micF* antisense RNA.[87–92] Outer membrane analysis of the *D. dadantii marA* mutant revealed increased expression of the OmpF protein (detected using *E. coli OmpF* antibody) compared to wild type OmpF levels. However, the analysis of the *D. dandantii* genome does not reveal the presence of a *micF* gene. This suggests that OmpF expression may be mediated by a different mechanism that is still to be elucidated in this bacterium (Ravirala et al., unpublished data).

4.3 Regulation by Two-Component System

Bacteria are known to harbor many two-component systems to respond to a wide range of environmental signals.[93] In general, TCSs consist of a sensor-histidine kinase domain, located in the membrane and a response regulator

(RR) domain, located in the cytoplasm. Following sensing the alterations in the environments, signaling initiates through autophosphorylation of the sensor protein at a conserved histidine residue. The phosphoryl group is then transferred to a conserved aspartate located in the response regulator. Phosphorylation of a response regulator alters the biochemical characteristics of its output domain as well as DNA binding abilities, moderating gene expression in bacteria.[93] The importance of TCSs in the regulation of antibiotic resistance have been studied in many bacterial species.[94–96]

4.3.1 BaeSR Two-Component Regulatory System

It is well known that MdtABC efflux pump gene expression in *E. coli* is moderated by stress response systems including BaeSR and CpxARP two-component systems. In addition to sensor kinase BaeS and response regulator BaeR, the BaeSR system regulates expression of the RND-type transporters including AcrD and MdtABC, and the periplasmic chaperone Spy.[53] Several studies demonstrated that BaeSR system is able to respond to a variety of environmental signals such as spheroplast formation and exposure to various chemicals including indole, tannins, flavonoids, tungstate, and zinc[50,97–99]

The Cpx system includes proteins such as sensor histidine kinase CpxA, the response regulator CpxR, and CpxP, a periplasmic inhibitor of CpxA.[100] Various stresses caused by disturbances in the bacterial cell envelope including alkaline pH, changes in the inner membrane composition, accumulation of misfolded envelope proteins, and attachment to hydrophobic surfaces by the outer membrane lipoprotein NlpE are responsible for triggering theCpx pathway.[101]

It has been demonstrated that overexpression of response regulators such as BaeR and CpxR result in upregulation of AcrD and MdtABC efflux pumps in *E. coli.*[102] Interestingly, induction studies with indole showed that the BaeSR two-component system is the major driver for CpxR- facilitated induction of multidrug transporter genes including acrD and mdtABC and that CpxR modulates the outcome of BaeR.

A recent study in *D. dadantii* showed the involvement of BaeSR system in the induction of MdtABC efflux pump gene expression (Thekkiniath et al., unpublished data). It has been shown that in *E. amylovora*, BaeSR but not CpxR plays a key role in regulation of MdtABC efflux pump gene expression and cell envelope stress response.[52,103] Additionally, Pletzer et al. demonstrated a link between BaeR and CpxR in the regulation of MdtABC efflux pumps and transcriptional cross talk with exopolysaccharide synthesis in *E. amylovora.*[104]

5. FUTURE DIRECTIONS

In summary, considerable advances on the physiological and functional roles of multidrug efflux pumps in several bacterial species have been made in the past decades. Our current understanding of multidrug efflux pumps suggests that efflux pump inhibitors play a key role in the elimination of bacterial infections. These efflux pump inhibitors serve two-fold functions including the re-establishment of the action of compounds to which efflux pumps confer resistance as well as decrease in the capacity of bacteria to colonize their host. Multidrug efflux pumps, in addition to serve as antibiotic resistance determinants, exhibit diverse functions and are important in plant–bacteria interactions. Understanding these functions is imperative to determining systems that connect antibiotic resistance with the physiology of the bacterium[105,106] both during infections and in natural environments. Thus, there is a need for extensive research on *Erwinia* and other pathogen efflux pumps to gain comprehensive knowledge about their regulation and natural functions.

ACKNOWLEDGEMENTS

We are grateful to Polrit Viravathana for his valuable comments and suggestions on the manuscript.

REFERENCES

1. Reverchon S, Nasser W, Robert-Baudouy J. pecS: a locus controlling pectinase, cellulase and blue pigment production in *Erwinia chrysanthemi*. *Mol Microbiol.* 1994;11: 1127–1139.
2. Levine A, Tenhaken R, Dixon R, Lamb C. H_2O_2 from the oxidative burst orchestrates the plant hypersensitive disease resistance response. *Cell.* 1994;79:583–593.
3. Dixon RA. Natural products and plant disease resistance. *Nature.* 2001;411:843–847.
4. Poole K. Efflux-mediated resistance to Fluoroquinolones in Gram-negative Bacteria. *Antimicrob Agents Chemother.* 2000;44:2233–2241.
5. Tegos GP, Haynes M, Strouse JJ, et al. Microbial efflux pump inhibition: tactics and strategies. *Curr Pharm Des.* 2011;17:1291–1302.
6. Putman M, van Veen HW, Konings WN. Molecular properties of bacterial multidrug transporters. *Microbiol Mol Biol Rev.* 2000;64:672–693.
7. Lubelski J, Konings WN, Driessen AJM. Distribution and physiology of ABC-type transporters contributing to multidrug resistance in bacteria. *Microbiol Mol Biol Rev.* 2007;71:463–476.
8. Law CJ, Maloney PC, Wang D-N. Ins and outs of major facilitator superfamily antiporters. *Annu Rev Microbiol.* 2008;62:289–305.
9. Tseng TT, Gratwick KS, Kollman J, et al. The RND permease superfamily: an ancient, ubiquitous and diverse family that includes human disease and development proteins. *J Mol Microbiol Biotechnol.* 1999;1:107–125.

10. Chung YJ, Saier Jr MH. SMR-type multidrug resistance pumps. *Curr Opin Drug Discov Devel.* 2001;4:237–245.
11. Moriyama Y, Hiasa M, Matsumoto T, Omote H. Multidrug and toxic compound extrusion (MATE)-type proteins as anchor transporters for the excretion of metabolic waste products and xenobiotics. *Xenobiotica.* 2008;38:1107–1118.
12. Piddock LJV. Multidrug-resistance efflux pumps?. not just for resistance. *Nat Rev Micro.* 2006;4:629–636.
13. Barabote RD, Thekkiniath J, Strauss RE, Vediyappan G, Fralick JA, San Francisco MJ. Xenobiotic efflux in bacteria and fungi: a genomics update. *Adv Enzymol Relat Areas Mol Biol.* 2011;77:237.
14. Li X-Z, Nikaido H. Efflux-mediated drug resistance in bacteria: an update. *Drugs.* 2009;69:1555–1623.
15. Ma D, Cook DN, Alberti M, Pon NG, Nikaido H, Hearst JE. Molecular cloning and characterization of acrA and acrE genes of *Escherichia coli. J Bacteriol.* 1993;175: 6299–6313.
16. Gotoh N, Tsujimoto H, Poole K, Yamagishi J, Nishino T. The outer membrane protein OprM of *Pseudomonas aeruginosa* is encoded by oprK of the mexA-mexB-oprK multidrug resistance operon. *Antimicrob Agents Chemother.* 1995;39:2567–2569.
17. González-Pasayo R, Martínez-Romero E. Multiresistance genes of *Rhizobium etli* CFN42. *Mol Plant Microbe Interact.* 2000;13:572–577.
18. Krummenacher P, Narberhaus F. Two genes encoding a putative multidrug efflux pump of the RND/MFP family are cotranscribed with an rpoH gene in *Bradyrhizobium japonicum. Gene.* 2000;241:247–254.
19. Palumbo JD, Kado CI, Phillips DA. An isoflavonoid-inducible efflux pump in *Agrobacterium tumefaciens* is involved in competitive colonization of roots. *J Bacteriol.* 1998;180:3107–3113.
20. Peng W-T, Nester EW. Characterization of a putative RND-type efflux system in *Agrobacterium tumefaciens. Gene.* 2001;270:245–252.
21. Martinez JL, Sánchez MB, Martínez-Solano L, et al. Functional role of bacterial multidrug efflux pumps in microbial natural ecosystems. *FEMS Microbiol Rev.* 2009;33: 430–449.
22. Walsh C. Molecular mechanisms that confer antibacterial drug resistance. *Nature.* 2000;406:775–781.
23. Murakami S, Nakashima R, Yamashita E, Yamaguchi A. Crystal structure of bacterial multidrug efflux transporter AcrB. *Nature.* 2002;419:587–593.
24. Fralick JA. Evidence that TolC is required for functioning of the Mar/AcrAB efflux pump of *Escherichia coli. J Bacteriol.* 1996;178:5803–5805.
25. Zgurskaya HI, Nikaido H. AcrA is a highly asymmetric protein capable of spanning the periplasm. *J Mol Biol.* 1999;285:409–420.
26. Nikaido H. How do exported proteins and antibiotics bypass the periplasm in Gram-negative bacterial cells? *Trends Microbiol.* 2000;8:481–483.
27. Koronakis V, Shaff A, Koronakis E, Luisi B, Hughes C. Crystal structure of the bacterial membrane protein TolC central to multidrug efflux and protein export. *Nature.* 2000;405:914–919.
28. Borges-Walmsley MI, Walmsley AR. The structure and function of drug pumps. *Trends Microbiol.* 2001;9:71–79.
29. Nikaido H, Basina M, Nguyen VY, Rosenberg EY. Multidrug efflux pump AcrAB of *Salmonella typhimurium* excretes only those [beta]-lactam antibiotics containing lipophilic side chains. *J Bacteriol.* 1998;180:4686–4692.
30. Nikaido H. Multidrug efflux pumps of gram-negative bacteria. *J Bacteriol.* 1996;178:5853.
31. Llama-Palacios A, López-Solanilla E, Rodríguez-Palenzuela P. The ybiT gene of *Erwinia chrysanthemi* codes for a putative ABC transporter and is involved in

competitiveness against endophytic bacteria during infection. *Appl Environ Microbiol.* 2002;68:1624–1630.

32. Barabote RD, Johnson OL, Zetina E, San Francisco SK, Fralick JA, San Francisco MJD. Erwinia chrysanthemi tolC is involved in resistance to antimicrobial plant chemicals and is essential for phytopathogenesis. *J Bacteriol.* 2003;185:5772–5778.
33. Valecillos AM, Palenzuela PR, López-Solanilla E. The role of several multidrug resistance systems in *Erwinia chrysanthemi* pathogenesis. *Mol Plant Microbe Interact.* 2006;19:607–613.
34. Burse A, Weingart H, Ullrich MS. The phytoalexin-inducible multidrug efflux pump AcrAB contributes to virulence in the fire blight pathogen, *Erwinia amylovora*. *Mol Plant Microbe Interact.* 2004;17:43–54.
35. Burse A, Weingart H, Ullrich MS. NorM, an *Erwinia amylovora* multidrug efflux pump involved in in vitro competition with other epiphytic bacteria. *Appl Environ Microbiol.* 2004;70:693–703.
36. Loake G, Grant M. Salicylic acid in plant defence—the players and protagonists. *Curr Opin Plant Biol.* 2007;10:466–472.
37. Ravirala RS, Barabote RD, Wheeler DM, et al. Efflux pump gene expression in *Erwinia chrysanthemi* is induced by exposure to phenolic acids. *Mol Plant Microbe Interact.* 2007;20:313–320.
38. Nair BM, Cheung KJ, Griffith A, Burns JL. Salicylate induces an antibiotic efflux pump in *Burkholderia cepacia* complex genomovar III (*B. cenocepacia*). *J Clin Invest.* 2004;113: 464–473.
39. Fralick JA, Burns-Keliher LL. Additive effect of tolC and rfa mutations on the hydrophobic barrier of the outer membrane of *Escherichia coli* K-12. *J Bacteriol.* 1994;176: 6404–6406.
40. Sharff A, Fanutti C, Shi J, Calladine C, Luisi B. The role of the TolC family in protein transport and multidrug efflux. *Eur J Biochem.* 2001;268:5011–5026.
41. Bavro VN, Pietras Z, Furnham N, et al. Assembly and channel opening in a bacterial drug efflux machine. *Mol Cell.* 2008;30:114–121.
42. Baucheron S, Chaslus-Dancla E, Cloeckaert A. Role of TolC and parC mutation in high-level fluoroquinolone resistance in *Salmonella enterica* serotype Typhimurium DT204. *J Antimicrob Chemother.* 2004;53:657–659.
43. Baucheron S, Mouline C, Praud K, Chaslus-Dancla E, Cloeckaert A. TolC but not AcrB is essential for multidrug-resistant *Salmonella enterica* serotype Typhimurium colonization of chicks. *J Antimicrob Chemother.* 2005;55:707–712.
44. Bunikis J, Tsao J, Luke CJ, Luna MG, Fish D, Barbour AG. *Borrelia burgdorferi* infection in a natural population of peromyscus leucopus mice: a longitudinal study in an area where lyme borreliosis is highly endemic. *J Infect Dis.* 2004;189: 1515–1523.
45. Al-Karablieh N, Weingart H, Ullrich MS. The outer membrane protein TolC is required for phytoalexin resistance and virulence of the fire blight pathogen *Erwinia amylovora*. *Microb Biotechnol.* 2009;2:465–475.
46. Elkins CA, Nikaido H. Substrate specificity of the RND-type multidrug efflux pumps AcrB and AcrD of *Escherichia coli* is determined predominately by two large periplasmic loops. *J Bacteriol.* 2002;184:6490–6498.
47. Nagakubo S, Nishino K, Hirata T, Yamaguchi A. The putative response regulator BaeR stimulates multidrug resistance of *Escherichia coli* via a novel multidrug exporter system, MdtABC. *J Bacteriol.* 2002;184:4161–4167.
48. Baranova N, Nikaido H. The BaeSR two-component regulatory system activates transcription of the yegMNOB (mdtABCD) transporter gene cluster in *Escherichia coli* and increases its resistance to novobiocin and deoxycholate. *J Bacteriol.* 2002;184: 4168–4176.

49. Nishino K, Nikaido E, Yamaguchi A. Regulation of multidrug efflux systems involved in multidrug and metal resistance of *Salmonellaenterica* Serovar Typhimurium. *JBacteriol.* 2007;189:9066–9075.
50. Wang D, Fierke CA. The BaeSR regulon is involved in defense against zinc toxicity in *E. coli*. *Metallomics.* 2013;5:372–383.
51. Appia-Ayme C, Patrick E, Sullivan MJ, et al. Novel inducers of the envelope stress response BaeSR in Salmonella Typhimurium: BaeR is critically required for tungstate waste disposal. *PLoS One.* 2011;6(8):e23713.
52. Pletzer D, Weingart H. Characterization and regulation of the resistance-nodulation-cell division-type multidrug efflux pumps MdtABC and MdtUVW from the fire blight pathogen *Erwinia amylovora*. *BMC Microbiol.* 2014;14:185.
53. Leblanc SKD, Oates CW, Raivio TL. Characterization of the induction and cellular role of the BaeSR two-component envelope stress response of *Escherichia coli*. *J Bacteriol.* 2011;193:3367–3375.
54. Nishino K, Latifi T, Groisman EA. Virulence and drug resistance roles of multidrug efflux systems of *Salmonella enterica* serovar Typhimurium. *Mol Microbiol.* 2006;59: 126–141.
55. Marger MD, Saier MH. A major superfamily of transmembrane facilitators that catalyse uniport, symport and antiport. *Trends Biochem Sci.* 1993;18:13–20.
56. Lomovskaya O, Lewis K. Emr, an *Escherichiacoli* locus for multidrug resistance. *Proc Natl Acad Sci USA.* 1992;89:8938–8942.
57. Colmer JA, Fralick JA, Hamood AN. Isolation and characterization of a putative multidrug resistance pump from *Vibrio cholerae*. *Mol Microbiol.* 1998;27:63–72.
58. Lee EH, Shafer WM. The farAB-encoded efflux pump mediates resistance of gonococci to long-chained antibacterial fatty acids. *Mol Microbiol.* 1999;33:839–845.
59. Okinaka Y, Yang C-H, Perna NT, Keen NT. Microarray profiling of *Erwinia chrysanthemi* 3937 genes that are regulated during plant infection. *Mol Plant Microbe Interact.* 2002;15:619–629.
60. Nishino K, Yamaguchi A. Analysis of a complete library of putative drug transporter genes in *Escherichia coli*. *JBacteriol.* 2001;183:5803–5812.
61. Condemine G. Characterization of SotA and SotB, two *Erwinia chrysanthemi* proteins which modify isopropyl-β-D-thiogalactopyranoside and lactose induction of the *Escherichia coli* lac promoter. *JBacteriol.* 2000;182:1340–1345.
62. Joko T, Hirata H, Tsuyumu S. Sugar transporter (MfsX) of the major facilitator superfamily is required for flagella-mediated pathogenesis in *Dickeya dadantii* 3937. *J General Plant Pathol.* 2007;73:266–273.
63. Haseloff BJ, Freeman TL, Valmeekam V, et al. The exuT gene of *Erwinia chrysanthemi* EC16: nucleotide sequence, expression, localization, and relevance of the gene product. *Mol Plant Microbe Interact.* 1998;11:270–276.
64. Higgins CF. ABC transporters: from microorganisms to man. *Annu Rev Cell Biol.* 1992;8:67–113.
65. Higgins CF. ABC transporters: physiology, structure and mechanism—an overview. *Res Microbiol.* 2001;152:205–210.
66. Jones PM, George AM. The ABC transporter structure and mechanism: perspectives on recent research. *Cell Mol Life Sci CMLS.* 2004;61:682–699.
67. Schoner B, Geistlich M, Rosteck P, et al. Sequence similarity between macrolide-resistance determinants and ATP-binding transport proteins. *Gene.* 1992;115: 93–96.
68. Brown MH, Paulsen IT, Skurray RA. The multidrug efflux protein NorM is a prototype of a new family of transporters. *Mol Microbiol.* 1999;31:394–395.
69. Morita Y, Kataoka A, Shiota S, Mizushima T, Tsuchiya T. NorM of *Vibrio parahaemolyticus* is an Na+-driven multidrug efflux pump. *JBacteriol.* 2000;182:6694–6697.

70. Huda MN, Morita Y, Kuroda T, Mizushima T, Tsuchiya T. Na+-driven multidrug efflux pump VcmA from *Vibrio cholerae* non-O1, a non-halophilic bacterium. *FEMS Microbiol Lett.* 2001;203:235–239.
71. Martínez JL, Baquero F. Interactions among strategies associated with bacterial infection: pathogenicity, epidemicity, and antibiotic resistance. *Clin Microbiol Rev.* 2002;15: 647–679.
72. Sun J, Deng Z, Yan A. Bacterial multidrug efflux pumps: mechanisms, physiology and pharmacological exploitations. *Biochem Biophys Res Commun.* 2014;453:254–267.
73. Alekshun MN, Levy SB, Mealy TR, Seaton BA, Head JF. The crystal structure of MarR, a regulator of multiple antibiotic resistance, at 2.3 A resolution. *Nat Struct Mol Biol.* 2001;8:710–714.
74. George AM, Levy SB. Gene in the major cotransduction gap of the Escherichia coli K-12 linkage map required for the expression of chromosomal resistance to tetracycline and other antibiotics. *J Bacteriol.* 1983;155:541–548.
75. George AM, Levy SB. Amplifiable resistance to tetracycline, chloramphenicol, and other antibiotics in *Escherichia coli*: involvement of a non-plasmid-determined efflux of tetracycline. *J Bacteriol.* 1983;155:531–540.
76. Praillet T, Reverchon S, Nasser W. Mutual control of the PecS/PecM couple, two proteins regulating virulence-factor synthesis in *Erwinia chrysanthemi*. *Mol Microbiol.* 1997;24:803–814.
77. Rouanet C, Nomura K, Tsuyumu S, Nasser W. Regulation of pelD and pelE, encoding major alkaline pectate lyases in *Erwinia chrysanthemi*: involvement of the main transcriptional factors. *J Bacteriol.* 1999;181:5948–5957.
78. Reverchon S, Rouanet C, Expert D, Nasser W. Characterization of indigoidine biosynthetic genes in *Erwinia chrysanthemi* and role of this blue pigment in pathogenicity. *J Bacteriol.* 2002;184:654–665.
79. Nasser W, Reverchon S, Vedel R, Boccara M. PecS and PecT coregulate the synthesis of HrpN and pectate lyases, two virulence determinants in *Erwinia chrysanthemi* 3937. *Mol Plant Microbe Interact.* 2005;18:1205–1214.
80. Hommais F, Oger-Desfeux C, Van Gijsegem F, et al. PecS is a global regulator of the symptomatic phase in the thytopathogenic bacterium *Erwinia chrysanthemi* 3937. *J Bacteriol.* 2008;190:7508–7522.
81. Haque MM, Kabir MS, Aini LQ, Hirata H, Tsuyumu S. SlyA, a MarR family transcriptional regulator, is essential for virulence in *Dickeya dadantii* 3937. *J Bacteriol.* 2009;191:5409–5418.
82. Buchmeier N, Bossie S, Chen CY, Fang FC, Guiney DG, Libby SJ. SlyA, a transcriptional regulator of *Salmonella typhimurium*, is required for resistance to oxidative stress and is expressed in the intracellular environment of macrophages. *Infect Immun.* 1997;65:3725–3730.
83. Stapleton MR, Norte VA, Read RC, Green J. Interaction of the *Salmonella typhimurium* transcription and virulence factor SlyA with target DNA and identification of members of the SlyA regulon. *J Biol Chem.* 2002;277:17630–17637.
84. Wyborn NR, Stapleton MR, Norte VA, Roberts RE, Grafton J, Green J. Regulation of *Escherichia coli* hemolysin E expression by H-NS and Salmonella SlyA. *J Bacteriol.* 2004;186:1620–1628.
85. Wilkinson SP, Grove A. Ligand-responsive transcriptional regulation by members of the MarR family of winged helix proteins. *Curr Issues Mol Biol.* 2006;8:51.
86. Duval V, Lister IM. MarA, SoxS and Rob of *Escherichia coli*—global regulators of multidrug resistance, virulence and stress response. *Int J Biotechnol Wellness Ind.* 2013;2:101.
87. Cohen SP, McMurry LM, Levy SB. marA locus causes decreased expression of OmpF porin in multiple-antibiotic-resistant (Mar) mutants of *Escherichia coli*. *J Bacteriol.* 1988;170:5416–5422.

88. Gambino L, Gracheck SJ, Miller PF. Overexpression of the MarA positive regulator is sufficient to confer multiple antibiotic resistance in *Escherichia coli*. *J Bacteriol.* 1993;175: 2888–2894.
89. Okusu H, Ma D, Nikaido H. AcrAB efflux pump plays a major role in the antibiotic resistance phenotype of *Escherichia coli* multiple-antibiotic-resistance (Mar) mutants. *J Bacteriol.* 1996;178:306–308.
90. Rosner JL. Nonheritable resistance to chloramphenicol and other antibiotics induced by salicylates and other chemotactic repellents in *Escherichia coli* K-12. *Proc Natl Acad Sci USA.* 1985;82:8771–8774.
91. Alekshun MN, Levy SB. The mar regulon: multiple resistance to antibiotics and other toxic chemicals. *Trends Microbiol.* 1999;7:410–413.
92. Randall LP, Woodward MJ. Multiple Antibiotic Resistance (mar) Locus in Salmonella enterica Serovar Typhimurium DT104. *Appl Environ Microbiol.* 2001;67:1190–1197.
93. West AH, Stock AM. Histidine kinases and response regulator proteins in two-component signaling systems. *Trends Biochem Sci.* 2001;26:369–376.
94. Sun S, Negrea A, Rhen M, Andersson DI. Genetic Analysis of Colistin Resistance in Salmonella enterica Serovar Typhimurium. *Antimicrob Agents Chemother.* 2009;53: 2298–2305.
95. Kishii R, Takei M. Relationship between the expression of ompF and quinolone resistance in *Escherichia coli*. *J Infect Chemother.* 2009;15:361–366.
96. Barrow K, Kwon DH. Alterations in two-component regulatory systems of phoPQ and pmrAB are associated with Polymyxin B resistance in clinical isolates of *Pseudomonas aeruginosa*. *Antimicrob Agents Chemother.* 2009;53:5150–5154.
97. Raffa RG, Raivio TL. A third envelope stress signal transduction pathway in *Escherichia coli*. *Mol Microbiol.* 2002;45:1599–1611.
98. Leblanc SK, Oates CW, Raivio TL. Characterization of the induction and cellular role of the BaeSR two-component envelope stress response of *Escherichia coli*. *J Bacteriol.* 2011;193:3367–3375.
99. Zoetendal EG, Smith AH, Sundset MA, Mackie RI. The BaeSR two-component regulatory system mediates resistance to condensed tannins in *Escherichia coli*. *Appl Environ Microbiol.* 2008;74:535–539.
100. Raivio TL, Popkin DL, Silhavy TJ. The Cpx envelope stress response is controlled by amplification and feedback inhibition. *J Bacteriol.* 1999;181:5263–5272.
101. Vogt SL, Raivio TL. Just scratching the surface: an expanding view of the Cpx envelope stress response. *FEMS Microbiol Lett.* 2012;326:2–11.
102. Hirakawa H, Nishino K, Hirata T, Yamaguchi A. Comprehensive studies of drug resistance mediated by overexpression of response regulators of two-component signal transduction systems in *Escherichia coli*. *J Bacteriol.* 2003;185:1851–1856.
103. Pletzer D, Weingart H. Characterization of AcrD, a resistance-nodulation-cell division-type multidrug efflux pump from the fire blight pathogen *Erwinia amylovora*. *BMC Microbiol.* 2014;14:13.
104. Pletzer D, Stahl A, Oja AE, Weingart H. Role of the cell envelope stress regulators BaeR and CpxR in control of RND-type multidrug efflux pumps and transcriptional cross talk with exopolysaccharide synthesis in *Erwinia amylovora*. *Arch Microbiol.* 2015;197:761–772.
105. Linares JF, Moreno R, Fajardo A, et al. The global regulator Crc modulates metabolism, susceptibility to antibiotics and virulence in *Pseudomonas aeruginosa*. *Environ Microbiol.* 2010;12:3196–3212.
106. Martínez JL, Rojo F. Metabolic regulation of antibiotic resistance. *FEMS Microbiol Rev.* 2011;35(5):768–789.

CHAPTER SIX

In Vitro Analysis of *Pseudomonas aeruginosa* Virulence Using Conditions That Mimic the Environment at Specific Infection Sites

J.A. Colmer-Hamood[1,2,*], N. Dzvova[1], C. Kruczek[3], A.N. Hamood[1,4]

[1] Department of Immunology and Molecular Microbiology, Texas Tech University Health Sciences Center, Lubbock, TX, United States
[2] Department of Medical Education, Texas Tech University Health Sciences Center, Lubbock, TX, United States
[3] Honors College, Texas Tech University, Lubbock, TX, United States
[4] Department of Surgery, Texas Tech University Health Sciences Center, Lubbock, TX, United States
Corresponding author. E-mail address: jane.colmer@ttuhsc.edu

Contents

Progress in Molecular Biology and Translational Science, Volume 142
ISSN 1877-1173
http://dx.doi.org/10.1016/bs.pmbts.2016.05.003

Abstract

Pseudomonas aeruginosa is a Gram-negative opportunistic pathogen that causes chronic lung infection in patients with cystic fibrosis (CF) and acute systemic infections in severely burned patients and immunocompromised patients including cancer patients undergoing chemotherapy and HIV infected individuals. In response to the environmental conditions at specific infection sites, *P. aeruginosa* expresses certain sets of cell-associated and extracellular virulence factors that produce tissue damage. Analyzing the mechanisms that govern the production of these virulence factors in vitro requires media that closely mimic the environmental conditions within the infection sites. In this chapter, we review studies based on media that closely resemble three in vivo conditions, the thick mucus accumulated within the lung alveoli of CF patients, the serum-rich wound bed and the bloodstream. Media resembling the CF alveolar mucus include standard laboratory media supplemented with sputum obtained from CF patients as well as prepared synthetic mucus media formulated to contain the individual components of CF sputum. Media supplemented with serum or individual serum components have served as surrogates for the soluble host components of wound infections, while whole blood has been used to investigate the adaptation of pathogens to the bloodstream. Studies using these media have provided valuable information regarding *P. aeruginosa* gene expression in different host environments as varying sets of genes were differentially regulated during growth in each medium. The unique effects observed indicate the essential role of these in vitro media that closely mimic the in vivo conditions in providing accurate information regarding the pathogenesis of *P. aeruginosa* infections.

1. INTRODUCTION

1.1 Rationale for Using Media That Mimic Host Infection Sites

Artificial media have long been used for the growth and characterization of bacteria. In the clinical setting, these media are designed to provide nutrients required by pathogens expected to be isolated from clinical specimens. Many of these media contain blood or serum, usually animal but sometimes human. In the research laboratory, rich media are used to support the general growth of the bacteria and specially developed media are used to analyze the production of various virulence factors elaborated by the bacterium under study. Only rarely do these media contain blood or serum. However, neither setting reproduces the environment of the host in which the infection is taking place. Thus, not all clinically relevant organisms can be grown on "standard" laboratory media, nor does the production of virulence factors under optimum conditions necessarily reflect their production at the infection site within the host. In this chapter, we review studies based on media that closely resemble three in vivo conditions: the thick accumulated mucus within the lung alveoli of CF patients; blood components such as serum, albumin, and transferrin present in wound exudate; and whole blood for bacteremia and sepsis.

1.2 *Pseudomonas aeruginosa*

Pseudomonas aeruginosa is a Gram-negative opportunistic pathogen that causes severe acute and chronic infections at different sites within the body such as urinary tract, skin (burn or surgical wounds), and the respiratory tract.[1–3] Patients susceptible to *P. aeruginosa* infection include severely burned patients, individuals with cystic fibrosis (CF), patients in intensive care units, and intubated patients on mechanical ventilators.[1,4–6] Immunocompromised patients including cancer patients undergoing chemotherapy and HIV-infected patients are also susceptible to *P. aeruginosa* infection.[2,4] The extensive tissue damage associated with these infections is due to the ability of *P. aeruginosa* to produce numerous cell-associated and extracellular factors.[3,5,6] Cell-associated factors include organelles associated with motility (pili and flagellum), alginate, and lipopolysaccharide (LPS), while extracellular factors include numerous toxins, proteases, and siderophores.

2. VIRULENCE FACTORS PRODUCED BY *P. AERUGINOSA*

2.1 Cell-Associated Virulence Factors

Cell-associated virulence factors include the flagellum, type IV pili, exopolysaccharide (EPS), and lipopolysaccharide (LPS).[3,5,6] In order for *P. aeruginosa* to establish infection, it must first attach to and then colonize host tissues. This initial adherence and colonization is mediated by the type IV pili and flagella in some cases.[6,7] Both pili, which mediate twitching motility, and the flagellum, which mediates swimming and swarming motilities, are also important in the cell motility observed in *P. aeruginosa.*[7–9] Additionally, certain strains of *P. aeruginosa* produce an alginate layer that surrounds the outside of the cell.[3,7] The alginate aids in adherence and inhibits phagocytosis.[7,10] The alginate layer is responsible for the mucoid phenotype that prevails in many chronic lung infections.[7,11,12] As a Gram-negative bacterium, *P. aeruginosa* also contains LPS in its outer membrane.[3] The lipid A component of LPS acts as an endotoxin if it is released from the cell upon lysis.[3,7] The release of this endotoxin causes a proinflammatory response that can result in septic shock within the host.[7,13,14]

2.2 Extracellular Virulence Factors

Extracellular (secreted) virulence factors include exotoxin A (ETA), type III secretion system (TTSS) effector molecules, elastases (LasB, LasA, alkaline protease, and protease IV), siderophores, and pyocyanin.[3,5–7] These factors contribute to tissue damage, bloodstream invasion, and bacterial dissemination throughout the host.[15] ETA catalyzes the ADP-ribosylation of host elongation factor-2, inhibiting host protein synthesis resulting in cell death.[3,7] ETA also depresses host responses to infection and influences *P. aeruginosa* virulence.[16,17] The TTSS effector molecules include exoenzyme S (ExoS), exoenzyme T (ExoT), exoenzyme U (ExoU), and exoenzyme Y (ExoY).[7,18] ExoS is an ADP-ribosyl transferase that causes cytotoxicity by disrupting normal cytoskeletal organization.[19–21] ExoT is also an ADP-ribosyl transferase that affects the eukaryotic cytoskeleton.[21] ExoU is a major cytotoxin secreted through the TTSS apparatus that is 100 times more cytotoxic than ExoS.[22,23] In an animal model of acute lung infection, ExoU caused decompartmentalization of the inflammatory response leading to sepsis.[24] ExoY is an adenylate cyclase that increases microvascular intercellular gap formation leading to increased lung permeability.[25,26] In addition to ETA and exoenzymes, *P. aeruginosa* secretes four main proteases; LasA (staphylolytic protease), LasB (elastase), alkaline protease, and protease IV.[3,7,27] The LasA and LasB proteases both exhibit elastolytic activity; however, most of the elastolytic activity within the host may be attributed to LasB.[28] Elastase plays an important role in multiple *P. aeruginosa* infections by promoting degradation of both blood vessel and lung alveoli tissues.[29–31] Production of both the LasA and LasB proteases is controlled by cell-to-cell quorum sensing signaling mechanisms.[32] Alkaline protease is a metalloprotease which is capable of degrading host complement proteins as well as cytokines and is secreted during bacteremic infections.[33,34] Protease IV is a serine protease which is also capable of degrading host complement proteins as well as host clotting factors.[35] Under iron-limited conditions, *P. aeruginosa* also secretes iron-scavenging molecules known as siderophores.[7] Two main siderophore systems are present in *P. aeruginosa*, the pyoverdine and pyochelin systems.[36] These molecules are able to strip Fe^{3+} from host Fe-binding proteins, such as transferrin and lactoferrin, and transport it back to the cell for utilization in several metabolic pathways.[36] Pyocyanin is the factor responsible for the blue-green pigmentation often observed with *P. aeruginosa* infections. It increases IL-8 production, depresses host responses, and induces neutrophils apoptosis.[37–39] Studies using animal models of acute and chronic lung infections confirmed the contribution of pyocyanin to the pathogenesis of *P. aeruginosa*.[40]

3. SPECIFIC VIRULENCE FACTORS AFFECTED BY HOST CONDITIONS

The production of virulence factors is stringently regulated and often depends upon conditions at specific infection sites. Numerous environmental conditions such as pH, temperature, iron availability, oxygen levels, cell density, and host factors determine which suite of virulence factors are produced at any given time.[36,41,42]

3.1 Alginate

The *P. aeruginosa* alginate consists of an anionic linear polymer composed of β-1,4-linked mannuronic acids and its epimer, α-L-guluronic acid.[43] Alginate is considered an important virulence factor for pathogenic *Pseudomonas* species as it contributes to the development of *P. aeruginosa* biofilm and the differentiation of microcolonies.[44,45] The involvement of alginate in biofilm formation is thought to be limited to the presence of *P. aeruginosa* within certain dehydrating conditions such as the lungs of CF patients,[45,46] as the alginate capsule increases the resistance of *P. aeruginosa* to desiccation.[47] Li et al.[48] showed that desiccating conditions induce alginate production by *P. aeruginosa*. In support of this finding, *P. aeruginosa* isolates obtained from the CF lung exhibit mucoidy, an alginate over-producing phenotype.[46,49]

The alginate capsule provides *P. aeruginosa* with several advantages that allow it to cause further tissue damage. For example, in response to the chemotactic inflammatory response within the infected lung, neutrophils and macrophages are recruited to the infection site.[50,51] However, due to the presence of the alginate capsule, these immune cells fail to eliminate *P. aeruginosa*. Instead, their excessive "frustrated" phagocytic activity produces further lung damage. Alginate also scavenges reactive oxygen species such as superoxide radicals that macrophages and neutrophils generate to eliminate bacterial pathogens.[50,51]

3.2 Iron Chelation

Within the host, iron is very limited as it is sequestered within iron binding proteins such as lactoferrin and transferrin.[52,53] Iron also exists within heme molecules such as hemoglobin and cytochromes.[54,55] Thus, to survive these conditions and depending on the type of infection, *P. aeruginosa* uses different strategies to obtain iron from the environment. One major strategy is the production of siderophores, low molecular weight iron-chelating compounds that bind iron at high affinity and that are then internalized into *P. aeruginosa* through specific receptors.[56–59] Depending on the availability of iron (Fe^{3+}), *P. aeruginosa* switches from production of low affinity iron-binding pyochelin to the energy-dependent pyoverdine that has a high affinity for iron and displaces iron from transferrin.[60] Experiments using the murine model of lung infection and thermal injury clearly showed that pyoverdine production is critical for the survival of *P. aeruginosa* during those infections.[61–63] Pyochelin is thought to play a role in sustaining the inflammatory response during chronic *P. aeruginosa* lung infection of CF patients.[64] Another *P. aeruginosa* strategy to obtain iron involves the uptake of heme molecules from the host hemoproteins hemoglobin, hemopexin, and myoglobin.[54,55] *P. aeruginosa* utilizes the *phu* system to extract heme directly from hemoglobin using a TonB-dependent outer membrane protein receptor.[65] In its second heme acquisition system, the Has system, *P. aeruginosa* extracts heme from hemoglobin by the hemophore HasA and transfers it to the membrane receptor HasR, through which the heme is internalized.[55] A third system that *P. aeruginosa* utilizes is the Feo system for the transport of ferrous iron (Fe^{2+}).[66] In this system, Fe^{2+} found in microaerobic or anaerobic conditions diffuses through the outer membrane and is transported to the cytoplasm through the Feo system.[66] The *P. aeruginosa* pyocyanin and its precursor, phenazine carboxylic acid, facilitate this process by reducing Fe^{3+} bound to the different host proteins to the Fe^{2+} state.[67] As the condition of the CF patients deteriorates, Fe^{2+} and phenazines accumulate in their lungs.[68]

3.3 ExoU Cytotoxin

P. aeruginosa and other Gram-negative pathogens possess TTSS, through which toxins (effectors) are translocated directly from the bacterial cytoplasm into the host cytosol.[69] *P. aeruginosa* secretes at least four effector molecules through the TTSS—ExoS, ExoT, ExoU, ExoY.[69] ExoU is a 74-kDa cytotoxin produced by cytotoxic strains of *P. aeruginosa*.[70,71] ExoU production is associated with the ability of *P. aeruginosa* to cause acute lung infection. In the murine model of acute pneumonia, ExoU production was shown to be associated with increased virulence, as well as systemic spread of *P. aeruginosa*.[70] Hauser et al.[72] suggested a correlation between ExoU production by *P. aeruginosa* and ventilator-associated pneumonia (VAP). About 29% of *P. aeruginosa* isolates obtained from patients with VAP produced detectable levels of ExoU.[72] Additional studies showed a close correlation between ExoU production by *P. aeruginosa* and severity of lung injury and dissemination of the bacteria into the bloodstream.[73,74]

3.4 Quorum Sensing

P. aeruginosa employs its three cell-to-cell communication systems, or quorum sensing (QS) systems, to coordinate the production of different virulence factors in response to environmental stimuli.[75] With QS, *P. aeruginosa* can sense population density and coordinate gene expression among the population through signaling molecules called autoinducers that bind their cognate receptors and activate gene expression directly or indirectly.[75] *P. aeruginosa* contains two well-characterized *N*-acyl-homoserine lactone based QS systems, the *las* and *rhl* systems, each consisting of a dedicated autoinducer synthase and its cognate transcriptional activator, LasI/LasR and RhlI/RhlR, respectively.[32,75,76] LasI is required for the synthesis of *N*-(3-oxododecanoyl)-homoserine lactone (C12-HSL), while RhlI is required for the synthesis of *N*-butyryl-L-homoserine lactone (C4-HSL).[32,76] These two systems, *las* and *rhl*, regulate about 6% of the *P. aeruginosa* genome with considerable overlap between the two systems.[77] Besides the *rhl* and *las* systems, *P. aeruginosa* possesses a third, 2-alkyl-4-quinolone-based QS system (*pqs*) that involves several quinolone signaling molecules, the most important of which are 2-heptyl-3-hydroxy-4-quinolone (*Pseudomonas* quinolone signal or PQS) and its immediate precursor 2-heptyl-4-quinolone (HHQ).[78] The HHQ PQS molecule is synthesized by proteins encoded by *pqsABCD* and *pqsH*, rather than by a single synthase.[79] Upon binding with its cognate receptor MvfR (PqsR), the complex regulates at least 140 *P. aeruginosa* genes, many of which are coregulated by the *rhl* system.[79,80]

The three QS systems are interlinked in a hierarchical manner, with the *las* system positively regulating *rhl* and *pqs*, *rhl* negatively regulating *pqs*, and *pqs* positively regulating *rhl*.[32,75,79,81] Additionally, numerous regulators further control the QS systems at both transcriptional and posttranscriptional levels.[75] Among these regulators, the transcriptional regulator RsaL represses the *lasI* in a negative feedback loop[82]; transcription of *rpoS* that encodes the stationary phase sigma factor RpoS is modestly activated by *rhl* and, in turn, subtly activates *lasR* and *rhlR* expression[83]; while MvaT controls the magnitude and timing of QS-dependent gene expression through a silencing mechanism that affects *lasI, rsaL, mvfR*, and *rpoS* and links all three systems.[84,85] Together, the QS systems play major roles in the regulation of virulence and survival factor production, biofilm maturation, motility phenotypes, and the transition to acute invasive infection.[32,75,76] Specific to the virulence factors discussed earlier: the *las* and *rhl* systems are activated by limited iron availability as found in the CF lung, while PQS chelates iron[86];

the secretion of TTSS cytotoxins has been shown to depend on PQS concentration[87]; and the switch to mucoidy phenotype has been shown to repress the *P. aeruginosa* QS systems leading to changes in virulence factor regulation,[88] while other authors showed a correlation of β-lactam resistance, with alginate production and QS in *P. aeruginosa.*[89]

4. SYNTHETIC MEDIA THAT MIMIC SPECIFIC HOST ENVIRONMENTS

Within the last decade, the role of different virulence factors in the pathogenesis of *P. aeruginosa* has been examined using laboratory media, several animal models, or tissue (biopsies) and secretions obtained from *P. aeruginosa* infected patients.[90–99] Most laboratory media were designed to induce maximum production of specific virulence factors. For example, the iron-deficient medium, chelated trypticase dialysate, or TSB-DC, is designed to induce maximum production of exotoxin A, siderophores and other iron-repressed factors.[95,100,101] Similarly the calcium-deficient medium, S-defined medium, is designed for maximum production of different proteins that constitute the TTSS including the TTSS effector molecules.[94] Another medium that was designed for maximum production of TTSS is TSB-DC treated with nitriloacetic acid to chelate calcium.[21] Additionally, phosphate-deficient medium is utilized for maximum production of phospholipase C and other phosphate repressed factors.[90] These media have been relied upon extensively to examine the mechanisms that govern the expression of different genes or operons that code for the machinery needed for synthesis, transport and regulation of various *P. aeruginosa* virulence factors.[18,102,103] Despite their tremendous contributions, however, these media do not reflect the exact environmental conditions at infection sites. In addition to iron, calcium and phosphate, other signals are likely to regulate the production of virulence factors at these infection sites.

4.1 Media That Mimic the CF Lung Environment for Analysis of Gene Expression

4.1.1 Cystic Fibrosis

CF is an autosomal recessive disorder that results from a mutation in the *CFTR* gene that codes for the CF transmembrane conductor regulatory protein (CFTR).[104,105] CF affects about 90,000 individuals worldwide and 30,000 in the United States.[106] Among the protean effects caused by *CFTR* mutations throughout the body, such mutations disrupt the function of the CFTR within lung epithelial cells.[104] This disruption varies from a total loss of function to the surface expression of a protein with poor chloride conductance.[105,107] Within the lung, *CFTR* mutations cause a defect in epithelial ion transport, which leads to the accumulation of thick, sticky mucus (sputum) that obstructs the airways and prevents normal mucociliary clearance.[108] Normal mucus consists primarily of water (95%), secreted mucins (MUC5AC and MUC5AB), free protein, and salts.[109] The most studied *CFTR* mutation is ΔF508, which accounts for about 70% of cases of CF. Due to oversecretion of mucin and the dehydration of the airway surface liquid, mucin concentration can reach greater than 8% (normal 1%) in respiratory secretions of these individuals (http://www.ncbi.nlm.nih.gov/pmc/articles/PMC2430075/). This can vary depending on the type of specific *CFTR* mutation, treatment the patient is receiving, as well as to infectious exacerbations. The amount of DNA present also contributes to viscosity and is also related to the amount of cellular destruction and the numbers of infectious organisms present. The ranges of mucin, DNA, and other components have been derived by studies that analyzed pooled CF sputum samples. This thin mucus is cleared via the cilia within the respiratory tract.[110] In CF, mucin over-secretion occurs, and extracellular DNA, filamentous actin, lipids, and proteoglycans as well as mucin degradation products are present in the mucus, which now must be expectorated as sputum.[110] Failure of the host defense mechanisms within the respiratory airways to clear bacterial infections leads to persistent bacterial infections.[111,112] It has been established that the loss of mucus viscosity has multiple effects on the ability of the host to clear infectious agents from the lungs. In 2005, Matsui et al. reported that dehydrated airway mucus prevents neutrophil capture and killing of bacteria on airway surfaces (J Immunol. 2005;175:1090-9). The elastic behavior and shear thickening lead to compaction of the mucus and increased adhesiveness, which makes it hard to clear. The stasis and adhesion are believed to initiate CF airway disease pathogenesis. Individuals with CF also have less beta-defensing hBD-1; originally thought to be due to salt-dependent inactivation, there is evidence from Dauletbaev et al. in 2001 that defensing expression in CF patients was not upregulated in response to inflammation (http://www.ncbi.nlm.nih.gov/pubmed/11844962). Another study showed that hBDs are lower in those CF patients with more advanced disease (http://www.sciencedirect.com/science/article/pii/S1569199303001280).

We were not trying to present a complete overview of cystic fibrosis, but rather discuss a few specific points that indicate the research we presented is tied to the pathophysiology of the disease at some level. The decline in host defense as well as the nutrients present within the accumulated mucus facilitate chronic colonization of the lung airways by different bacterial pathogens.[64,111–114]

Initially, the colonizing bacteria are Gram-positive organisms such as *Staphylococcus aureus* and fastidious Gram-negative bacteria like *Haemophilus influenzae*, with *P. aeruginosa* entering the bacterial population later on.[64,112,114] Once established, *P. aeruginosa* displaces the other pathogens becoming the main pathogen within the lungs and inaugurating persistent infection.[64,115] Besides bacterial competition, *P. aeruginosa* faces additional challenges within the lungs of CF patients. Among these challenges are constant exposure to antibiotics, as a result of treatment,[116] and the continuous attack from the host innate immune system including antimicrobial peptides secreted by epithelial cells and reactive oxygen species and antimicrobial compounds produced and secreted by neutrophils and macrophages recruited to the infection site.[117] To adapt to these stressful conditions, *P. aeruginosa* responds by initiating a series of responses leading to chronic infection.[11,64,115]

4.1.2 Chronic *P. aeruginosa* Lung Infection in CF

A distinctive feature of the establishment of a chronic infection is that the *P. aeruginosa* strains produce excessive amounts of alginate, or mucoid exopolysaccharide,[118,119] which provides the producing strain with the mucoidy phenotype.[118,120–122] This phenotype is seen almost exclusively in *P. aeruginosa* isolated from CF patients, although normal sputum can also induce alginate production.[118] Chronic *P. aeruginosa* lung infections are associated with deteriorating lung function and a worsening prognosis for CF patients.[111,123] The microaerophilic and/or anaerobic conditions generated by mucus hyper secretion within the CF lung, as well as the DNA present from the necrosis of recruited neutrophils, further facilitate establishment of chronic *P. aeruginosa* infection.[124,125] Indeed, DNA has been shown to be necessary for *P. aeruginosa* biofilm formation within the CF lung.[126] CF sputum, therefore, is critical for the survival and proliferation of the infecting *P. aeruginosa* during chronic infection of the CF lung. As it supports the growth of *P. aeruginosa*, CF sputum also likely influences *P. aeruginosa* pathogenesis by inducing the production of different virulence factors, including the alginate capsule, and promoting resistance to different antibiotics. Several investigators have utilized different laboratory media to address the effect of CF sputum on the growth and virulence of *P. aeruginosa*.

4.1.3 CF Respiratory Mucus Medium

Wang *et al.*[127] developed an in vivo expression technology (IVET) selection system for *P. aeruginosa* by constructing a *purEF*Δ in *P. aeruginosa* strain PAK that was used to generate cointegrates carrying random insertions of the promoterless *purEF* gene. The PAK-AR2 cointegrates were introduced into neutropenic mice to identify promoters of *P. aeruginosa* genes whose expression was enhanced by in vivo growth.[127] They then adapted the IVET system to identify *P. aeruginosa* promoters that were specifically induced by respiratory mucus derived from CF patients.[128] Sputum obtained from two CF patients during physical therapy was dialyzed against distilled water to remove potential low molecular weight substances that promoted the growth of PAK-AR2 alone.[128] Initial titration experiments using minimal medium A (MMA)[129] supplemented with 50, 25, 10 and 5% respiratory mucus (RM) showed that optimum growth was obtained using MMA with 10% RM.[128] Promoters inducible by the CF respiratory mucus—that is, cointegrates that expressed *purEF*—were isolated by growing the strains carrying the cointegrates in either MMA or MMA/10% RM at 37°C for 16 h and enriched using three cycles of growth in MMA/10% RM.[128] DNA analysis revealed that 22 isolates carried the promoterless *purEF* fusion within three different *P. aeruginosa* genetic loci.[128] One gene recovered was *fptA*, which codes for the outer membrane protein pyochelin receptor, FptA.[130,131] Based on these results, the authors suggested that mucus serves as an iron chelator and generates iron-deficient conditions within the CF lung.[128] It is interesting, however, that the expression of the pyoverdine receptor FpvA was not enhanced,[128] since the pyoverdine system has been shown to be far more efficient than the pyochelin system in binding iron.[56] Another locus whose expression was enhanced by the CF sputum was *np20*, which codes for the zinc responsive transcriptional regulator, Np20.[132] The *np20* gene, which is required for pyochelin production, regulates the expression of *dksA2* in a zinc-dependent manner.[133] Ellison et al.[132] suggested that Np20 is the *P. aeruginosa* zinc uptake regulator (Zur) that regulates the transcription of genes encoding the high affinity ZnuABC transport system. The third locus recovered encodes MigA, a α-1,6-rhamnosyltransferase required for lipo-saccharide synthesis in *P. aeruginosa*.[134] MigA and another rhamnosyltransferase, WapR, are differentially regulated by the Rhl quorum sensing system and affect the architecture of biofilm formation.[135,136]

4.1.4 Mucopurulent Material Medium

Wolfgang et al.[137] used a whole genome microarray to examine the response of *P. aeruginosa* PAK to mucopurulent material (MPM) collected from two adult, chronically infected CF patients. The MPM was sterilized by UV irradiation, stored at −80°C, and used to supplement minimal medium M63.[129,137] Compared with the growth of PAK at 37°C in M63 with 40 mM glucose, the growth of PAK in M63 with 10% MPM significantly altered the expression of 61 genes; the expression of 47 genes was activated while that of 14 genes was repressed.[137] Among the genes whose expression was significantly induced was *rhlI* which codes for the autoinducer synthase enzyme that synthesizes the C4-HSL.[138] The expression of other genes that constitute the Rhl-dependent network were also activated.[32,137] In contrast, the expression of the flagellar gene, *fliC*, was repressed.[137,139] The repression of *fliC* expression was rapid, such that within 2 h of exposure to 10% MPM, *fliC* mRNA was reduced by 80%, and by 4 h postexposure, almost no *fliC* mRNA was detected.[137] Based on these results, Wolfgang et al.[137] suggested that these changes are part of the *P. aeruginosa* strategy to adapt to the environment and persist within the CF lung during chronic infection. The *P. aeruginosa* flagellum is highly immunogenic. Therefore, during chronic infection, *P. aeruginosa* represses the synthesis of flagellar protein to prevent its detection by the host defense system.[137] The increase in the expression of QS and QS-related genes, which constitutes another part of *P. aeruginosa* adaptation, supported the results of a previous study that demonstrated detectable levels of the autoinducers C12-HSL and C4-HSL in sputum samples obtained from CF patients colonized with *P. aeruginosa.*[140]

4.1.5 Ten Percent CF Sputum Medium

Palmer et al.[141] also utilized microarray analysis to examine the effect of CF sputum as a sole source of carbon and energy on the physiology of *P. aeruginosa.* They grew *P. aeruginosa*, PA14 in 3-(*N*-morpholino) propanesulfonic acid (MOPS) buffered medium supplemented with either 20% glucose or 10% CF sputum (10% CF sputum medium).[141,142] Once obtained from the CF patients, sputum samples were immediately frozen on dry ice and stored at −80°C. The samples were then thawed and lyophilized, and the lyophilized powder was stored at −20°C under desiccation. Compared with the growth in the glucose supplemented MOPs, the growth of PAK in 10% CF sputum medium differentially regulated the expression of 147 genes.[141] Most of these genes (113) were upregulated, while 34 were repressed.[141] Genes involved in branched chain and aromatic amino acid catabolism were highly upregulated while those involved in the biosynthesis of these amino acids were repressed. Similar to the analysis of Wolfgang et al.,[137] Palmer et al.[141] showed that *fliC* expression was repressed. More than 95% of *P. aeruginosa* grown in 10% CF sputum medium were nonmotile. Further analysis confirmed that the repression of motility is not due to mucus only, but due to a heat labile factor within the sputum. In contrast to the findings of Wolfgang et al.,[137] no changes were observed in the *las* or *rhl* QS genes while the expression of the *pqsA-E* operon, which is involved in biosynthesis of the quinolone signaling molecule PQS,[143] was strongly enhanced by growth in 10% CF sputum medium.[141] Detection of very low levels of C12-HSL and C4-HSL excluded the possibility that endogenous *P. aeruginosa* autoinducers within the CF sputum had induced the production of PQS.[141] Besides the effect on PQS production, the growth of PA14 in 10% CF sputum medium increased the expression of *hcnABC* and *phzABCDE*, two operons that are positively regulated by PQS.[141–143] Finally, similar to the study by Wang et al.,[128] the growth of PA14 in 10% CF sputum medium enhanced the expression of the pyochelin genes dramatically.[141] However, in this study, the expression of the pyoverdine genes was also increased, although at lower levels (3- to 15-fold less) than that of the pyochelin genes.[141]

4.1.6 Synthetic CF Sputum Medium

The CF sputum is complex.[109,110] Therefore, to identify the specific components of sputum that influence the growth of *P. aeruginosa* as well as the expression of different genes within it, Palmer et al.[144] developed synthetic CF sputum medium (SCFM). To do that, these authors analyzed sputum samples from 12 CF patients for the levels of free amino acids, cations, anions, glucose, and lactate.[144] Samples were desiccated and the powdered CF sputum was suspended in distilled water to final concentrations of 10, 20, and 25%.[141,144] The suspension was homogenized, the insoluble materials were removed by centrifugation, and the clear solution was utilized in chromatography and enzymatic analysis. Based on the analyses, Palmer et al.[144] synthesized SCFM which contained ions, amino acids and lactones at levels comparable to their levels in the CF sputum. From findings of other studies, SCFM was supplemented with ferric sulfate.[145,146] To compensate for the relative buffering capacity of the lung, SCFM was also supplemented with 10% MOPS.[144] Results showed that 54% of the genes in the *P. aeruginosa* strain PA14 that had previously been shown to be differentially expressed by growth in CF sputum, were differentially expressed when PA14 was grown in SCFM.[141,144] These genes included those involved in amino acid catabolism, PQS production, iron acquisition, and chemotaxis/flagellar motility.[144] Since carbon source catabolism affects *P. aeruginosa* virulence, the authors determined the carbon utilization profiles of PA14 in SCFM and 10% CF sputum medium. They found that PA14 consumed the same six carbon sources first in both types of media: proline, alanine, arginine, glutamate, aspartate, and lactate; suggesting that SCFM is an in vivo relevant medium with which to examine carbon preference.[144] As the aromatic amino acids tryptophan, tyrosine, and phenylalanine are known to influence PQS biosynthesis,[141,147] further analysis was done using SCFM with serine substituted for these amino acids. Results from these experiments revealed that aromatic amino acids are important for PQS production.[144] Additionally, aromatic amino acids in SCFM influenced the production of pyocyanin whose synthesis depends on PQS-dependent antimicrobial activity as measured by the ability of PA14 to lyse the Gram-positive bacterium, *S. aureus*.[144,148]

4.2 Media to Analyze Biofilm Formation by CF Lung Isolates

4.2.1 Biofilms

Under hazardous conditions, microorganisms attach to either biotic or abiotic surfaces and form protective structures termed biofilms.[149] Mature biofilms are mushroom shaped multicellular structures in which microorganisms are surrounded by a protective layer termed the glycocalyx. The glycocalyx is formed by the extrapolysaccharide matrix (EPS) produced by the microorganisms and the host surrounding tissue.[149,150] Within the biofilm the resistance of microorganisms to host immune response and various antibiotics is significantly enhanced.[149] *P. aeruginosa* forms biofilms at different infection sites including chronic wounds,[91] burn wounds,[92] and the lung alveoli of CF patients.[149,151] Since the recognition of the clinical importance of biofilms, many different investigators have utilized both static (stationary) and continuous flow-through systems and a variety of minimal media to investigate the formation of *P. aeruginosa* biofilms.[152–159] Similar to the experience with gene expression, investigators began adding host components to the biofilm medium to more closely simulate in vivo conditions.

4.2.2 Use of Mucin as a Substrate for Biofilm Formation

Landry et al.[160] investigated the effect of mucin, the main component of CF sputum, on the development of *P. aeruginosa* biofilm on glass and mucin-coated surfaces. Using the once-through continuous culture biofilm system[161] and the chemically defined EPRI medium,[154] Landry et al.[160]

showed that compared with *P. aeruginosa* biofilms developed on a glass surface, those developed on a mucin-coated surface formed larger aggregates that were more resistant to tobramycin. Analysis of specific *P. aeruginosa* mutants revealed that the mucin-coated surface blocks *P. aeruginosa* surface-associated motility.[160] The authors further suggested that *P. aeruginosa* adheres to the mucin-coated surface through both FliD-dependent and FliD-independent mechanisms.[160] FliD is a mucin-adhesin located at the distal end of the helical flagellar filament.[162] The FliD-mucin interaction immobilizes the bacteria on the mucin-coated surfaces allowing the immobilized bacteria to multiply and form large aggregates.[160]

4.2.3 Artificial Sputum Medium

Previous studies indicated that within the thick alveolar mucus of the CF lung, *P. aeruginosa* grows under either low oxygen tension (10% O_2, microaerophilic) or in an anoxic environment (0% O_2, anaerobic).[112,125,163] Under these conditions *P. aeruginosa* forms biofilms. However, unlike the typical microbial biofilms, *P. aeruginosa* forms microcolonies in which bacteria adhere to the mucus within the CF sputum and to each other, rather than to a solid surface.[164,165] To examine the development of these biofilm/microcolonies, Sriramulu et al.[166] developed a specific medium, artificial sputum medium (ASM+), that contains the essential ingredients of CF sputum: mucin, DNA, sodium chloride, potassium chloride, lecithin (from egg yolk emulsion), and amino acids, plus diethylenetriaminepentaacetic acid (as a cation chelator). The concentrations of these ingredients are similar to those found in an average sputum of a patient with CF.[167] Upon growth in ASM+, *P. aeruginosa* formed clumps visible by microscopy, that were difficult to disrupt even by vigorous pipetting.[166] Bacterial cells were attached to each other but not to the polystyrene or glass surfaces. The removal of amino acids from the ASM+ reduced the number of microcolonies and increased number of planktonic cells. The removal of mucin reduced *P. aeruginosa* growth and caused the cells to attach to the polystyrene surfaces similar to typical biofilms. Lecithin, DNA, and salts were required to variable extents for the formation of microcolonies, while tryptophan contributed to the formation of tight microcolonies.[166] Among the *P. aeruginosa* virulence factors examined, alginate and pili, but not flagellar-mediated motility played a role in forming tight microcolonies; additionally, the QS regulator LasR, but not RhlR, influenced microcolony development.[166] Two-dimensional gel analysis revealed that the growth of *P. aeruginosa* in ASM+ also increased the production of OprF outer membrane protein.[166,168,169]

4.2.4 Modified Artificial Sputum Medium

To examine the effect of CF sputum on the expression of *P. aeruginosa* genes, Fung et al.[170] modified ASM+[166] by adding bovine serum albumin (BSA), increasing the concentration of porcine stomach mucin, and lowering the DNA concentration. They called their modified medium ASMDM.[170] The addition of BSA was based on a previous study that showed high albumin concentration in the sputum of CF patients compared with the albumin level in sputum obtained from non-CF patients.[171] Fung et al.[170] grew *P. aeruginosa* strain PA14 in MOPS/glucose medium or in ASMDM and examined the expression of PA14 genes using microarray experiments. They compared these results with those of Palmer et al.[141] in which PA14 was grown in SCFM. The growth in MSMDM or SCFM enhanced the expression of some of the same nutrient-controlled genes.[170] However, several virulence genes were upregulated upon growth in ASMDM only; 24 QS genes, 23 TTSS genes, and several genes involved in anaerobic metabolism.[170] Enhanced expression of different QS and TTSS genes during acute infection in vivo and in vitro conditions had been previously documented.[172–175] The increase in the expression of TTSS was not due to low calcium content of ASMDM as this medium, as well as SCFM, contain the same amount of calcium.[141,170] The increased expression of genes involved in anaerobic metabolism suggests that ASMDM reflects the previously described hypoxic condition in the CF airway mucus more faithfully than SCFM.[112,170] The authors suggested that the increased expression of certain anaerobic metabolism genes may have contributed to the observed increase in the expression of TTSS genes as nitrous oxide produced by anaerobic metabolism is critical for the assembly of the entire TTSS.[176] Another major difference between ASMDM and SCFM is the expression of iron-regulated genes. Several iron-regulated genes including pyochelin and pyoverdine genes were down regulated upon the growth of PA14 in ASMDM but were upregulated when the strain was grown in SCFM.[141,170] The iron-regulated genes were repressed even though ASMDM contained the iron-chelator diethylenetriaminepentaacetic acid.[166,170] Fung et al.[170] suggested that despite the chelator, ASMDM contained sufficient iron to repress these genes, although the source of this iron was not identified.

4.2.5 Use of Artificial Sputum Medium to Examine Formation of Biofilm-Like Structures

Haley et al.[177] utilized ASM+ [166] and confocal laser scanning microscopy (CLSM) to examine the development of *P. aeruginosa* microcolonies

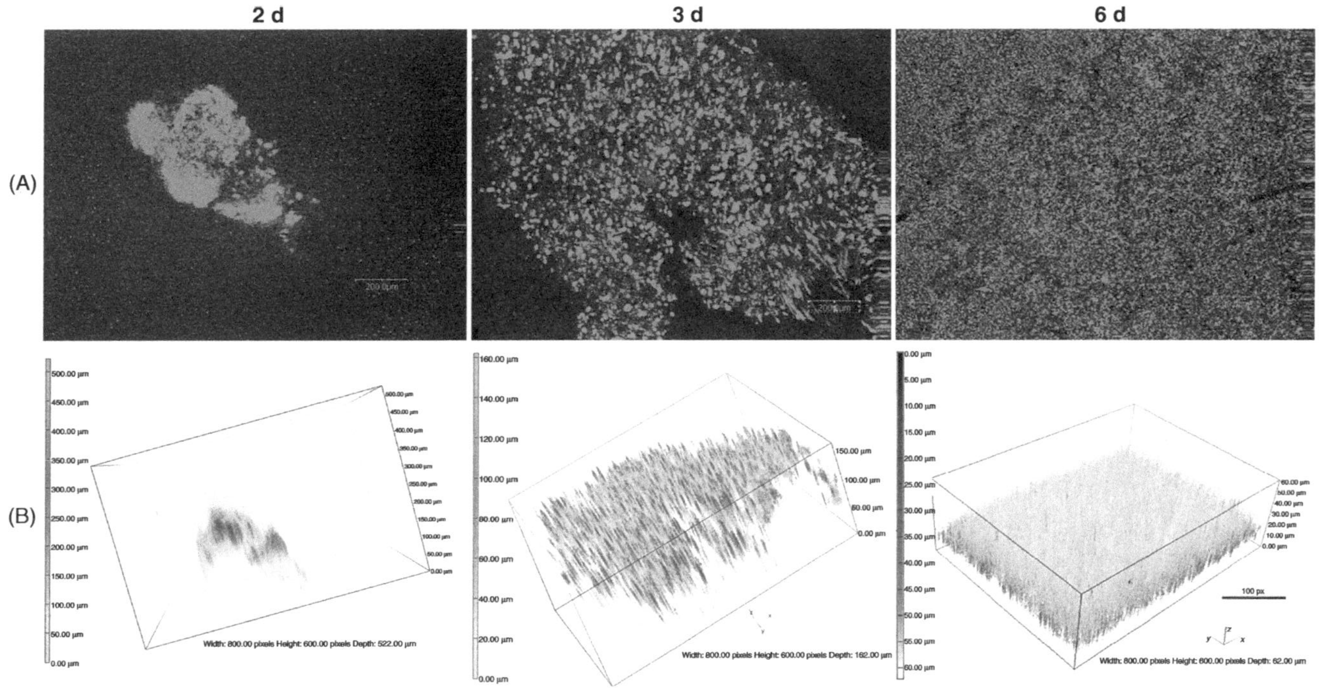

Figure 1 PAO1 BLS vary structurally over time. PAO1/pMRP9-1, a plasmid carrying the gene for green fluorescent protein (appears gray here, green in the web version), was inoculated into ASM+ to an initial OD_{600} of 0.02–0.03. One-milliliter aliquots were distributed into wells of 24-well microtiter plate and the plate was incubated at 37°C in static conditions under 20% EO_2 for up to 6 days without changing the medium. (A) CLSM micrographs of BLS at 2, 3, and 6 days postinoculation; magnification, 10X; bars, 200.00 nm. (B) The 3-D architecture of the BLS shown in (A). Boxes, 800.00 px W × 600.00 px H; bars, 100 px. *Haley et al.; licensee BioMed Central Ltd. 2012.*[177] *Figure used without changes as permitted under the terms of the Creative Commons Attribution License (http://creativecommons.org/licenses/by/2.0).*

described previously[164,165] and the conditions that influence their development. They also addressed the interaction between *P. aeruginosa* and *S. aureus* within the lung alveoli of CF patients.[177] Strains carrying plasmids encoding either green fluorescent or red fluorescent proteins were used for image analyses, and the computer program COMSTAT was utilized to quantify the architectural features of the microcolonies.[177,178] Initial analyses revealed that upon growth in ASM+, the *P. aeruginosa* strain PAO1 produced floating structures, termed biofilm-like structures (BLS), that have all the characteristics of well-developed biofilms.[177] Extending the incubation of PAO1 in ASM+ from 2 to 6 days increased the size and thickness of BLS considerably (Fig. 1). Adjusting the mucin concentration affected the architecture of the BLS: compared with BLS produced at a mucin concentration of 5 mg/mL (standard for ASM+), reducing the amount of mucin to 2.5 mg/mL or increasing it to 10 mg/mL resulted in the formation of thicker, more homogeneous BLS.[177] This variation was not due to a difference in the growth of PAO1, as comparable number of microorganisms were recovered at each of the mucin concentrations.[177] In contrast, increasing or decreasing the DNA concentration led to production of BLS that were thinner, covered less surface area, and contained less biomass.[177] To examine the effect of environmental oxygen (EO_2), PAO1 was grown in ASM+ under aerobic (20% EO_2), microaerobic (10% EO_2) or anaerobic (0% EO_2). Under 0% EO_2, the culture was supplemented with potassium nitrate as a terminal electron acceptor.[112,177] Compared with the BLS formed under 20% EO_2, those formed under 10% EO_2 were more fully developed, more dense, and increased in their depth within the medium.[177] In contrast, BLS developed under 0% EO_2 were smaller and less developed.

The architecture of the BLS also varied with the *P. aeruginosa* strain tested; compared with the BLS produced by PAO1, those produced by strain PAK were similar, while strain PA103 and a *P. aeruginosa* isolate obtained from a patient with a lower respiratory tract infection were significantly smaller.[177] To examine the contribution of the QS systems to BLS development, Haley et al.[177] used PAO1 mutants defective in *lasR, lasI, rhlR, rhlI*, and *pqsA*.[179–182] The *las* and *pqs* mutants produced BLS that visually and architecturally resembled those produced by the PAO1, while those produced by the *rhl* mutants appeared smaller and less dense and were significantly different in most parameters measured (Fig. 2) suggesting that the *rhl* QS system plays a role in the development of BLS.[177]

Using the same experimental application, Haley et al.[177] showed that similar to *P. aeruginosa, S. aureus* forms BLS that were more developed under

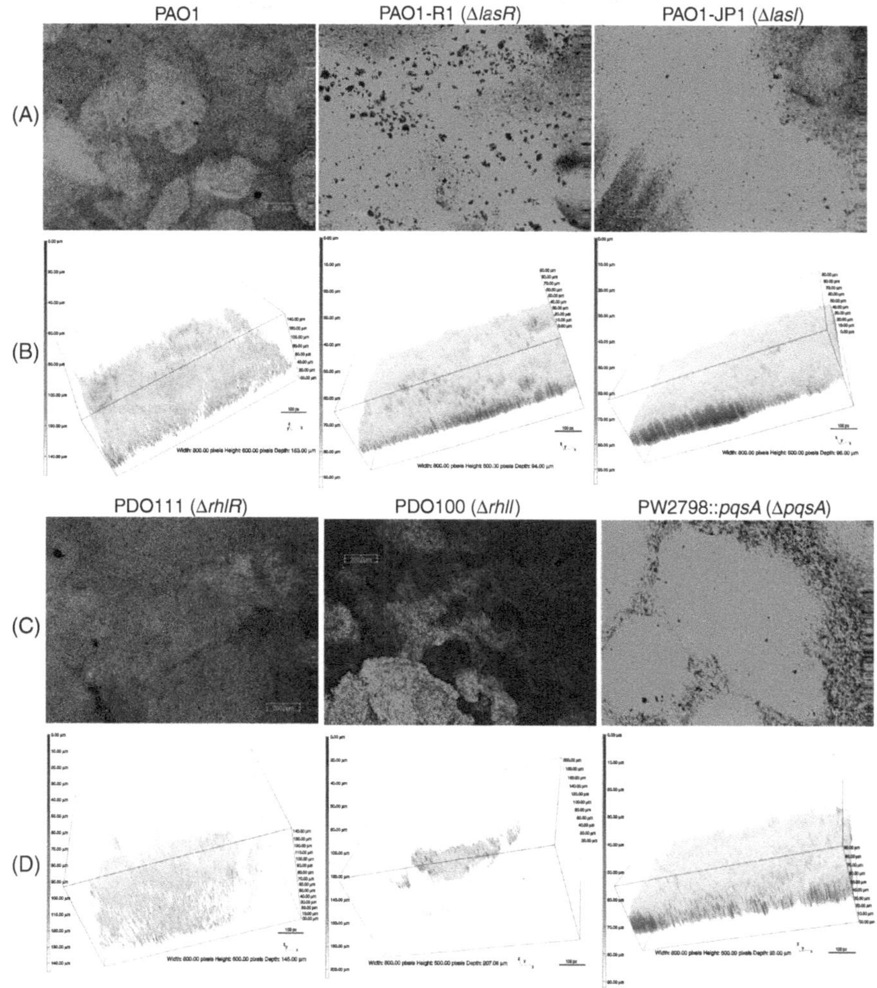

Figure 2 Loss of individual QS genes affects BLS formation. PAO1 strains defective in the *lasR* (PAO-R1), *lasI* (PAO-JP1), *rhlR* (PDO-111), *rhlI* (PDO100), or *pqsA* (PW2798::*pqsA-lacZ*) genes were transformed with pMRP9-1 and the transformants plus PAO1/pMRP9-1 as a control were grown in ASM+ under 10% EO_2 without shaking for 3 days. The BLS were analyzed by CLSM and the 3-D architecture constructed. (A and C) Representative micrographs of the BLS; magnification, 10×; bar, 200.00 nm. (B and D) Respective 3-D images constructed from the CLSM micrographs. Boxes, 800.00 px W × 600 px H; bars, 100 px. *Haley et al.; licensee BioMed Central Ltd. 2012.*[177] *Figure used without changes as permitted under the terms of the Creative Commons Attribution License (http://creativecommons.org/licenses/by/2.0).*

20% EO_2 than 10% EO_2. They then examined the interaction of *P. aeruginosa* and *S. aureus* when grown together. When cocultured in ASM+ under 20% EO_2 for 48 h, *P. aeruginosa* inhibited the growth of *S. aureus.*[177] To examine the influence of *P. aeruginosa* on partially developed *S. aureus* BLS, PAO1 was inoculated into ASM+ containing 8-h *S. aureus* BLS. Analysis of the structures post-*P. aeruginosa* inoculation revealed that the CFU of *S. aureus* began decreasing at 8 h, that by 40 h only remnants of the *S. aureus* BLS could be detected, and that by 56 h, the *S. aureus* BLS had been completely replaced by those of *P. aeruginosa* (Fig. 3).[177]

4.3 Media for Analysis of *P. aeruginosa* Virulence in Wound and Systemic Infections

4.3.1 Acute and Chronic Wounds

Acute wounds represent external damage to intact skin and may occur surgically or from accidental trauma—bites, burns, and other trauma, from minor cuts and abrasions to severe crush or gunshot injuries. While the time varies with the extent and type of the wound, healing of an acute wound is expected to occur within 30 days.[183] Acute wound infections are caused by numerous types of bacteria including *P. aeruginosa* (about 1%).[183] Chronic wounds, caused by compromise of the integrity of dermal/epidermal tissue, are frequently due to vascular disorders (peripheral vascular disease or venous hypertension) and metabolic diseases (such as diabetes).[183] These wounds often exhibit a marked failure to heal, providing a site where multiple species infect the tissues.[183] Burn wounds present a unique circumstance; infections begin acutely but, depending on the type of burn and the necessary treatment, acute infections often progress to the chronic state, leading to failure of the wound to heal.[184] Within chronic burn wounds, *P. aeruginosa* is often predominant, although other pathogens are frequently found as well.[183–185] These pathogens are often present in biofilms, from which they may emerge in planktonic form to invade the surrounding tissues and the bloodstream.[186,187] Whether the infection is acute or chronic, systemic responses such as the systemic inflammatory response syndrome and severe sepsis (see Section 4.3.4) may occur.[92,183–185,187]

4.3.2 Effect of Individual Blood Components on P. aeruginosa Biofilm Formation

Since biofilm formation within an infected wound site or on indwelling devices has been shown to precede *P. aeruginosa* systemic infections,[92,149,184,188] investigators have examined the effect of specific blood

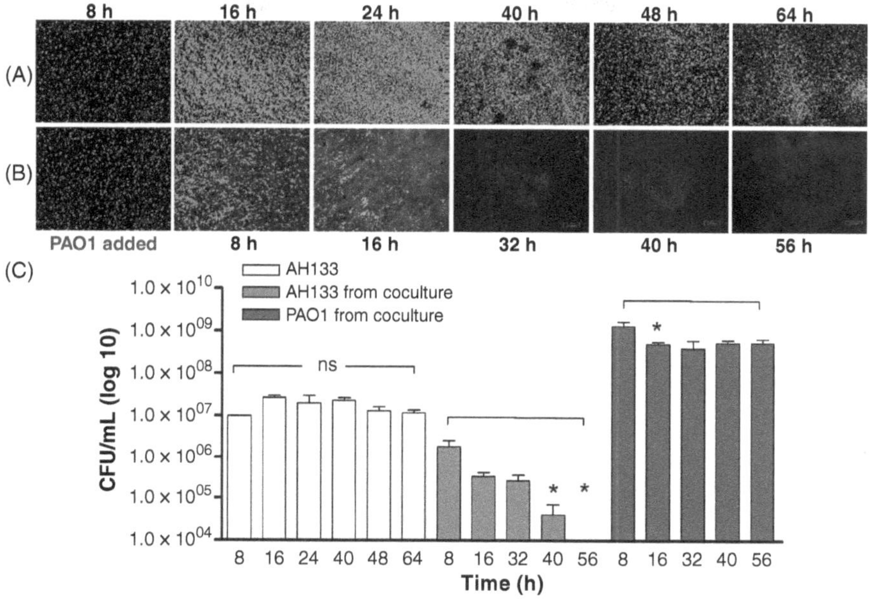

Figure 3 Elimination of *S. aureus* AH133 BLS is due to the bactericidal effect of PAO1. PAO1/p7605 [dark gray (red in the web version)] gradually eliminates previously formed AH133 [light gray (green) in the web version] BLS. ASM+ was inoculated with AH133 and the cultures were grown for 8 h to allow for the partial development of AH133 BLS. (A) One culture was continued without addition of PAO1 for a total of 64 h. (B) The other culture was inoculated with PAO1/p7605 (starting density similar to that used to initiate the AH133 culture). Incubation was continued for an additional 56 h and cultures were analyzed at 8- to 16-h intervals by CLSM (exact time points indicated earlier and below micrographs). (C) PAO1 is bactericidal to AH133. Two sets of wells containing ASM+ were inoculated with AH133 and incubated for 8 h. PAO1/pMP7605 was added to one set of wells and incubation of both sets was continued for 56 h. At the specified time points, the gelatinous mass was obtained and the CFU/mL of each species was determined using selective media (methods). White bars: AH133 CFU/mL in single culture; light gray (green in web version) bars, CFU/mL of AH133 in the coculture; dark gray (red in web version) bars, CFU/mL of PAO1/pMP7605 from the coculture. Values represent the means of at least three independent experiments ± SEM. *Haley et al.; licensee BioMed Central Ltd. 2012.[177] Figure used without changes as permitted under the terms of the Creative Commons Attribution License (http://creativecommons.org/licenses/by/2.0).*

components on the attachment stage of biofilm formation by *P. aeruginosa*. Ardehali et al.[189] showed that, at low concentrations, the blood protein transferrin interferes with the ability of *P. aeruginosa* to bind to a polyurethane surface. Similarly, Singh[190] demonstrated that human lactoferrin blocks *P. aeruginosa* biofilm formation by sequestering iron and enhancing *P. aeruginosa* surface motility. Using M9 minimal medium supplemented with 10%

adult bovine serum (ABS), Hammond et al.[191] examined the effect of serum on biofilm development by *P. aeruginosa* in both static and continuous flow-through culture systems.[11,192] In both systems, ABS significantly inhibited the development of *P. aeruginosa* biofilms on a plastic surface as well as eliminating *P. aeruginosa* partially formed biofilms.[191] Further analysis revealed that ABS eliminated *P. aeruginosa* biofilms by significantly enhancing the pilus-mediated twitching motility without affecting flagellum-mediated swimming motility.[191] Although twitching motility is essential for early stages of biofilm development, it is detrimental at later stages.[155] Similar results were obtained when *P. aeruginosa* biofilms were developed using small pieces of intravenous catheter and human serum.[191] Based on these results, Hammond et al.[191] suggested that serum binds to the catheter surface and physically hinders the attachment of *P. aeruginosa*. Additionally, serum eliminates *P. aeruginosa* biofilm formation by significantly enhancing the twitching motility.[191]

4.3.3 Effect of Blood Components on P. aeruginosa Gene Expression

Kruczek et al.[193] utilized microarray experiments to examine the effect of 10% ABS on the expression of different genes of *P. aeruginosa* strain PAO1. PAO1 was grown in either the iron-deficient medium TSB-DC or in TSB-DC containing 10% ABS and the cultures were harvested at either the early exponential phase or the early stationary phase of growth.[193] Compared with TSB-DC, the growth of PAO1 to early exponential phase in TSB-DC/10% ABS significantly altered the expression of 178 genes; expression of 138 genes was enhanced, while expression of 40 genes was repressed.[193] In contrast to the growth at early exponential phase, the growth at early stationery phase influenced the expression of few genes.[193] Most of the genes whose expression was significantly induced at early exponential phase are negatively regulated by iron including the *toxA*-positive regulatory gene *regA*[194] and the pyoverdine and pyochelin genes.[193] These genes carry specific binding sequences for either the ferric uptake regulator Fur (Fur-binding box)[36] or the pyoverdine regulatory factor PvdS (iron-starvation box)[195] within their upstream regions.[193] Kruczek et al.[193] further analyzed the mechanism of ABS-induced enhancement in the expression of iron-regulated genes using *regA* and *pvdS* genes as representatives. Experimental evidence showed that serum enhancement of the expression of these genes occurs through a mechanism that is not related to iron. Despite the depletion of the intracellular pool of iron by several growth passages in TSB-DC, serum still significantly enhanced *pvdS* expression in PAO1.[193]

Additionally, the observed increase in gene expression was shown to be Fur-independent, since serum significantly enhanced *pvdS* expression in both PAO1 and its Fur-deficient mutant.[193] Initial experiments indicated that the potential serum factor that enhances the expression of these genes is a protein.[193] Results of further analysis using membrane fractionation experiments suggested that the potential factor is more than 50 kDa in molecular weight.[193] Since albumin, which is a major serum protein,[196] has been previously shown to influence bacterial virulence,[197,198] Kruczek et al.[193] examined the effect of human serum, albumin-depleted human serum, and human serum albumin on the expression of *regA* and *pvdS* and obtained results similar to those with ABS and albumin-depleted ABS. Depletion of albumin from human or bovine serum abrogated the enhancement of *regA* and *pvdS* expression.[193] Further experiments showed that another serum protein, apotransferrin also increased *regA* and *pvdS* expression. Based on the results of these different experiments, Kruczek et al.[193] suggested that the *P. aeruginosa* iron-regulated genes are induced through two iron-independent mechanisms, one of which is albumin-related while the other is apotransferrin-related.

Kruczek et al.[199] examined the effect of adult bovine serum (serum) on the expression of QS and QS-related genes in the *P. aeruginosa* strain PAO1. The strain was grown in either Luria Bertani broth (LB) or LB broth containing 10% serum (LBS) to an early or late phase of growth. The expression of PAO1 genes was examined using either qRT-PCR or transcriptional fusion systems. At early stages of growth, LBS-grown PAO1 exhibited significant decreases in expression of *lasI*, *rhlI*, and *pqsA* but showed no changes on the expression of *lasR* and *rhlR*.[199] In contrast, the expression of all of these genes was significantly increased during late stages of growth. A similar pattern was observed for the expression of the *lasB* and *rhlA* genes, which are tightly controlled by the *las* and *rhl* systems, respectively.[199] Additionally, growth in the presence of serum repressed the production of all three autoinducers (C12-HSL, C4-HSL, and PQS) at early stages of growth, but enhanced their production at the later stages of growth.[199] Using qRT-PCR, Kruczek et al.[199] further showed that the influence of serum on the expression of several QS-controlled genes—*lasA*, *hcB*, *plcD*, and *phzC*—is similar to the effect on the QS systems. Serum also produced comparable effects on LasB elastolytic activity, LasA staphylolytic activity, and phospholipase C activity produced by PAO1.[199]

The *P. aeruginosa* QS systems are regulated by numerous positive and negative regulators. The positive regulators include Vfr,VqsR, and

GacA[200–202]; whereas the negative regulators include MvaT, RsmA, QscA, RpoS and RpoN.[85,203–206] Therefore using qRT-PCR, Kruczek et al.[199] examined the possibility that serum affects the QS systems through one or more of these regulators. Results suggested that potential regulation may occur through *vqsR* or *mvaT* at early stages of growth and *vfr* at late stages of growth. Using *vqsR* and *mvaT* mutants Kruczek et al.[199] ruled out the possibility that serum influences QS systems through either gene product during early stages of growth. Serum significantly repressed *lasI* expression in each mutant and in their parent strain PAO1.[199] During late stage of growth, however, serum may regulate the *las* system through *vfr*. At this stage of growth, serum enhanced the expression of *lasI* and *lasR* in PAO1, but not in the *vfr* isogenic mutant.[199] Based on the results of these experiments, Kruczek et al.[199] suggested that, depending on the stage of growth, serum differentially regulates the expression of QS and QS-controlled genes; and that at the late stage of growth, serum induced the expression of *las* genes through *vfr* (Fig. 4).

4.3.4 Bacteremia and Sepsis

Despite improvements in healthcare, bacteremia, or bloodstream infection (BSI), is still a major cause of morbidity and mortality.[207,208] Both Gram-negative and Gram-positive bacteria may invade the blood from the respiratory tract, the urinary tract, intraabdominal infection, or other sites within the body.[207] Bacterial factors such as LPS (endotoxin) of Gram-negative bacteria or lipoteichoic acid from the cell walls of Gram-positive bacteria are recognized by the innate immune system pattern recognition receptors, especially Toll-like receptors. This recognition triggers a cascade of proinflammatory cytokines that elicit a strong host response critical for clearance of the bacterial infection, first at the site of the infection and then systemically.[207,209] This response is called sepsis, which was defined in 1992 by an international consensus panel as a systemic inflammatory response to infection.[207,209] While frequently beneficial, the sepsis response can also lead to organ damage (severe sepsis) and/or hypotension (septic shock), either of which can be fatal.[7,13,14,207,209] Numerous risk factors for BSI with severe sepsis have been defined, including chronic diseases such as AIDS, chronic obstructive pulmonary disease, many cancers, and use of immunosuppressive drugs; severe burns; the very young and the elderly; and those with neutropenia for any reason.[4,5,208,209]

Among the few pathogens that cause significant numbers of BSI is *P. aeruginosa*, which causes 3-7% of all BSI and 23–26% of Gram-negative

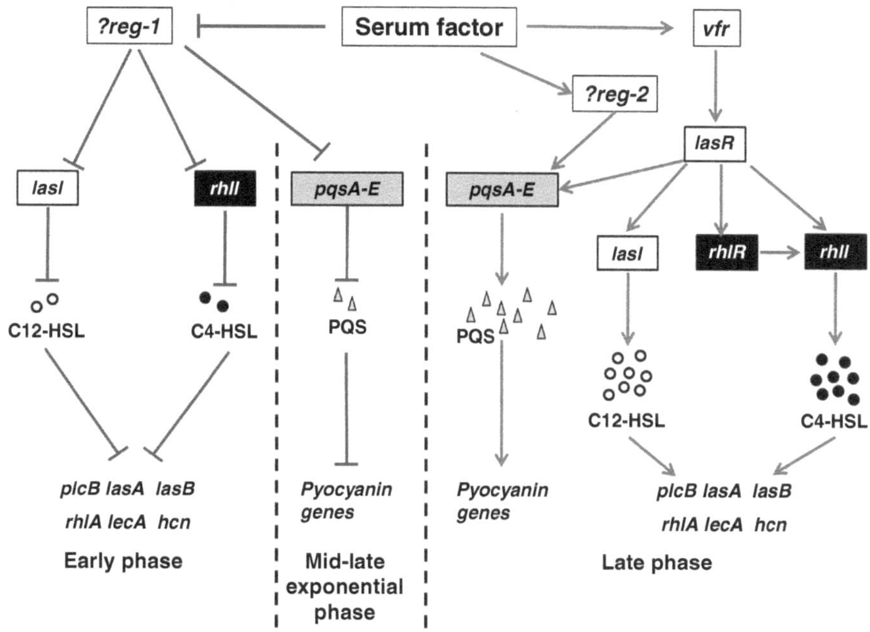

Figure 4 Diagram illustrating the possible mechanisms described by Kruczek et al.[199] through which serum represses and then enhances the expression of PAO1 QS and QS-controlled virulence genes. The specific gene(s) through which serum represses the QS systems at early and mid-late exponential phases is not yet known *(?reg-1)*. Low levels of C12-HSL, C4-HSL, and PQS are insufficient to activate LasR, RhlR, and MvfR with subsequent reduction in QS-related gene expression. At the late phase of growth, the serum factor enhances *vfr* expression, which enhances *lasR* expression with subsequent enhancement of C12-HSL, C4-HSL, and PQS synthesis leading to increased expression of QS-related genes. Additionally, the serum factor may enhance the expression of another yet unknown regulatory gene (*?reg-2*) that affects expression of *pqsA-E*. − −|, indicates repression of gene expression; →, indicates enhancement of gene expression. Experimental evidence presented by Kruczek et al.[199] showed that the QS regulatory genes *vqsR*, *rsmA*, *mvaT*, *gacA*, and *rpoN* were not involved in this regulatory process.

BSI.[210,211] The prognosis for *P. aeruginosa* BSI is very poor, with the mortality rate for patients with *P. aeruginosa* BSI ranging from 27 to 48%.[2,208,212] One of the major factors that predisposes patients to *P. aeruginosa* BSI is the disruption of the physical barriers of the skin and mucosal surfaces.[5] Severely burned patients are immunocompromised due to the loss of the skin barrier as well as changes within the innate and adaptive immune systems.[4] Bacterial pathogens, such as *P. aeruginosa*, colonize the exposed underlying tissue, establish an infection, and invade the bloodstream, leading to sepsis. Another route of entry to the bloodstream is via introduction of foreign bodies such as urinary catheters, vascular catheters, and drainage tubes, which may disrupt the integrity of the mucosal surfaces.[210] *P. aeruginosa* colonizes the surface of these devices and spreads to the bloodstream.[192,210] Depending on the location of *P. aeruginosa* infection in the host, certain virulence factors may play a more prominent role in the pathogenesis of *P. aeruginosa* infection than others.[72,73,153,212] Previous studies have extensively analyzed the role of specific virulence factors in the pathogenesis of acute and chronic *P. aeruginosa* infections; however, the role of virulence factors during BSI has not been completely defined.[153,155,190] A major contributing factor to this problem is the lack of in vitro conditions that closely mimic the in vivo conditions during the growth of *P. aeruginosa* in blood.

4.3.5 Effect of Transfer to Whole Blood on Bacterial Gene Expression

Another approach to examine the effect of blood or blood components on the virulence of *P. aeruginosa* is by transferring growing bacteria to whole blood to mimic their transition from an extravascular site into the bloodstream. Investigators utilized this approach to examine the effect of blood on the virulence of Gram-positive bacterial pathogens. Graham et al.[213] utilized this approach to examine the mechanism that the bacterial pathogen *Streptococcus pyogenes*, or Group A *Streptococcus* (GAS), uses to adapt to the environment of blood during systemic infection. GAS causes skin and mucosal infections in humans.[214,215] However as it transits to blood, it causes life threatening infections, including puerperal sepsis, necrotizing fasciitis, and streptococcal toxic shock syndrome.[214,215] GAS strains were grown in Todd–Hewitt broth to a late exponential phase, harvested and suspended in the same volume of whole blood from human volunteers.[213] Blood cultures were grown at 37°C, samples were obtained at 30, 60, and 90 min, and the global gene expression was determined. The expression of many genes that may enhance bacterial survival, including those coding for super antigens and host evasion proteins were enhanced.[213] Additionally, as blood is a protein-rich environment, an increase in the expression of genes involved in proteolysis, transport and catabolism of oligopeptides was detected.[213]

Similarly, Mereghetti et al.[216] examined the effect of blood on the expression of different genes of *Streptococcus agalactiae*, group B *Streptococcus* (GBS), which causes serious late onset neonatal infections such as bacteremia and meningitis.[217] The GBS was grown in a rich laboratory medium and shifted to blood from healthy volunteers.[216] Blood cultures were incubated at either 37°C or 40°C (to mimic high fever associated with severe infections) and samples were obtained at 30 and 90 min. Microarray analysis revealed a significant and sudden shift in the expression of various genes, including those involved in carbohydrate metabolism. At 30 min, 83% of the genes were down regulated.[216] However, genes encoding proteins involved in interaction with the host coagulation and fibrinolysis systems and bacteria-host interaction were upregulated.[216] Compared with the growth at 37°C, growth at 40°C produced additional transcriptome variations.[216]

4.3.6 Use of Whole Blood in an Ex Vivo Culture System

While serum is present in wounds as part of the wound exudate, once *P. aeruginosa* enters the bloodstream, the organism finds itself in the presence of whole blood, exposed to numerous proteins, sugars, lipids, and ions in addition to the cellular constituents. We investigated the changes in global gene expression that *P. aeruginosa* strain PA14 undergoes as it adapts to the environment of whole blood during systemic infection using the closed loop flow-through system (CLFT). The CLFT is an ex vivo model that mimics blood circulating in arteries.[218] In this model, blood is injected into a continuous loop made of silicone tubing (16-mm platinum-cured silicone tubing to allow oxygen diffusion), the loop is submerged in a 37°C water bath, and the blood within the tubing is propelled at 176 mL/min by a peristaltic pump.[218] To mimic natural infections, PA14 was grown overnight in LB and then inoculated into 7 mL of blood from healthy volunteers at a low inoculum (10^2–10^3 CFU/mL) and the blood cultures were incubated at 37°C in the CLFT (N.D., unpublished data, 2015). Initial experiments confirmed that the growth of PA14 in blood was comparable to its growth in LBB. Additional experiments showed that incubating the culture for more than 8 h resulted in extensive lysis of red blood cells. The cultures were harvested at 8 h post inoculation and RNA was extracted. As a control, PA14 was grown under the same conditions in LB. Transcriptome profiles were analyzed by RNA-seq. Compared with LB, the growth of PA14 in blood altered the expression of more than 2000 genes (N.D., unpublished data, 2015). Most genes of the TTSS, as well as the pyoverdine iron-acquisition system, were upregulated; however many QS, and QS-controlled genes were down regulated (N.D., unpublished data, 2015).

P. aeruginosa causes serious systemic infections in severely burned patients.[184,219–221] Therefore, we investigated the possibility that burn-induced changes in host blood components facilitate *P. aeruginosa* virulence by altering the expression of different genes. We grew PA14 individually in whole blood from either three healthy volunteers or three severely burned

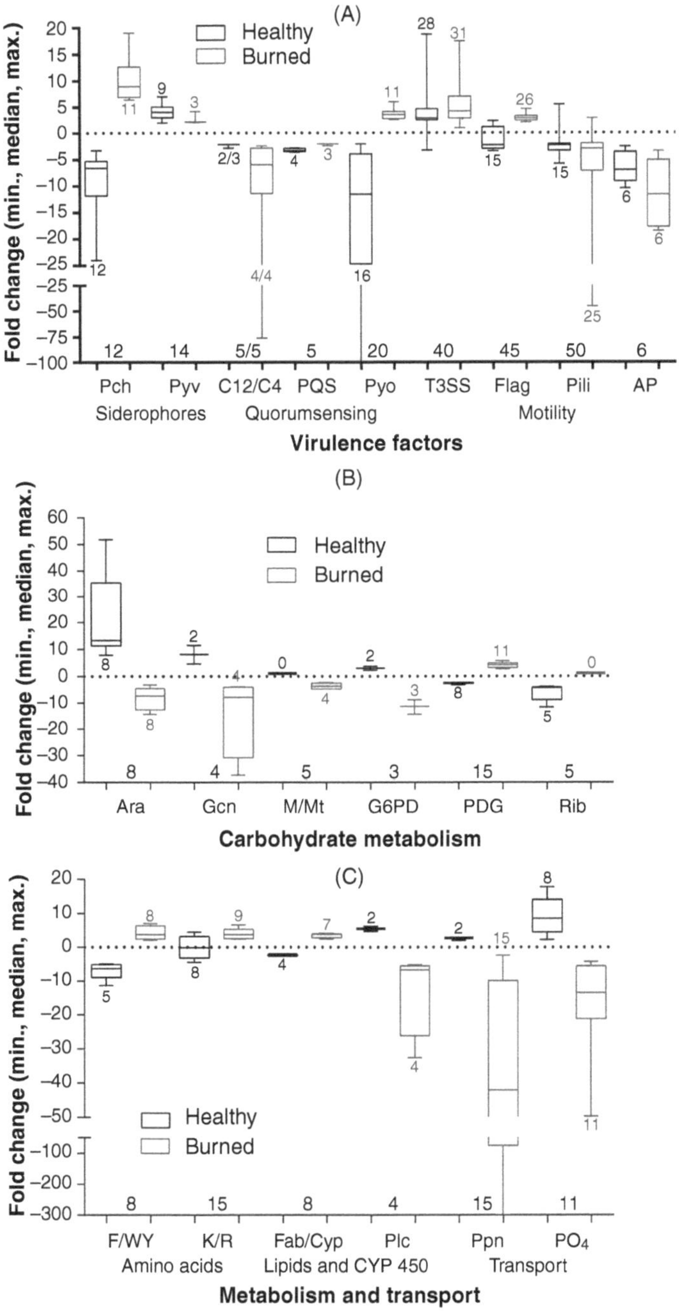

Figure 5 Comparison of *P. aeruginosa* gene expression when grown in whole blood obtained from healthy volunteers or burned patients. Functions of gene products were found using the "Gene Ontology (GO) Term Annotations" search feature of the

patients in the CLFT system described earlier, and examined global gene expression by RNA-Seq.[222] Compared with growth in blood from healthy volunteers, the growth of PA14 strain in the blood from severely burned patients significantly altered the expression of 2596 genes; 1060 genes were enhanced, while 1536 genes were repressed. Genes whose expression was significantly reduced included *lasR*, *rhlR*, and *pqsA-E* related to *las* and *rhl* QS, the virulence factors controlled by the *las* and *rhl* systems, and the transport of heme, phosphate and phosphonate (Fig. 5).[222] Genes whose expression was enhanced include genes related to the TTSS, the pyochelin iron-acquisition system, flagellum synthesis, and pyocyanin production (despite reduction in expression of the *pqs* genes) (Fig. 5).[222] Results of RNA-Seq experiments were confirmed by qRT-PCR. At this time the mechanism through which burn-induced changes increased or decreased the expression of different genes is not known. Analysis of different ions within blood samples from the three burn patients failed to support the possibility that the observed variations in gene expression were due to the alterations in levels of these ions.[222] For example, the expression of the TTSS genes is known to be repressed by calcium and maximum expression of these genes occurs when *P. aeruginosa* is grown in calcium-deficient medium.[223] While calcium levels in the blood from all three patients were low, the levels were still sufficient to repress the TTSS genes. Based on this, we suggested

◀ *Pseudomonas* Genome Database (http://www.pseudomonas.com/). Box and whisker plot of gene expression in blood from a healthy volunteer *(healthy)* was compared to expression in LB as a control while that of gene expression in blood from 3 burned patients *(burned)* was compared to expression in blood from a healthy volunteer. Expression of virulence factor genes (A), genes for carbohydrate metabolism (B), protein and lipid metabolism (C), and transport of phosphonate and phosphate (C) was considered differentially regulated if expression varied by two-fold or more from relevant control. The numbers of genes found for each GO annotation are noted above the X axis; the number of genes differentially regulated under each condition are indicated above or below each box. (A) *Pch*, pyochelin synthesis; *Pyv*, pyoverdine synthesis; *C12/C4*, *las* and *rhl* QS systems and two representative genes controlled by each system; *PQS*, *Pseudomonas* quinolone signal synthesis genes; *Pyo*, pyocyanin synthesis (controlled by PQS); *T3SS*, type III secretion system; *Flag*, flagellum synthesis; *Pili*, type IV pili/fimbriae synthesis; *AP*, alkaline protease synthesis and secretion. (B) *Ara*, arabinose utilization; *Glc*, gluconate utilization; *Ml/Mt*, utilization of mannitol and maltose; *G6PD*, glucose-6-phosphate dehydrogenase operon; *PDG*, peptidoglycan synthesis and cell division; *Rib*, ribose utilization operon. (C) *F/W/Y*, aromatic amino acid utilization and transport; *K/R*, lysine and arginine utilization and transport; *Fab/Cyp*, fatty acid metabolism and CYP 450; *Plc*, phophospholipase C; *Ppn*, phosphonate transport; *PO_4*, phosphate transport.

that severe burn induces the expression of different genes of the TTSS in blood through a calcium-independent mechanism.[222]

5. CONCLUSIONS

In this chapter, we have reviewed studies based on media that closely resemble three in vivo conditions: the thick mucus accumulated within the lung alveoli of CF patients, the serum-rich wound bed and the bloodstream. Different sets of *P. aeruginosa* virulence factors were expressed or repressed under each infection condition. In SCFM, expression of *P. aeruginosa* genes for the synthesis of branched-chain amino acid, PQS, and pyochelin were enhanced while genes for catabolism of aromatic amino acids and flagellum synthesis were repressed.[144] A similar analysis using a more refined medium, ASMDM, revealed similar expression patterns as SCFM for amino acid synthesis/catabolism and PQS synthesis, but found major differences in that pyochelin synthesis gene expression was reduced while genes for the T3SS and anaerobic metabolism were enhanced.[170] Using media containing 10% serum as a surrogate for wound exudate, Kruczek et al.[193,199] found that genes for pyoverdine and pyochelin synthesis were enhanced at early stages of growth, while QS genes (*las, rhl*, and *pqs*) were repressed at early stages of growth but enhanced at late exponential phase of growth. These studies revealed changes in expression of up to 200 genes. Additionally, differences between genes expressed or repressed in healthy blood versus blood from burn patients were elucidated. In these studies, thousands of genes were affected. Similar to the studies in CF sputum medium, genes for metabolism were also affected (Fig. 5). These genes, necessary for survival under the new conditions faced by the bacterium, may represent targets for future therapies for infection. The unique effects observed indicate the essential role of these in vitro media that closely mimic in vivo conditions in providing accurate information regarding the pathogenesis of *P. aeruginosa* infections.

REFERENCES

1. Doring G, Pier GB. Vaccines and immunotherapy against *Pseudomonas aeruginosa*. *Vaccine*. 2008;26:1011–1024.
2. Driscoll JA, Brody SL, Kollef MH. The epidemiology, pathogenesis and treatment of *Pseudomonas aeruginosa* infections. *Drugs*. 2007;67:351–368.
3. Pier GB, Ramphal R. *Pseudomonas aeruginosa*. In: Mandell GL, Bennett JE, Dolin R, eds. In: *Mandell, Douglas, and Bennett's Principles and Practice of Infectious Diseases 7th ed. Philadelphia: Churchill Livingstone; 2010:.*

4. Branski LK, Al-Mousawi A, Rivero H, Jeschke MG, Sanford AP, Herndon DN. Emerging infections in burns. *Surg Infect.* 2009;10:389–397.
5. Sadikot RT, Blackwell TS, Christman JW, Prince AS. Pathogen–host interactions in *Pseudomonas aeruginosa* pneumonia. *Am J Respir Crit Care Med.* Jun 1 2005;171 (11):1209–1223.
6. van Delden C. Virulence factors in *Pseudomonas aeruginosa.* In: Ramos J-L, ed. In: *Pseudomonas: Virulence and Gene Regulation. New York, NY: Kluwer Academic/Plenum Publishers; 2004:.*
7. Kipnis E, Sawa T, Wiener-Kronish J. Targeting mechanisms of *Pseudomonas aeruginosa* pathogenesis. *Med Mal Infect.* 2006;36:78–91.
8. Feldman M, Bryan R, Rajan S, et al. Role of flagella in pathogenesis of *Pseudomonas aeruginosa* pulmonary infection. *Infect Immun.* 1998;66:43–51.
9. Mattick JS. Type IV pili and twitching motility. *Annu Rev Microbiol.* 2002;56:289–314.
10. Vasil ML. *Pseudomonas aeruginosa*: biology, mechanisms of virulence, epidemiology. *J Pediatr.* 1986;108:800–805.
11. Davies JC. *Pseudomonas aeruginosa* in cystic fibrosis: pathogenesis and persistence. *Paediatr Respir Rev.* 2002;3:128–134.
12. Mathee K, Ciofu O, Sternberg C, et al. Mucoid conversion of *Pseudomonas aeruginosa* by hydrogen peroxide: a mechanism for virulence activation in the cystic fibrosis lung. *Microbiology.* 1999;145:1349–1357.
13. Backhed F, Normark S, Schweda EK, Oscarson S, Richter-Dahlfors A. Structural requirements for TLR4-mediated LPS signalling: a biological role for LPS modifications. *Microbes Infect.* 2003;5:1057–1063.
14. Wieland CW, Siegmund B, Senaldi G, Vasil ML, Dinarello CA, Fantuzzi G. Pulmonary inflammation induced by *Pseudomonas aeruginosa* lipopolysaccharide, phospholipase C, and exotoxin A: role of interferon regulatory factor 1. *Infect Immun.* 2002;70: 1352–1358.
15. Woods DE, Iglewski BH. Toxins of *Pseudomonas aeruginosa*: new perspectives. *Rev Infect Dis.* 1983;5(suppl 4):S715–S722.
16. Miyazaki S, Matsumoto T, Tateda K, Ohno A, Yamaguchi K. Role of exotoxin A in inducing severe *Pseudomonas aeruginosa* infections in mice. *J Med Microbiol.* 1995;43:169–175.
17. Schultz MJ, Speelman P, Zaat SA, Hack CE, van Deventer SJ, van der Poll T. The effect of *Pseudomonas* exotoxin A on cytokine production in whole blood exposed to *Pseudomonas aeruginosa. FEMS Immunol Med Microbiol.* 2000;29:227–232.
18. Frank DW. The exoenzyme S regulon of *Pseudomonas aeruginosa. Mol Microbiol.* 1997;26:621–629.
19. Fu H, Coburn J, Collier RJ. The eukaryotic host factor that activates exoenzyme S of *Pseudomonas aeruginosa* is a member of the 14-3-3 protein family. *Proc Natl Acad Sci USA.* 1993;90:2320–2324.
20. Krall R, Sun J, Pederson KJ, Barbieri JT. In vivo Rho GTPase-activating protein activity of *Pseudomonas aeruginosa* cytotoxin ExoS. *Infect Immun.* 2002;70:360–367.
21. Yahr TL, Barbieri JT, Frank DW. Genetic relationship between the 53- and 49-kilodalton forms of exoenzyme S from *Pseudomonas aeruginosa. J Bacteriol.* 1996;178: 1412–1419.
22. Ader F, Le Berre R, Faure K, et al. Alveolar response to *Pseudomonas aeruginosa*: role of the type III secretion system. *Infect Immun.* 2005;73:4263–4271.
23. Lee VT, Smith RS, Tümmler B, Lory S. Activities of *Pseudomonas aeruginosa* effectors secreted by the type iii secretion system in vitro and during infection. *Infect Immun.* 2005;73:1695–1705.
24. Sawa T, Yahr TL, Ohara M, et al. Active and passive immunization with the *Pseudomonas* V antigen protects against type III intoxication and lung injury. *Nat Med.* 1999;5: 392–398.

25. Sayner SL, Frank DW, King J, Chen H, VandeWaa J, Stevens T. Paradoxical cAMP-induced lung endothelial hyperpermeability revealed by *Pseudomonas aeruginosa* ExoY. *Circ Res.* 2004;95:196–203.
26. Yahr TL, Vallis AJ, Hancock MK, Barbieri JT, Frank DW. ExoY, an adenylate cyclase secreted by the *Pseudomonas aeruginosa* type III system. *Proc Natl Acad Sci USA.* 1998;95:13899–13904.
27. Caballero AR, Moreau JM, Engel LS, Marquart ME, Hill JM, O'Callaghan RJ. *Pseudomonas aeruginosa* protease IV enzyme assays and comparison to other *Pseudomonas* proteases. *Anal Biochem.* 2001;290:330–337.
28. Matsumoto K. Role of bacterial proteases in pseudomonal and serratial keratitis. *Biol Biochem.* 2004;385:1007–1016.
29. Kessler E, Safrin M, Gustin JK, Ohman DE. Elastase and the LasA protease of *Pseudomonas aeruginosa* are secreted with their propeptides. *J Biol Biochem.* 1998;273:30225–30231.
30. Morihara K, Homma JY. *Pseudomonas* proteases. In: Holder IA, ed. In: *BacterialEnzymes andVirulence. Boca Raton: CRC Press; 1985:.*
31. Schultz DR, Miller KD. Elastase of *Pseudomonas aeruginosa*: inactivation of complement components and complement-derived chemotactic and phagocytic factors. *Infect Immun.* 1974;10:128–135.
32. Rumbaugh KP, Griswold JA, Hamood AN. The role of quorum sensing in the in vivo virulence of *Pseudomonas aeruginosa. Microbes Infect.* 2000;2:1721–1731.
33. Guzzo J, Pages JM, Duong F, Lazdunski A, Murgier M. *Pseudomonas aeruginosa* alkaline protease: evidence for secretion genes and study of secretion mechanism. *J Bacteriol.* 1991;173:5290–5297.
34. Kipnis E, Guery BP, Tournoys A, et al. Massive alveolar thrombin activation in *Pseudomonas aeruginosa*-induced acute lung injury. *Shock.* 2004;21:444–451.
35. Malloy JL, Veldhuizen RA, Thibodeaux BA, O'Callaghan RJ, Wright JR. *Pseudomonas aeruginosa* protease IV degrades surfactant proteins and inhibits surfactant host defense and biophysical functions. *Am J Physiol Lung Cell Mol Physiol.* 2005;288:L409–L418.
36. Vasil ML, Ochsner UA. The response of *Pseudomonas aeruginosa* to iron: genetics, biochemistry and virulence. *Mol Microbiol.* 1999;34:399–413.
37. Allen L, Dockrell DH, Pattery T, et al. Pyocyanin production by *Pseudomonas aeruginosa* induces neutrophil apoptosis and impairs neutrophil-mediated host defenses in vivo. *J Immunol.* 2005;174:3643–3649.
38. Denning GM, Railsback MA, Rasmussen GT, Cox CD, Britigan BE. *Pseudomonas* pyocyanine alters calcium signaling in human airway epithelial cells. *Am J Physiol.* 1998;274:L893–L900.
39. Leidal KG, Munson KL, Denning GM. Small molecular weight secretory factors from *Pseudomonas aeruginosa* have opposite effects on IL-8 and RANTES expression by human airway epithelial cells. *Am J Respir Cell Mol Biol.* 2001;25:186–195.
40. Lau GW, Ran H, Kong F, Hassett DJ, Mavrodi D. *Pseudomonas aeruginosa* pyocyanin is critical for lung infection in mice. *Infect Immun.* 2004;72:4275–4278.
41. Miller AF, Brudvig GW. Manganese and calcium requirements for reconstitution of oxygen-evolution activity in manganese-depleted photosystem II membranes. *Biochemistry.* 1989;28:8181–8190.
42. Ochsner UA, Vasil ML. Gene repression by the ferric uptake regulator in *Pseudomonas aeruginosa:* cycle selection of iron-regulated genes. *Proc Natl Acad Sci USA.* 1996;93:4409–4414.
43. Hay ID, Gatland K, Campisano A, Jordens JZ, Rehm BHA. Impact of alginate overproduction on attachment and biofilm architecture of a supermucoid *Pseudomonas aeruginosa* strain. *Appl Environ Microbiol.* 2009;75:6022–6025.

44. Halverson LJ. Role of alginate in bacterial biofilms. In: Rehm BHA, ed. In: *Alginates: Biology and Applications. Berlin: Springer Berlin Heidelberg; 2009:.*
45. Hay ID, Wang Y, Moradali MF, Rehman ZU, Rehm BH. Genetics and regulation of bacterial alginate production. *Environ Microbiol.* 2014;16:2997–3011.
46. Boucher RC. Airway surface dehydration in cystic fibrosis: pathogenesis and therapy. *Annu Rev Med.* 2007;58:157–170.
47. Chang WS, van de Mortel M, Nielsen L, Nino de Guzman G, Li X, Halverson LJ. Alginate production by *Pseudomonas putida* creates a hydrated microenvironment and contributes to biofilm architecture and stress tolerance under water-limiting conditions. *J Bacteriol.* 2007;189:8290–8299.
48. Li X, Nielsen L, Nolan C, Halverson LJ. Transient alginate gene expression by *Pseudomonas putida* biofilm residents under water-limiting conditions reflects adaptation to the local environment. *Environ Microbiol.* 2010;12:1578–1590.
49. Costerton JW. Cystic fibrosis pathogenesis and the role of biofilms in persistent infection. *Trends Microbiol.* 2001;9:50–52.
50. Pedersen SS, Hoiby N, Espersen F, Koch C. Role of alginate in infection with mucoid *Pseudomonas aeruginosa* in cystic fibrosis. *Thorax.* 1992;47:6–13.
51. Simpson JA, Smith SE, Dean RT. Scavenging by alginate of free radicals released by macrophages. *Free Rad Biol Med.* 1989;6:347–353.
52. Cornelissen CN, Sparling PF. Iron piracy: acquisition of transferrin-bound iron by bacterial pathogens. *Mol Microbiol.* 1994;14:843–850.
53. Wang J, Pantopoulos K. Regulation of cellular iron metabolism. *Biochemical J.* 2011;434:365–381.
54. Cornelis P, Dingemans J. *Pseudomonas aeruginosa* adapts its iron uptake strategies in function of the type of infections. *Front Cell Infect Microbiol.* 2013;3(3):75.
55. Wandersman C, Delepelaire P. Haemophore functions revisited. *Mol Microbiol.* 2012;85:618–631.
56. Dumas Z, Ross-Gillespie A, Kümmerli R. Switching between apparently redundant iron-uptake mechanisms benefits bacteria in changeable environments. *Proc Biol Soc.* 2013;280:20131055.
57. Hider RC, Kong XL. Glutathione: a key component of the cytoplasmic labile iron pool. *Biometals.* 2011;24:1179–1187.
58. Schalk IJ, Guillon L. Fate of ferrisiderophores after import across bacterial outer membranes: different iron release strategies are observed in the cytoplasm or periplasm depending on the siderophore pathways. *Amino Acids.* 2013;44:1267–1277.
59. Schalk IJ, Mislin GL, Brillet K. Structure, function and binding selectivity and stereoselectivity of siderophore-iron outer membrane transporters. *Curr Top Membr.* 2012;69:37–66.
60. Cornelis P. Iron uptake and metabolism in pseudomonads. *Appl Microbiol Biotechnol.* 2010;86:1637–1645.
61. Meyer JM. Pyoverdines: pigments, siderophores and potential taxonomic markers of fluorescent *Pseudomonas* species. *Arch Microbiol.* 2000;174:135–142.
62. Meyer JM, Neely A, Stintzi A, Georges C, Holder IA. Pyoverdin is essential for virulence of *Pseudomonas aeruginosa. Infect Immun.* 1996;64:518–523.
63. Takase H, Nitanai H, Hoshino K, Otani T. Impact of siderophore production on *Pseudomonas aeruginosa* infections in immunosuppressed mice. *Infect Immun.* 2000;68:1834–1839.
64. Lyczak JB, Cannon CL, Pier GB. Lung infections associated with cystic fibrosis. *Clin Microbiol Rev.* 2002;15:194–222.
65. Ochsner UA, Johnson Z, Vasil ML. Genetics and regulation of two distinct haem-uptake systems, *phu* and *has*, in *Pseudomonas aeruginosa. Microbiology.* 2000;146:185–198.

66. Cartron ML, Maddocks S, Gillingham P, Craven CJ, Andrews SC. Feo—transport of ferrous iron into bacteria. *Biometals.* 2006;19:143–157.
67. Wang Y, Wilks JC, Danhorn T, Ramos I, Croal L, Newman DK. Phenazine-1-carboxylic acid promotes bacterial biofilm development via ferrous iron acquisition. *J Bacteriol.* 2011;193:3606–3617.
68. Hunter RC, Asfour F, Dingemans J, et al. Ferrous iron is a significant component of bioavailable iron in cystic fibrosis airways. *mBio.* 2013;4:e00557–e00613.
69. Hauser AR. The type III secretion system of *Pseudomonas aeruginosa*: infection by injection. *Nat Rev Microbiol.* 2009;7:654–665.
70. Finck-Barbancon V, Goranson J, Zhu L, et al. ExoU expression by *Pseudomonas aeruginosa* correlates with acute cytotoxicity and epithelial injury. *Mol Microbiol.* 1997;25:547–557.
71. Hauser AR, Fleiszig S, Kang PJ, Mostov K, Engel JN. Defects in type III secretion correlate with internalization of *Pseudomonas aeruginosa* by epithelial cells. *Infect Immun.* 1998;66:1413–1420.
72. Hauser AR, Cobb E, Bodi M, et al. Type III protein secretion is associated with poor clinical outcomes in patients with ventilator-associated pneumonia caused by *Pseudomonas aeruginosa. Crit Care Med.* 2002;30:521–528.
73. Allewelt M, Coleman FT, Grout M, Priebe GP, Pier GB. Acquisition of expression of the *Pseudomonas aeruginosa* ExoU cytotoxin leads to increased bacterial virulence in a murine model of acute pneumonia and systemic spread. *Infect Immun.* 2000;68: 3998–4004.
74. Kurahashi K, Kajikawa O, Sawa T, et al. Pathogenesis of septic shock in *Pseudomonas aeruginosa* pneumonia. *J Clin Invest.* 1999;104:743–750.
75. Coggan KA, Wolfgang MC. Global regulatory pathways and cross-talk control *Pseudomonas aeruginosa* environmental lifestyle and virulence phenotype. *Curr Issues Mol Biol.* 2012;14:47–70.
76. de Kievit TR, Iglewski BH. Bacterial quorum sensing in pathogenic relationships. *Infect Immun.* 2000;68:4839–4849.
77. Schuster M, Lostroh CP, Ogi T, Greenberg EP. Identification, timing, and signal specificity of *Pseudomonas aeruginosa* quorum-controlled genes: a transcriptome analysis. *J Bacteriol.* 2003;185:2066–2079.
78. Pesci EC, Milbank JB, Pearson JP, et al. Quinolone signaling in the cell-to-cell communication system of *Pseudomonas aeruginosa. Proc Natl Acad Sci USA.* 1999;96: 11229–11234.
79. Wade DS, Calfee MW, Rocha ER, et al. Regulation of *Pseudomonas* quinolone signal synthesis in *Pseudomonas aeruginosa. J Bacteriol.* 2005;187:4372–4380.
80. Deziel E, Gopalan S, Tampakaki AP, et al. The contribution of MvfR to *Pseudomonas aeruginosa* pathogenesis and quorum sensing circuitry regulation: multiple quorum sensing-regulated genes are modulated without affecting *lasRI, rhlRI* or the production of *N*-acyl-L-homoserine lactones. *Mol Microbiol.* 2005;55:998–1014.
81. McGrath S, Wade DS, Pesci EC. Dueling quorum sensing systems in *Pseudomonas aeruginosa* control the production of the *Pseudomonas* quinolone signal (PQS). *FEMS Microbiol Lett.* 2004;230:27–34.
82. de Kievit T, Seed PC, Nezezon J, Passador L, Iglewski BH. RsaL, a novel repressor of virulence gene expression in *Pseudomonas aeruginosa. J Bacteriol.* 1999;181:2175–2184.
83. Whiteley M, Parsek MR, Greenberg EP. Regulation of quorum sensing by RpoS in *Pseudomonas aeruginosa. J Bacteriol.* 2000;182:4356–4360.
84. Castang S, McManus HR, Turner KH, Dove SL. H-NS family members function coordinately in an opportunistic pathogen. *Proc Natl Acad Sci USA.* 2008;105: 18947–18952.

85. Diggle SP, Winzer K, Lazdunski A, Williams P, Cámara M. Advancing the quorum in *Pseudomonas aeruginosa*: MvaT and the regulation of *N*-acylhomoserine lactone production and virulence gene expression. *J Bacteriol.* 2002;184:2576–2586.
86. Smith DJ, Lamont IL, Anderson GJ, Reid DW. Targeting iron uptake to control *Pseudomonas aeruginosa* infections in cystic fibrosis. *Eur Respir J.* 2013;42:1723–1736.
87. Singh G, Wu B, Baek MS, et al. Secretion of *Pseudomonas aeruginosa* type III cytotoxins is dependent on *Pseudomonas* quinolone signal concentration. *Microbial Pathog.* 2010;49:196–203.
88. Ryall B, Carrara M, Zlosnik JE, et al. The mucoid switch in *Pseudomonas aeruginosa* represses quorum sensing systems and leads to complex changes to stationary phase virulence factor regulation. *PLoS One.* 2014;9:e96166.
89. Balasubramanian D, Kong KF, Jayawardena SR, Leal SM, Sautter RT, Mathee K. Co-regulation of β-lactam resistance, alginate production and quorum sensing in *Pseudomonas aeruginosa. J Med Microbiol.* 2011;60:147–156.
90. Berka RM, Vasil ML. Phospholipase C (heat-labile hemolysin) of *Pseudomonas aeruginosa*: purification and preliminary characterization. *J Bacteriol.* 1982;152:239–245.
91. James GA, Swogger E, Wolcott R, et al. Biofilms in chronic wounds. *Wound Rep Regen.* 2008;16:37–44.
92. Kennedy P, Brammah S, Wills E. Burns, biofilm and a new appraisal of burn wound sepsis. *Burns.* 2010;36:49–56.
93. Lim YW, Evangelista JS, Schmieder R, et al. Clinical insights from metagenomic analysis of sputum samples from patients with cystic fibrosis. *J Clin Microbiol.* 2014;52:425–437.
94. Nicas TI, Iglewski BH. Isolation and characterization of transposon-induced mutants of *Pseudomonas aeruginosa* deficient in production of exoenzyme S. *Infect Immun.* 1984;45:470–474.
95. Ohman DE, Sadoff JC, Iglewski BH. Toxin A-deficient mutants of *Pseudomonas aeruginosa* PA103: isolation and characterization. *Infect Immun.* 1980;28:899–908.
96. Rumbaugh KP, Griswold JA, Iglewski BH, Hamood AN. Contribution of quorum sensing to the virulence of *Pseudomonas aeruginosa* in burn wound infections. *Infect Immun.* 1999;67:5854–5862.
97. Storey DG, Ujack EE, Rabin HR, Mitchell I. *Pseudomonas aeruginosa lasR* transcription correlates with the transcription of *lasA*, *lasB*, and *toxA* in chronic lung infections associated with cystic fibrosis. *Infect Immun.* 1998;66:2521–2528.
98. Sy A, Srinivasan M, Mascarenhas J, et al. *Pseudomonas aeruginosa* keratitis: outcomes and response to corticosteroid treatment. *Invest Ophthalmol Vis Sci.* 2012;53:267–272.
99. Wargo MJ, Gross MJ, Rajamani S, et al. Hemolytic phospholipase C inhibition protects lung function during *Pseudomonas aeruginosa* infection. *Am J Respir Crit Care Med.* 2011;184:345–354.
100. Oglesby-Sherrouse AG, Djapgne L, Nguyen AT, Vasil AI, Vasil ML. The complex interplay of iron, biofilm formation, and mucoidy affecting antimicrobial resistance of *Pseudomonas aeruginosa. Pathog Dis.* 2014;70:307–320.
101. Wilderman PJ, Vasil AI, Johnson Z, et al. Characterization of an endoprotease (PrpL) encoded by a PvdS-regulated gene in *Pseudomonas aeruginosa. Infect Immun.* 2001;69:5385–5394.
102. Ochsner UA, Wilderman PJ, Vasil AI, Vasil ML. GeneChip expression analysis of the iron starvation response in *Pseudomonas aeruginosa*: identification of novel pyoverdine biosynthesis genes. *Mol Microbiol.* 2002;45:1277–1287.
103. Sage AE, Vasil ML. Osmoprotectant-dependent expression of *plcH*, encoding the hemolytic phospholipase C, is subject to novel catabolite repression control in *Pseudomonas aeruginosa* PAO1. *J Bacteriol.* 1997;179:4874–4881.

104. Davis PB, Drumm M, Konstan MW. Cystic fibrosis. *Am J Respir Crit Care Med.* 1996;154:1229–1256.
105. Gibson RL, Burns JL, Ramsey BW. Pathophysiology and management of pulmonary infections in cystic fibrosis. *Am J Respir Crit Care Med.* 2003;168:918–951.
106. Hamosh A, FitzSimmons SC, Macek Jr M, Knowles MR, Rosenstein BJ, Cutting GR. Comparison of the clinical manifestations of cystic fibrosis in black and white patients. *J Pediatr.* 1998;132:255–259.
107. Welsh MJ, Smith AE. Molecular mechanisms of CFTR chloride channel dysfunction in cystic fibrosis. *Cell.* 1993;73:1251–1254.
108. Wang Y, Wrennall JA, Cai Z, Li H, Sheppard DN. Understanding how cystic fibrosis mutations disrupt CFTR function: from single molecules to animal models. *Int J Biochem Cell Biol.* 2014;52:47–57.
109. Creeth JM. Constituents of mucus and their separation. *Br Med Bull.* 1978;34:17–24.
110. Voynow JA, Rubin BK. Mucins, mucus, and sputum. *Chest.* 2009;135:505–512.
111. Ciofu O, Hansen CR, Hoiby N. Respiratory bacterial infections in cystic fibrosis. *Curr Opin Pulm Med.* 2013;19:251–258.
112. Hassett DJ, Cuppoletti J, Trapnell B, et al. Anaerobic metabolism and quorum sensing by *Pseudomonas aeruginosa* biofilms in chronically infected cystic fibrosis airways: rethinking antibiotic treatment strategies and drug targets. *Adv Drug Del Rev.* 2002;54: 1425–1443.
113. Johansen HK, Kovesi TA, Koch C, Corey M, Hoiby N, Levison H. *Pseudomonas aeruginosa* and *Burkholderia cepacia* infection in cystic fibrosis patients treated in Toronto and Copenhagen. *Pediatr Pulmonol.* 1998;26:89–96.
114. Rowe SM, Miller S, Sorscher EJ. Cystic fibrosis. *N Engl J Med.* 2005;352:1992–2001.
115. Heijerman H. Infection and inflammation in cystic fibrosis: a short review. *J Cyst Fibos.* 2005;4(suppl 2):3–5.
116. Doring G, Flume P, Heijerman H, Elborn JS. Consensus Study G. Treatment of lung infection in patients with cystic fibrosis: current and future strategies. *J Cyst Fibos.* 2012;11:461–479.
117. Hartl D, Gaggar A, Bruscia E, et al. Innate immunity in cystic fibrosis lung disease. *J Cyst Fibos.* 2012;11:363–382.
118. Ohman DE, Chakrabarty AM. Utilization of human respiratory secretions by mucoid *Pseudomonas aeruginosa* of cystic fibrosis origin. *Infect Immun.* 1982;37:662–669.
119. Pier GB, Matthews Jr WJ, Eardley DD. Immunochemical characterization of the mucoid exopolysaccharide of *Pseudomonas aeruginosa*. *J Infect Dis.* 1983;147:494–503.
120. Hauser AR, Jain M, Bar-Meir M, McColley SA. Clinical significance of microbial infection and adaptation in cystic fibrosis. *Clin Microbiol Rev.* 2011;24:29–70.
121. Mayer-Hamblett N, Ramsey BW, Kulasekara HD, et al. *Pseudomonas aeruginosa* phenotypes associated with eradication failure in children with cystic fibrosis. *Clin Infect Dis.* 2014;59:624–631.
122. Mayer-Hamblett N, Rosenfeld M, Gibson RL, et al. *Pseudomonas aeruginosa* in vitro phenotypes distinguish cystic fibrosis infection stages and outcomes. *Am J Respir Crit Care Med.* 2014;190:289–297.
123. Romling U, Fiedler B, Bosshammer J, et al. Epidemiology of chronic *Pseudomonas aeruginosa* infections in cystic fibrosis. *J Infect Dis.* 1994;170:1616–1621.
124. Boucher RC. New concepts of the pathogenesis of cystic fibrosis lung disease. *Eur Respir J.* 2004;23:146–158.
125. Worlitzsch D, Tarran R, Ulrich M, et al. Effects of reduced mucus oxygen concentration in airway *Pseudomonas* infections of cystic fibrosis patients. *J Clin Invest.* 2002;109:317–325.
126. Whitchurch CB, Tolker-Nielsen T, Ragas PC, Mattick JS. Extracellular DNA required for bacterial biofilm formation. *Science.* 2002;295:1487.

127. Wang J, Mushegian A, Lory S, Jin S. Large-scale isolation of candidate virulence genes of *Pseudomonas aeruginosa* by in vivo selection. *Proc Natl Acad Sci USA.* 1996;93: 10434–10439.
128. Wang J, Lory S, Ramphal R, Jin S. Isolation and characterization of *Pseudomonas aeruginosa* genes inducible by respiratory mucus derived from cystic fibrosis patients. *Mol Microbiol.* 1996;22:1005–1012.
129. Miller JH. *Experiments in molecular genetics.* Cold Spring Harbor, NY: CHSL Press; 1972.
130. Ankenbauer RG. Cloning of the outer membrane high-affinity Fe(III)-pyochelin receptor of *Pseudomonas aeruginosa. J Bacteriol.* 1992;174:4401–4409.
131. Ankenbauer RG, Quan HN. FptA, the Fe(III)-pyochelin receptor of *Pseudomonas aeruginosa*: a phenolate siderophore receptor homologous to hydroxamate siderophore receptors. *J Bacteriol.* 1994;176:307–319.
132. Ellison ML, Farrow JM, Parrish W, Danell AS, Pesci EC. The transcriptional regulator Np20 is the zinc uptake regulator in *Pseudomonas aeruginosa. PLoS One.* 2013;8: e75389.
133. Blaby-Haas CE, Furman R, Rodionov DA, Artsimovitch I, de Crecy-Lagard V. Role of a Zn-independent DksA in Zn homeostasis and stringent response. *Mol Microbiol.* 2011;79:700–715.
134. Poon KK, Westman EL, Vinogradov E, Jin S, Lam JS. Functional characterization of MigA and WapR: putative rhamnosyltransferases involved in outer core oligosaccharide biosynthesis of *Pseudomonas aeruginosa. J Bacteriol.* 2008;190:1857–1865.
135. Kocincova D, Ostler SL, Anderson EM, Lam JS. Rhamnosyltransferase genes *migA* and *wapR* are regulated in a differential manner to modulate the quantities of core oligosaccharide glycoforms produced by *Pseudomonas aeruginosa. J Bacteriol.* 2012;194: 4295–4300.
136. Lau PC, Lindhout T, Beveridge TJ, Dutcher JR, Lam JS. Differential lipopolysaccharide core capping leads to quantitative and correlated modifications of mechanical and structural properties in *Pseudomonas aeruginosa* biofilms. *J Bacteriol.* 2009;191: 6618–6631.
137. Wolfgang MC, Jyot J, Goodman AL, Ramphal R, Lory S. *Pseudomonas aeruginosa* regulates flagellin expression as part of a global response to airway fluid from cystic fibrosis patients. *Proc Natl Acad Sci USA.* 2004;101:6664–6668.
138. Smith RS, Iglewski BH. *P. aeruginosa* quorum-sensing systems and virulence. *Curr Opin Microbiol.* 2003;6:56–60.
139. Brimer CD, Montie TC. Cloning and comparison of *fliC* genes and identification of glycosylation in the flagellin of *Pseudomonas aeruginosa* a-type strains. *J Bacteriol.* 1998;180:3209–3217.
140. Singh PK, Schaefer AL, Parsek MR, Moninger TO, Welsh MJ, Greenberg EP. Quorum-sensing signals indicate that cystic fibrosis lungs are infected with bacterial biofilms. *Nature.* 2000;407:762–764.
141. Palmer KL, Mashburn LM, Singh PK, Whiteley M. Cystic fibrosis sputum supports growth and cues key aspects of *Pseudomonas aeruginosa* physiology. *J Bacteriol.* 2005;187:5267–5277.
142. Dubern JF, Diggle SP. Quorum sensing by 2-alkyl-4-quinolones in *Pseudomonas aeruginosa* and other bacterial species. *Mol Biosys.* 2008;4:882–888.
143. Diggle SP, Cornelis P, Williams P, Cámara M. 4-quinolone signalling in *Pseudomonas aeruginosa*: old molecules, new perspectives. *Int J Med Microbiol.* 2006;296:83–91.
144. Palmer KL, Aye LM, Whiteley M. Nutritional cues control *Pseudomonas aeruginosa* multicellular behavior in cystic fibrosis sputum. *J Bacteriol.* 2007;189:8079–8087.
145. Reid DW, Withers NJ, Francis L, Wilson JW, Kotsimbos TC. Iron deficiency in cystic fibrosis: relationship to lung disease severity and chronic *Pseudomonas aeruginosa* infection. *Chest.* 2002;121:48–54.

146. Stites SW, Plautz MW, Bailey K, O'Brien-Ladner AR, Wesselius LJ. Increased concentrations of iron and isoferritins in the lower respiratory tract of patients with stable cystic fibrosis. *Am J Respir Crit Care Med.* 1999;160:796–801.
147. Farrow 3rd JM, Pesci EC. Two distinct pathways supply anthranilate as a precursor of the *Pseudomonas* quinolone signal. *J Bacteriol.* 2007;189:3425–3433.
148. Mashburn LM, Jett AM, Akins DR, Whiteley M. *Staphylococcus aureus* serves as an iron source for *Pseudomonas aeruginosa* during in vivo coculture. *J Bacteriol.* 2005;187: 554–566.
149. Donlan RM, Costerton JW. Biofilms: survival mechanisms of clinically relevant microorganisms. *Clin Microbiol Rev.* 2002;15:167–193.
150. Ghannoum MA, O'Toole GA. *Microbial biofilms.* Washington, D.C: ASM Press; 2004.
151. Prince AS. Biofilms, antimicrobial resistance, and airway infection. *N Engl J Med.* 2002;347:1110–1111.
152. Davies DG, Geesey GG. Regulation of the alginate biosynthesis gene *algC* in *Pseudomonas aeruginosa* during biofilm development in continuous culture. *Appl Environ Microbiol.* 1995;61:860–867.
153. Davies DG, Parsek MR, Pearson JP, Iglewski BH, Costerton JW, Greenberg EP. The involvement of cell-to-cell signals in the development of a bacterial biofilm. *Science.* 1998;280:295–298.
154. Hentzer M, Teitzel GM, Balzer GJ, et al. Alginate overproduction affects *Pseudomonas aeruginosa* biofilm structure and function. *J Bacteriol.* 2001;183:5395–5401.
155. O'Toole GA, Kolter R. Initiation of biofilm formation in *Pseudomonas fluorescens* WCS365 proceeds via multiple, convergent signalling pathways: a genetic analysis. *Mol Microbiol.* 1998;28:449–461.
156. Pereira FD, Bonatto CC, Lopes CA, Pereira AL, Silva LP. Use of MALDI-TOF mass spectrometry to analyze the molecular profile of *Pseudomonas aeruginosa* biofilms grown on glass and plastic surfaces. *Microbial Pathog.* 2015;86:32–37.
157. Sauer K, Cullen MC, Rickard AH, Zeef LA, Davies DG, Gilbert P. Characterization of nutrient-induced dispersion in *Pseudomonas aeruginosa* PAO1 biofilm. *J Bacteriol.* 2004;186:7312–7326.
158. Schaber JA, Carty NL, McDonald NA, et al. Analysis of quorum sensing-deficient clinical isolates of *Pseudomonas aeruginosa. J Med Microbiol.* 2004;53:841–853.
159. Singh PK, Parsek MR, Greenberg EP, Welsh MJ. A component of innate immunity prevents bacterial biofilm development. *Nature.* 2002;417:552–555.
160. Landry RM. Mucin-*Pseudomonas aeruginosa* interactions promote biofilm formation and antibiotic resistance. *Mol Microbiol.* 2006;59:142–151.
161. Christensen BB, Sternberg C, Andersen JB, et al. Molecular tools for study of biofilm physiology. *Methods Enzymol.* 1999;310:20–42.
162. Arora SK, Ritchings BW, Almira EC, Lory S, Ramphal R. The *Pseudomonas aeruginosa* flagellar cap protein, FliD, is responsible for mucin adhesion. *Infect Immun.* 1998;66:1000–1007.
163. Schobert M, Jahn D. Anaerobic physiology of *Pseudomonas aeruginosa* in the cystic fibrosis lung. *Int J Med Microbiol.* 2010;300:549–556.
164. Baltimore RS, Christie CD, Smith GJ. Immunohistopathologic localization of *Pseudomonas aeruginosa* in lungs from patients with cystic fibrosis. Implications for the pathogenesis of progressive lung deterioration. *Am Rev Respir Dis.* 1989;140:1650–1661.
165. Lam J, Chan R, Lam K, Costerton JW. Production of mucoid microcolonies by *Pseudomonas aeruginosa* within infected lungs in cystic fibrosis. *Infect Immun.* 1980;28:546–556.
166. Sriramulu DD, Lunsdorf H, Lam JS, Romling U. Microcolony formation: a novel biofilm model of *Pseudomonas aeruginosa* for the cystic fibrosis lung. *J Med Microbiol.* 2005;54:667–676.

167. Ghani M, Soothill JS. Ceftazidime, gentamicin, and rifampicin, in combination, kill biofilms of mucoid *Pseudomonas aeruginosa*. *Can J Microbiol.* 1997;43:999–1004.
168. Fito-Boncompte L, Chapalain A, Bouffartigues E, et al. Full virulence of *Pseudomonas aeruginosa* requires OprF. *Infect Immun.* 2011;79:1176–1186.
169. Woodruff WA, Hancock RE. *Pseudomonas aeruginosa* outer membrane protein F: structural role and relationship to the *Escherichia coli* OmpA protein. *J Bacteriol.* 1989;171:3304–3309.
170. Fung C, Naughton S, Turnbull L, et al. Gene expression of *Pseudomonas aeruginosa* in a mucin-containing synthetic growth medium mimicking cystic fibrosis lung sputum. *J Med Microbiol.* 2010;59:1089–1100.
171. Sagel SD, Kapsner R, Osberg I, Sontag MK, Accurso FJ. Airway inflammation in children with cystic fibrosis and healthy children assessed by sputum induction. *Am J Respir Crit Care Med.* 2001;164:1425–1431.
172. Berthelot P, Attree I, Plesiat P, et al. Genotypic and phenotypic analysis of type III secretion system in a cohort of *Pseudomonas aeruginosa* bacteremia isolates: evidence for a possible association between O serotypes and *exo* genes. *J Infect Dis.* 2003;188:512–518.
173. Christensen LD, Moser C, Jensen PO, et al. Impact of *Pseudomonas aeruginosa* quorum sensing on biofilm persistence in an in vivo intraperitoneal foreign-body infection model. *Microbiology.* 2007;153:2312–2320.
174. Roy-Burman A, Savel RH, Racine S, et al. Type III protein secretion is associated with death in lower respiratory and systemic *Pseudomonas aeruginosa* infections. *J Infect Dis.* 2001;183:1767–1774.
175. Wu H, Song Z, Givskov M, et al. *Pseudomonas aeruginosa* mutations in *lasI* and *rhlI* quorum sensing systems result in milder chronic lung infection. *Microbiology.* 2001;147:1105–1113.
176. Van Alst NE, Wellington M, Clark VL, Haidaris CG, Iglewski BH. Nitrite reductase NirS is required for type III secretion system expression and virulence in the human monocyte cell line THP-1 by *Pseudomonas aeruginosa*. *Infect Immun.* 2009;77: 4446–4454.
177. Haley CL, Colmer-Hamood JA, Hamood AN. Characterization of biofilm-like structures formed by *Pseudomonas aeruginosa* in a synthetic mucus medium. *BMC Microbiol.* 2012;12:181.
178. Heydorn A, Nielsen AT, Hentzer M, et al. Quantification of biofilm structures by the novel computer program COMSTAT. *Microbiology.* 2000;146:2395–2407.
179. Brint JM, Ohman DE. Synthesis of multiple exoproducts in *Pseudomonas aeruginosa* is under the control of RhlR-RhlI, another set of regulators in strain PAO1 with homology to the autoinducer-responsive LuxR-LuxI family. *J Bacteriol.* 1995;177:7155–7163.
180. Gambello MJ, Iglewski BH. Cloning and characterization of the *Pseudomonas aeruginosa lasR* gene, a transcriptional activator of elastase expression. *J Bacteriol.* 1991;173: 3000–3009.
181. Jacobs MA, Alwood A, Thaipisuttikul I, et al. Comprehensive transposon mutant library of *Pseudomonas aeruginosa*. *Proc Natl Acad Sci USA.* 2003;100:14339–14344.
182. Pearson JP, Pesci EC, Iglewski BH. Roles of *Pseudomonas aeruginosa las* and *rhl* quorum-sensing systems in control of elastase and rhamnolipid biosynthesis genes. *J Bacteriol.* 1997;179:5756–5767.
183. Bowler PG, Duerden BI, Armstrong DG. Wound microbiology and associated approaches to wound management. *Clin Microbiol Rev.* 2001;14:244–269.
184. Church D, Elsayed S, Reid O, Winston B, Lindsay R. Burn wound infections. *Clin Microbiol Rev.* 2006;19:403–434.
185. Pruitt Jr BA, McManus AT, Kim SH, Goodwin CW. Burn wound infections: current status. *World J Surg.* 1998;22:135–145.

186. Hall-Stoodley L, Stoodley P. Evolving concepts in biofilm infections. *Cell Microbiol.* 2009;11:1034–1043.
187. Parsek MR, Singh PK. Bacterial biofilms: an emerging link to disease pathogenesis. *Annu Rev Microbiol.* 2003;57:677–701.
188. Costerton JW, Stewart PS, Greenberg EP. Bacterial biofilms: a common cause of persistent infections. *Science.* 1999;284:1318–1322.
189. Ardehali R, Shi L, Janatova J, Mohammad SF, Burns GL. The effect of apo-transferrin on bacterial adhesion to biomaterials. *Artif Organs.* 2002;26:512–520.
190. Singh PK. Iron sequestration by human lactoferrin stimulates *P. aeruginosa* surface motility and blocks biofilm formation. *Biometals.* 2004;17:267–270.
191. Hammond A, Dertien J, Colmer-Hamood JA, Griswold JA, Hamood AN. Serum inhibits *P. aeruginosa* biofilm formation on plastic surfaces and intravenous catheters. *J Surg Res.* 2010;159:735–746.
192. Deziel E, Comeau Y, Villemur R. Initiation of biofilm formation by *Pseudomonas aeruginosa* 57RP correlates with emergence of hyperpiliated and highly adherent phenotypic variants deficient in swimming, swarming, and twitching motilities. *J Bacteriol.* 2001;183:1195–1204.
193. Kruczek C, Wachtel M, Alabady MS, Payton PR, Colmer-Hamood JA, Hamood AN. Serum albumin alters the expression of iron-controlled genes in *Pseudomonas aeruginosa. Microbiology.* 2012;158:353–367.
194. Hamood AN, Colmer-Hamood JA, Carty NL, Regulation of *Pseudomonas aeruginosa* exotoxin A synthesis.Ramos J-L, ed. *Pseudomonas Virulence and gene regulation.* 2. New York, NY: Kluwer Academic/Plenum Publishers; 2004: .
195. Cornelis P, Matthijs S, Van Oeffelen L. Iron uptake regulation in *Pseudomonas aeruginosa. Biometals.* 2009;22:15–22.
196. Rondeau P, Bourdon E. The glycation of albumin: structural and functional impacts. *Biochimie.* 2011;93:645–658.
197. Somarajan SR, La Rosa SL, Singh KV, Roh JH, Hook M, Murray BE. The fibronectin-binding protein Fnm contributes to adherence to extracellular matrix components and virulence of *Enterococcus faecium. Infect Immun.* 2015;83:4653–4661.
198. St John S, Blower R, Popova TG, et al. *Bacillus anthracis* co-opts nitric oxide and host serum albumin for pathogenicity in hypoxic conditions. *Front Cell Infect Microbiol.* 2013;3:16.
199. Kruczek C, Qaisar U, Colmer-Hamood JA, Hamood AN. Serum influences the expression of *Pseudomonas aeruginosa* quorum-sensing genes and QS-controlled virulence genes during early and late stages of growth. *Microbiologyopen.* 2014;3: 64–79.
200. Albus AM, Pesci EC, Runyen-Janecky LJ, West SE, Iglewski BH. Vfr controls quorum sensing in *Pseudomonas aeruginosa. J Bacteriol.* 1997;179:3928–3935.
201. Juhas M, Wiehlmann L, Huber B, et al. Global regulation of quorum sensing and virulence by VqsR in *Pseudomonas aeruginosa. Microbiology.* 2004;150:831–841.
202. Reimmann C, Beyeler M, Latifi A, et al. The global activator GacA of *Pseudomonas aeruginosa* PAO positively controls the production of the autoinducer N-butyryl-homoserine lactone and the formation of the virulence factors pyocyanin, cyanide, and lipase. *Mol Microbiol.* 1997;24:309–319.
203. Chugani SA, Whitely M, Lee KM, D'Argenio D, Manoil C, Green NM. QscR, a modulator of quorum-sensing signal synthesis and virulence in *Pseudomonas aeruginosa. Proc Natl Acad Sci USA.* 2001;98:2752–2757.
204. Heurlier K, Dénervaud V, Pessi G, Reimmann C, Haas D. Negative control of quorum sensing by RpoN (sigma54) in *Pseudomonas aeruginosa* PAO1. *J Bacteriol.* 2003;185: 2227–2235.

205. Pessi G, Williams F, Hindle Z, et al. The global posttranscriptional regulator RsmA modulates production of virulence determinants and N-acylhomoserine lactones in *Pseudomonas aeruginosa*. *J Bacteriol.* 2001;183:6676–6683.
206. Schuster M, Hawkins AC, Harwood CS, Greenberg EP. The *Pseudomonas aeruginosa* RpoS regulon and its relationship to quorum sensing. *Mol Microbiol.* 2004;51:973–985.
207. Angus DC, van der Poll T. Severe sepsis and septic shock. *N Engl J Med.* 2013;369:2063.
208. Hattemer A, Hauser A, Diaz M, et al. Bacterial and clinical characteristics of health care- and community-acquired bloodstream infections due to *Pseudomonas aeruginosa*. *Antimicrob Agents Chemother.* 2013;57:3969–3975.
209. Sandquist M, Wong HR. Biomarkers of sepsis and their potential value in diagnosis, prognosis and treatment. *Expert Rev Clin Immunol.* 2014;10:1349–1356.
210. El Amari EB, Chamot E, Auckenthaler R, Pechere JC, Van Delden C. Influence of previous exposure to antibiotic therapy on the susceptibility pattern of *Pseudomonas aeruginosa* bacteremic isolates. *Clin Infect Dis.* 2001;33:1859–1864.
211. El-Solh AA, Hattemer A, Hauser AR, Alhajhusain A, Vora H. Clinical outcomes of type III *Pseudomonas aeruginosa* bacteremia. *Crit Care Med.* 2012;40:1157–1163.
212. Wisplinghoff H, Bischoff T, Tallent SM, Seifert H, Wenzel RP, Edmond MB. Nosocomial bloodstream infections in US hospitals: analysis of 24,179 cases from a prospective nationwide surveillance study. *Clin Infect Dis.* 2004;39:309–317.
213. Graham MR, Virtaneva K, Porcella SF, et al. Group A *Streptococcus* transcriptome dynamics during growth in human blood reveals bacterial adaptive and survival strategies. *Am J Pathol.* 2005;166:455–465.
214. Cunningham MW. Pathogenesis of group A streptococcal infections. *Clin Microbiol Rev.* 2000;13:470–511.
215. Musser JM, Krause RM. The revival of group A streptococcal diseases, with a commentary on staphylococcal toxic shock syndrome. In: Krause RM, ed. In: *Emerging Infections. New York: Academic Press; 1998:.*
216. Mereghetti L, Sitkiewicz I, Green NM, Musser JM. Extensive adaptive changes occur in the transcriptome of *Streptococcus agalactiae* (group B *Streptococcus*) in response to incubation with human blood. *PLoS One.* 2008;3:e3143.
217. Phares CR, Lynfield R, Farley MM, et al. Epidemiology of invasive group B streptococcal disease in the United States, 1999–2005. *JAMA.* 2008;299:2056–2065.
218. Kealey CP, Whelan SA, Chun YJ, et al. In vitro hemocompatibility of thin film nitinol in stenotic flow conditions. *Biomaterials.* 2010;31:8864–8871.
219. Fitzwater J, Purdue GF, Hunt JL, O'Keefe GE. The risk factors and time course of sepsis and organ dysfunction after burn trauma. *J Trauma.* 2003;54:959–966.
220. Mason Jr AD, McManus AT, Pruitt Jr BA. Association of burn mortality and bacteremia. A 25-year review. *Arch Surg.* 1986;121:1027–1031.
221. Raz-Pasteur A, Hussein K, Finkelstein R, Ullmann Y, Egozi D. Blood stream infections (BSI) in severe burn patients—early and late BSI: a 9-year study. *Burns.* 2013;39:636–642.
222. Kruczek C, Kottapalli KR, Dissanaike S, Dzvova N, Griswold JA, Colmer-Hamood JA, Hamood AN. Major transcriptome changes accompany the growth of *Pseudomonas aeruginosa* in blood from patients with severe thermal injuries. *PLoS One.* 2016;11(3): e0149229.
223. Horsman SR, Moore RA, Lewenza S. Calcium chelation by alginate activates the type III secretion system in mucoid *Pseudomonas aeruginosa* biofilms. *PLoS One.* 2012;7: e46826.

CHAPTER SEVEN

Host Responses to Biofilm

C. Watters[1], D. Fleming[2,3], D. Bishop[1], K.P. Rumbaugh[2,3,*]

[1] Wound Infections Department, Naval Medical Research Center, Silver Spring, MD, United States
[2] Department of Surgery, Texas Tech University Health Sciences Center, Lubbock, TX, United States
[3] Department of Immunology and Molecular Microbiology, Texas Tech University Health Sciences Center, Lubbock, TX, United States
* Corresponding author. E-mail address: kendra.rumbaugh@ttuhsc.edu

Contents

Abstract

From birth to death the human host immune system interacts with bacterial cells. Biofilms are communities of microbes embedded in matrices composed of extracellular polymeric substance (EPS), and have been implicated in both the healthy microbiome and disease states. The immune system recognizes many different bacterial patterns, molecules, and antigens, but these components can be camouflaged in the biofilm mode of growth. Instead, immune cells come into contact with components of the EPS matrix, a diverse, hydrated mixture of extracellular DNA (bacterial and host), proteins, polysaccharides, and lipids. As bacterial cells transition from planktonic to biofilm-associated they produce small molecules, which can increase inflammation, induce cell

Progress in Molecular Biology and Translational Science, Volume 142
ISSN 1877-1173
http://dx.doi.org/10.1016/bs.pmbts.2016.05.007

death, and even cause necrosis. To survive, invading bacteria must overcome the epithelial barrier, host microbiome, complement, and a variety of leukocytes. If bacteria can evade these initial cell populations they have an increased chance at surviving and causing ongoing disease in the host. Planktonic cells are readily cleared, but biofilms reduce the effectiveness of both polymorphonuclear neutrophils and macrophages. In addition, in the presence of these cells, biofilm formation is actively enhanced, and components of host immune cells are assimilated into the EPS matrix. While pathogenic biofilms contribute to states of chronic inflammation, probiotic *Lactobacillus* biofilms cause a negligible immune response and, in states of inflammation, exhibit robust antiinflammatory properties. These probiotic biofilms colonize and protect the gut and vagina, and have been implicated in improved healing of damaged skin. Overall, biofilms stimulate a unique immune response that we are only beginning to understand.

1. HOST IMMUNE RESPONSES TO BIOFILM CONSTITUENTS

The immune response to biofilms is complex and often contradictory. Biofilms can both suppress and overstimulate the immune system, and it is likely that this differential response depends on several factors including the immune status of the host, the anatomical location of the biofilm, the species composition of the biofilm and the specific antigens immune cells encounter. The immune system recognizes many bacterial factors; some are cell-associated (eg, lipopolysaccharide, flagella, and cell wall) and some are secreted. Various secreted factors make up the biofilm matrix, and some are exoproducts that serve other functions. All of these proteins have the capacity to be immunogenic, but an important factor to consider is what the immune system "sees" of the biofilms. If bacteria are "buried" within their own extracellular polymeric substance (EPS), which exoproducts can penetrate and with which antigens do immune cells come into contact?

The biofilm EPS is a complex composition of water, exopolysaccharides, exoproteins, extracellular DNA, lipids, and biosurfactants, among other components[1] (Fig. 1). These molecules collectively comprise the structural, adhesive, protective, and metabolic properties unique to the biofilm mode of microbial life. Apart from the immune response mounted against the biofilm-associated organisms themselves, various components of the EPS have been shown to be important for the host immune response to biofilm infections, both in terms of immunogenesis and immunomodulation. In this section, we will summarize what is currently known about how the host immune system recognizes and responds to three major EPS constituents;

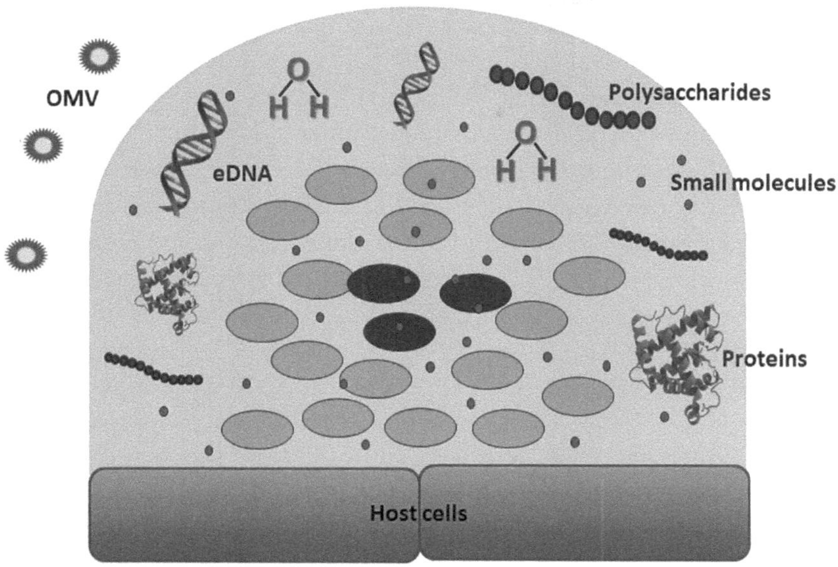

Figure 1 The biofilm EPS is a complex composition of water, exopolysaccharides, exoproteins, and extracellular DNA, among other components. These components collectively comprise the structural, adhesive, protective, and metabolic properties unique to the biofilm mode of microbial life. Apart from the immune response mounted against the biofilm-associated organisms themselves [gray ovals (green ovals in the web version), with persisters colored black (purple in the web version), various components of the EPS can elicit host immune responses, which potentially alter the course of infection. *OMV*, outer membrane vesicles.

exopolysaccharides, exoproteins, and extracellular DNA. Additionally, we will summarize the literature pertaining to how each may contribute to biofilm persistence.

1.1 The Immune Response to Exopolysaccharides

Extracellular polysaccharides, or exopolysaccharides, are major EPS constituents of nearly all biofilms.[1–3] Their presence provides a multitude of integral attributes to the biofilm. This includes, but is not limited to, mechanical stability, physical and chemical protection against antimicrobials and host defense cells, adhesion and aggregation of biofilm cells, tolerance to desiccation, and sorption of organic and inorganic compounds, as well as serving as a carbon source in times of nutrient limitation.[1,4]

How the host specifically deals with EPS exopolysaccharides is understudied at best, and what *has* been published paints a picture of great complexity, with both immunogenic and antiimmunogenic properties having been identified. Multiple studies have shown that certain bacterial exopolysaccharides can inhibit the general immune response toward biofilms. For example, Murofushi et al. demonstrated that certain exopolysaccharides in *Lactobacillus plantarum* biofilms decreased the production of the proinflammatory cytokines, IL-6, IL-8, and MCP-1 (monocyte chemotactic protein-1 also known as CCL2), by increasing the negative regulators of Toll-like receptor-4 (TLR4), an important pattern recognition receptor (PRR) for the detection of pathogen-associated molecular patterns (PAMPs).[5] They fractionated the EPS from a strain of *L. plantarum* isolated from pickled shallots into neutral (NPS) and acidic (APS) portions via anion exchange chromatography, and then stimulated cells from a porcine intestinal epithelial (PIE) cell line, in vitro, by adding a specific EPS fraction, or the bacterium itself, to the cultures. The expression of IL-6, IL-8, and MCP-1 was measured by quantitating mRNA via RT-qPCR, and the authors found that, upon challenging PIE cells with an enterotoxigenic *Escherichia coli* (ETEC) strain, pretreatment with APS led to significantly lower levels of IL-6, IL-8, and MCP-1 mRNAs than control cells. The investigators demonstrated the dependence on TLR4 by showing that anti-TLR4 antibodies significantly reduced the ability of APS to decrease the levels of the three cytokines. These results led the authors to conclude that *L. plantarum* APS may be useful as a probiotic for reducing the inflammatory damage resulting from ETEC infection.

Raffatellu et al. showed that capsular exopolysaccharides from *Salmonella typhi* can also reduce the expression of IL-8 by human intestinal epithelial cells, ex vivo.[6] The authors explanted tissue with either encapsulated *S. typhi*, or a Vi capsular antigen knockout mutant, and performed real-time PCR to quantify IL-8 mRNA. It was determined that the observed decrease in IL-8 production was due to reduced TLR5 and TLR4 activity [which recognize *S. typhi* flagellar proteins and lipopolysaccharides (LPS), respectively] by transfecting human embryonic kidney cells with TLR5 or TLR4, and comparing IL-8 expression to that of wild-type control cells. Their conclusion was that the *S. typhi* capsular exopolysaccharides were able to attenuate the activity of both TLRs. Similarly, Bylund et al. found that exopolysaccharides from a clinical strain of *Burkholderia cepacia* inhibited neutrophil chemotaxis and production of reactive oxygen species (ROS).[7] The authors discovered that purified *B. cepacia* significantly hindered neutrophil chemotaxis toward fMLF, a potent chemoattractant, in a dose-dependent fashion, and reduced neutrophil ROS production by inhibiting stimulation by the

NADPH-oxidase activators, PMA and fMLF (another function of fMLF, aside from being a chemoattractant).

Apart from inhibition of the innate immune cells, such as neutrophils and macrophages, certain biofilm exopolysaccharides, such as alginate, produced by *Pseudomonas aeruginosa*, have been shown to inhibit complement activation.[8] Interestingly, alginate is also known to inhibit macrophage killing of *P. aeruginosa*,[9] in vitro, resulting in increased resistance to phagocytosis.

Additional evidence that some exopolysaccharides may be able to indirectly inhibit the immune system came from Aslam et al., who showed that the exopolysaccharides from a variety of animal and plant pathogens are able to chelate calcium, which serves as a secondary messenger in PRR signal-transduction pathways.[10] By purifying the exopolysaccharides from a variety of pathogens, including *P. aeruginosa*, the authors found that each species bound calcium at a concentration between 5 and 43 μg Ca^{2+} per mg of EPS upon in vitro infection of a plant model. In summary, the previous studies suggest that EPS exopolysaccharides may play a vital role in immune evasion by biofilm-associated microbes.

Conversely, the literature shows that other exopolysaccharides can contain PAMPs that are recognized by the PRRs (especially toll-like receptor 2, TLR2) of innate immune cells, inducing an immune response. Lin et al. showed that TLR2 recognizes the biofilm exopolysaccharides produced by *Thermus aquaticus*, inducing the production of proinflammatory cytokines by the peritoneal macrophages of mice.[11] By treating murine macrophages with the purified *T. aquaticus* exopolysaccharide, TA-1, in vitro, and measuring cytokine levels, the authors determined that expression of both TNF-α and IL-6 were elevated in comparison to controls. It was also found that the macrophage response to TA-1 was TLR2-mediated by pretreating the cells with anti-TLR2 antibodies and observing significantly lower expression of proinflammatory cytokines. This indicates that TLR2 serves as a PRR for specific PAMP(s) on the exopolysaccharide, TA-1. Additionally, and through similar means, Graveline et al. demonstrated human monocyte TLR2s also recognize capsular polysaccharides produced by *Streptococcus suis*, triggering the release of the proinflammatory cytokines, IL-1, IL-6, IL-8, and MCP-1, in vitro.[12]

Taken together, these varied results illuminate the complex nature of the host response to exopolysaccharides. In general, it is assumed that polysaccharides, being a major contributor to the structural makeup of biofilm EPS, play a significant role in conferring physical and chemical protection to the associated microbes. Although numerous methods have elucidated how certain exopolysaccharides attenuate the immune response, some appear to elicit the opposite response. Activating the immune system isn't necessarily counterproductive for the biofilm, as long as it can compensate for host-derived, or antimicrobial-derived damage. In fact, chronic inflammation can be ideal for certain biofilm-associated infections, such as chronic wound infections, where autoimmune damage of the host-tissue is necessary for maintaining an optimal environment for persistence.[13,14]

Last, given that the general purpose of a biofilm seems to be the protection of the microbes from their environment, we hypothesize that there often exists a delicate balance differential in polysaccharide expression (coupled with differential expression of other EPS constituents) that work to counter any detrimental attributes, such as immunogenicity. In support of this theory, *P. aeruginosa* has been shown to differentially express its three main exopolysaccharides, Psl (associated with a nonmucoid EPS phenotype), Pel (associated with a mucoid EPS phenotype), and alginate (also associated with a mucoid EPS phenotype) in response to various types of immune system assault.[15]

Mishra et al. showed that Psl-deficient strains of *P. aeruginosa* were significantly more susceptible to serum deposition of complement components C3, C5, and C7, increasing the response of phagocytes toward the bacteria (ie, opsonophagocytosis and oxidative burst), in vitro.[16] Given that Psl is hypothesized to be important for initial colonization in cystic fibrosis (CF) biofilm infections, prior to later conversion to a more-mucoid phenotype, the authors also tested to see if Psl-deficient *P. aeruginosa* had significantly reduced fitness in a short-term mouse CF infection model when coinoculated, in a 1:1 ratio, with wild-type bacteria. It was determined that the wild-type strain had a 4.5-fold fitness advantage over the Psl-deficient strain after 12 h of infection, indicating the importance of Psl expression in the early stages of infection. Thus, *P. aeruginosa* biofilm infections exhibit varying polysaccharide compositions in order to alter their resulting phenotypes in response to stress from the host immune system.

1.2 The Immune Response to Extracellular DNA

It is widely accepted that extracellular DNA (eDNA) functions as a structural scaffolding within the biofilm EPS of many organisms, facilitates horizontal gene transfer between microbes, and aids in bacterial adhesion and aggregation.[17–21] The role of host response to bacterial DNA for immune clearance has been widely investigated. For more than 20 years, it has been known that oligodeoxynucleotides containing unmethylated 5'-Cytosine-phosphate-Guanine-3' (CpG) motifs, present in bacterial DNA, but not in mammalian DNA, are recognized as PAMPs by TLR9 on the surface of innate immune cells and B cells.[22–24] Therefore, bacterial DNA can trigger both innate and acquired immune responses. Furthermore, Trevani et al. showed that bacterial DNA can also activate neutrophils through CpG-independent mechanisms, in vitro, by binding to a TLR/Toll-IL-1R receptor complex associated with the MyD88 adaptor molecule.[25] They found that, through multiple pathways, including but likely not limited to, the p38 MAPK, ERK1/2, JNK, and PI3K/Akt pathways, neutrophil functions are activated by various means, including AP-1 and NF-κB stimulation of IL-8.

Like the exopolysaccharides, almost all of the work done on understanding the interaction between the host immune system and bacterial DNA has not been biofilm-specific. While it would seem likely that the known immunogenic PAMPs of bacterial DNA would also be present in the eDNA of biofilm EPS, several studies have reported that a significant portion of the structural matrix eDNA can come from lysed host cells, such as neutrophils, and thus would presumably be far less antigenic than bacterial DNA.[21,26,27] However, Bass et al. found that degradation of the eDNA from in vitro *P. aeruginosa* biofilms with DNAse1 greatly reduced the release of the neutrophil proinflammatory cytokines, IL-8 and IL-1b.[28] This suggests that eDNA may be a major immunogenic component of the EPS matrix.

Other research has shown that EPS-eDNA may exhibit immunomodulatory effects. For example, Mulcahy et al. showed in vitro that subinhibitory concentrations of eDNA present in *P. aeruginosa* biofilms can chelate cations, creating a cation-limited environment that induces the bacterial expression of resistance genes to cationic antimicrobial peptides, important host defense molecules of the innate immune system produced by neutrophils, platelets, and epithelia.[29] On the other hand, the authors also showed that higher concentrations of eDNA were antimicrobial for multiple bacterial species, leading to cell membrane disruption and lysis. Mulcahy et al. suggest the observed effect of eDNA at high concentration may be one of the underlying mechanisms of action of neutrophil extracellular traps (NETs), ejected "nets" of neutrophil webs of nuclear DNA coated in antimicrobial histones and proteases that trap and kill bacteria.[29,30]

In summary, as with other EPS components, the antiimmune properties of biofilm eDNA may compensate for any inherent immunogenesis arising from their presence. However, given the highly immunogenic nature of bacterial DNA, we predict that other mechanisms may limit the access of the immune system to the eDNA PAMPs. More work is needed to determine the specific host responses to EPS-eDNA, as the literature is currently limited.

1.3 The Immune Response to Exoproteins in the Biofilm

Extracellular proteins represent another important EPS component that can rival exopolysaccharides in terms of total biofilm mass.[31–34] Functionally, exoproteins are vital to the maintenance and modification of the EPS by performing enzymatic degradation of structural elements for dispersal events and compositional restructuring.[35,36] In addition to maintaining and modifying the matrix, certain EPS proteins, such as DNA-binding proteins, are known to be key factors in the physical stability of the biofilm architecture. Specifically targeting these proteins with antibodies has been investigated as a therapeutic avenue for biofilm dispersal.[37] Along that line, many other proteins have been found to be biofilm-specific, or at least upregulated in biofilm growth, and could represent additional antibody targets.[38–43] Also, exoproteins are known to function in the extracellular digestion of macromolecules for nutrient acquisition, surface and EPS-scaffolding adhesion, protection against host defense cells and molecules, and cell-to-cell communication[1,32,34]; thus, targeting any of these mechanisms clinically has the potential to hinder biofilm persistence and virulence.

Many groups have attempted to characterize the biofilm proteome, and it is certainly clear that biofilms display altered transcriptomic and proteomic phenotypes. For example, Passmore et al. compared the *P. aeruginosa* biofilm and planktonic "secretome" by 2D-DiGE revealing several more proteins that were highly expressed by *P. aeruginosa* biofilm-derived cells than planktonic cells.[39] Similarly, Zijnge and coworkers used LC-MS/MS to identify several new putative virulence factors produced by *Aggregatibacter actinomycetemcomitans* while in a biofilm.[44] While studies like these may identify biofilm-specific proteins, we do not yet know the roles these proteins may play in modulating the immune responses to biofilms.

In general, there are certain characteristics of bacterial proteins that act as PAMPs, or danger-associated molecular patterns (DAMPs, such as *n*-formyl methionine, the prokaryotic protein translation-initiator peptide), that inherently trigger an innate immune response.[45] Formyl peptide receptors (FPRs) are G-protein-coupled receptors present on the surface of host phagocytes that are induced upon detection and binding of prokaryotic formylated peptides, inflammation and bactericidal activity, such as neutrophil degranulation and superoxide production.[46–49] Other PRRs that have been shown to recognize microbial protein PAMPs include TLR2 and TLR5.[50] However, the immune response to EPS-specific exoproteins has not been sufficiently investigated, despite the growing understanding of their contribution to biofilm structure and function. Here we report on the current literature regarding the host response to extracellular proteins in the EPS, the most significant of which pertains to the ubiquitous opportunistic pathogen, *Staphylococcus aureus*.

It has been shown that *S. aureus* biofilms harbor an extensive exoproteome of more than 30 different proteins, with functions ranging from pathogenesis (ie, leukocidin and hemolysin), to immunomodulation (ie, lipoprotein, immunodominant antigens, protein A, and IgG-binding protein), carbohydrate metabolism (ie, phosphoglycerate mutase, enolase, glyceraldehyde-3-phosphate dehydrogenase, alcohol dehydrogenase, and fructose bisphosphate aldolase), DNA metabolism, and stress-response (ie, foldase, DNA binding protein 2, nuclease, and superoxide dismutase).[34,51–53] Brady et al. found that more than 20 of the biofilm-specific proteins in *S. aureus* osteomyelitis biofilms were immunogenic.[51] By using a clinical strain of methicillin-resistant *S. aureus* (MRSA) from a patient with osteomyelitis, the researchers induced biofilm formation in rabbit tibias, in vivo. The rabbit sera was then collected and utilized to probe for proteins in in vitro biofilms (to circumvent the problem of host-protein contamination from in vivo biofilm samples) via western blot and MALDI-TOF analysis, to identify staphylococcal immunogens in comparison to existing data generated from planktonic cultures.

In another study investigating the immunogenicity of *S. aureus* biofilms, mouse peritoneal leukocytes were stimulated with biofilm or planktonic staphylococcal cultures or their acellular filtrates, and then the production of cytokines (TNFα, IL-6, IL-10, MCP-1, and MIP-1α), hemolytic activity, and staphylokinase production was determined.[54] It was found that similar staphylococcal components and metabolites, which possessed immunomodulatory properties, were present in both biofilm and planktonic filtrates. This suggests that *S. aureus* biofilms possess an "open" character to some degree, and despite the cover of the bacterial microcolony by EPS, many *S. aureus* virulence factors (eg, PG, LTA, Hla) leak into the external environment and likely play a role in pathogenesis.

These data point to the inherent antigenic nature of many biofilm proteins, suggesting other components than the innate immune system are involved, given the humoral response. It should be noted that, despite the significant immune response, biofilms are still able to persist, further highlighting the physical and chemical protection afforded to pathogens living inside a biofilm. However, some researchers have attempted to use the antigenicity of bacterial exoproteins to a clinical advantage. For example, Gil et al. recently found that inoculation of mice with an exoproteome extract of a *S. aureus* biofilm significantly inhibited subsequent biofilm formation.[53] This would suggest that immature *S. aureus* biofilms are less able to compensate for the immunogenicity of EPS proteins, and that significant exoproteome expression is more likely to be seen in established biofilms. This hypothesis was supported by Brady et al., who found that some of the immunogenic proteins only elicited an antibody response following a lengthy infection period of more than 28, or even 42 days.[51]

Other experiments have been conducted to elucidate biofilm-specific, protective antigens by reacting biofilms (or planktonic cultures) with serum or other bodily fluids from diseased or nondiseased hosts. For example, Saniu et al. were able to identify specific *Streptococcus mutans* biofilm cellular proteins that were recognized by immunoglobulin A (IgA) in saliva from individuals who had dental caries, versus those who did not.[55] Interestingly, the vast majority of proteins that elicited IgA reactivity were intracellular proteins involved in intracellular processes such as amino acid biosynthesis, glycolysis, and cell division, providing more evidence that, despite biofilm cells being buried in EPS, their intercellular proteins are still accessible to the immune system.

Clearly the secretome of planktonic cell cultures is very different from that of biofilms; however, we know very little about how the biofilm secretome changes in mono versus polymicrobial biofilms, or if the secretome seen in vitro even resembles that seen in vivo. The population composition certainly makes a difference. For example, while the diversity of *P. aeruginosa* proteins secreted in polymicrobial conditions was lower than in monoculture conditions, some secreted proteins (such as the virulence factor ToxA and hemophore HasAp) were uniquely expressed in polymicrobial biofilms.[56] Not surprisingly, host cells respond differently to biofilms depending on their species composition. In a recent study multilayered human gingival epithelial cultures were exposed to biofilms made by up to 10 different oral species of bacteria in vitro. The investigators then characterized the epithelial secretome by liquid chromatography-tandem mass spectrometry.[57] The group identified a total of 192 secreted host proteins and noted that more were downregulated than upregulated following exposure to biofilms. Importantly, the host cell secretome was highly influenced by the biofilm species composition.

1.4 The Immune Challenge of Outer Membrane Vesicles

In recent years, the presence of outer membrane vesicles (OMVs) in the EPS of Gram negative bacterial biofilms, especially those of *P. aeruginosa*, has become apparent.[58,59] OMVs are spherical, bilayered phospholipids that are derived from the outer membrane of a variety of Gram-negative bacteria and consist of membrane lipids, LPS, proteins, and DNA.[60] They perform key roles in the pathogenesis and survival of bacteria, including facilitating the secretion of insoluble molecules and the distant, protected delivery of concentrated enzymes to specific targets.[61] Although their production is correlated with the bacterial stress response, which is also involved in the production of biofilm, their precise biological role within the EPS is unclear. However, given the fact that they are a common constituent of Gram-negative bacterial-derived biofilms, and that comparison of planktonic versus biofilm-derived OMVs has indicated functional roles in proteolysis and binding of antimicrobials,[61] it would appear that they are an important component of the EPS matrix.

Due to the concentrated and combined expression of immunogenic components that contain multiple PAMPs that are recognized by various PRRs, OMVs have been found to be highly immunogenic in comparison to soluble antigenic molecules.[60] Although exoproteins have been covered in detail thus far, Toyofuku et al. recently determined that up to 30% of the EPS proteins from *P. aeruginosa* biofilms grown in vitro were OMV-associated.[62] Given these results, we expect EPS-associated OMV research to become an increasingly prevalent area of study that will open doors to a greater understanding of biofilm structure and function, including host-EPS interaction.

1.5 Host Responses to Biofilm-Associated Small Molecules

Bacteria in biofilms secrete many small molecules that affect the immune response. These small molecules include secondary metabolites, pigmented cyanins (redox active secondary messengers) and antibiotics. However, some of the best-studied are those produced for cell-to-cell signaling or quorum sensing (QS), during which a "quorum," or threshold number of bacteria, are able to synchronize different behaviors or functions including the production of virulence factors or the formation of biofilms in some species of bacteria.[63] QS is typically mediated by small molecules in Gram-negative bacteria and small peptides in Gram-positive bacteria. While these signals bind to bacterial transmembrane or intracellular receptors, they can also modulate cell signaling in host cells, resulting in modified host responses.

Several different types of host cell effects, generated by QS signals, have been documented and thoroughly reviewed.[64–66] Most of these interkingdom effects have been described in response to the *N*-(3-oxododecanoyl)-homoserine lactone (3-oxo-C12-HSL) made by *P. aeruginosa*. 3-oxo-C12-HSL activates the intracellular LasR transcriptional regulator in *P. aeruginosa*, promoting the expression of many different bacterial genes; it also causes a variety of effects in different types in mammalian cells. For example, 3-oxo-C12-HSL can act as an attractant for neutrophils and stimulate the production of the chemokine IL-8.[67–69] It can also cause apoptosis in a number of different cell types including macrophages and neutrophils,[70] fibroblasts and vascular endothelial cells,[69] airway epithelial cells[71] and human mesenchymal stem cells,[72] although the mechanisms of action appear to differ. 3-oxo-C12-HSL-induced apoptosis in some cells is linked to elevated intracellular calcium release[69,73] and proceeds through the activation of the intrinsic mitochondrial pathway[74] or elevation of caspase 3 and 8 activities (extrinsic pathway) in other cell types.[70]

3-oxo-C12-HSL can either suppress immune responses or stimulate proinflammatory responses, depending on the host cell type, 3-oxo-C12-HSL concentrations, or other experimental parameters. While the exact mechanisms involved are not completely understood, these interactions seem to be independent of canonical pathogen-associated molecular PRR signaling pathways.[75] It has been proposed that over the course of more than a billion years of coevolution, prokaryotes and mammalian cells have developed dual-conduit signaling pathways over which bacterial QS signals and mammalian hormones travel.[76–79] However, while some host cell receptors can bind quorum signals and mediate signaling events,[80–83] there is not yet evidence that any specific mammalian receptors evolved specifically for quorum signals.

While 3-oxo-C12-HSL appears to have a wide spectrum of effects on the host response, it is likely that many other QS molecules can also alter host immune responses; however, it must be noted that most of the documented effects have been observed in vitro after cells in culture were stimulated with exogenous, purified QS molecules. It is unclear whether these effects occur during infection in a much more complex environment and heterogeneous population of both bacterial cells and host cells. As discussed previously for exoproteins, it is also unclear what concentration of QS signals are produced inside a biofilm under varying conditions, and the diffusion kinetics from biofilms to host cells.

Certain secondary messaging molecules produced by bacteria have been shown to illicit a host immune response, and may represent a means by which the host can specifically respond to biofilm-associated infections. Cyclic dinucleotides (c-di-NMPs), in particular cyclic diguanylate monophosphate (c-di-GMP), have emerged as important, universal secondary messaging molecules that trigger the biofilm mode of life in a multitude of bacterial species.[84] Of note, c-di-NMPs have been shown to be involved in the cessation of motility, cellular adhesion, regulation of cell-cycle control, biofilm formation and maintenance, cell wall metabolism, DNA repair, gene expression control, the regulation of virulence factors, and many other processes.[84–87] Recently, it has been shown that the host immune system can recognize c-di-NMPs by way of PRR signaling.[88] C-di-GMP and cyclic adenosine monophosphate (c-di-AMP) have been shown to be recognized as PAMPs, triggering a host type I interferon gamma (IFN-γ) innate immune response.[89,90] Given the heavy involvement and elevated levels of c-di-NMPs produced during biofilm development and persistence, this research represents one of the first examples of possible biofilm-associated danger signals that can be detected by the immune system.

Another secondary metabolite that has been shown to be involved in host/biofilm interactions is pyocyanin, a redox-active, membrane-permeable phenazine pigment produced by *P. aeruginosa* that has the ability to kill both host cells (including immune cells), and the cells of competing microorganisms.[91,92] Recent studies have shown that H_2O_2 generation via pyocyanin production is involved in the release of eDNA from lysed subpopulations of *P. aeruginosa*,[93] and that pyocyanin promotes the binding of eDNA to *P. aeruginosa*, facilitating aggregation and EPS formation.[94] Interestingly, it has also been found that pyocyanin-induced ROS production can lead to neutrophil NET production.[95] Given that certain chronic disease states, such as lupus erythematosus and CF, are associated with host tissue damage caused by aberrantly high NET production,[96–98] and that pyocyanin is known to be highly correlated with chronic *P. aeruginosa* biofilm infection in CF lungs,[99,100] it is possible that certain *P. aeruginosa* biofilms exhibit high levels of pyocyanin production in order to create excessive levels of NETosis, which is the release of NETs. Aberrant NET production then causes collateral host tissue damage, making chronic infection more likely. This scenario is representative of a pathogen which takes advantage of the physical protection afforded to it by the biofilm mode of life by triggering an immune response that would normally be detrimental to planktonic cells and instead utilizing the misdirected damage to the host tissue to facilitate prolonged infection.

In summary, how the host responds to biofilm EPS in general, as well as to the specific components, remains unclear. More research is needed to understand the complex interplay between immunogenesis and immunomodulation, as well as how the actions of the immune system may contribute to the disease-state of the various types of biofilm infections. It is also important to note a potential fundamental inadequacy in all of the previous research; nearly all studies were performed in vitro, and/or with limited, if any, polymicrobial interactions. Thus, understanding the differences in composition, function, and behavior between in vitro and in vivo biofilms is an avenue that must be explored before all biofilm and EPS-related studies, including how the host interacts with the various biofilm components, can be readily accepted. This is especially true when it is taken into consideration that many biofilm-associated infections are allowed to persist, in part, due to a general immune-incompetence of the host, or their location inside an immune-limited site, such as a urethral catheter or a diabetic foot ulcer. Thus, any in vitro models of these infection types must account for the depleted immune system, something that is lacking in the literature. Also, it has become increasingly apparent that single-species biofilms are rarely seen clinically, with most instead being polymicrobial. Thus, interspecies, and even interkingdom interactions only add to biofilm complexity.[101–106] Therefore, results ascertained from purely in vitro, single-species studies cannot be considered entirely clinically applicable.

2. PATHOGENIC BIOFILMS AND HOST CELLS

Human host immunity, which involves an intricate interplay of signals and cell populations, has traditionally been divided into the innate and adaptive (cell mediated and humoral) immune systems. Innate immunity is composed of the general epithelial barrier, host microbiome, complement, and a variety of leukocytes including both phagocytic (macrophages, neutrophils) and nonphagocytic (natural killer) cells. Cells of the innate immune system utilize PRRs to recognize broadly conserved molecules specific to microbes such as LPS and peptidoglycan.[107] These initial host cells are critical in quelling bacterial invaders and stimulating other facets of the immune system through the release of chemokines and cytokines (Fig. 2) Pathogenic responses establishing infection often use evasion strategies against these early immune responders. As such, there is a large body of research examining the interplay between bacterial and innate immune cells, and how bacteria have evolved to overcome immunity. Biofilms provide a safe haven to bacteria from environmental assaults, including antibiotics and immune cells, and there is growing evidence that biofilms frustrate specific components of the immune response, contributing to chronic inflammation. This section focuses on the primary innate immune cell populations associated with bacterial clearance and how they respond to biofilms versus planktonic cells.

2.1 Neutrophils and Pathogenic Biofilms

Polymorphonuclear neutrophils (PMNs) represent the body's first responders to inflammation, and make up the bulk of circulating leukocytes.[108] Tissue-specific cells such as epithelial, mast cells, and tissue macrophages recruit PMNs to the damaged or infected site by releasing the chemoattractants IL-8 and CXCL1/CXCL2,[109] and the activating cytokine IL-6.[110] Once activated, PMNs rapidly chemotax to the designated area, phagocytizing microbes and releasing the proinflammatory cytokines TNF-α, IL-8, and IL-1β.[111] This creates a positive loop attracting and priming more neutrophils and generating inflammation at the local tissue site. Bacteria are then inactivated by PMNs primarily through four mechanisms: releasing antimicrobial granules (eg, defensins, cathelicidins, lysozyme, and lactoferrin), phagocytosis, generation of ROS, and the release of NETs.[112] Successful phagocytosis of bacteria by PMNs is dependent upon opsonization. Opsonization is the process by which a microbe is labeled for removal via complement and/or antibodies, and subsequently phagocytosed.[113] When bacteria are opsonized with both IgG and C3b, they bind to the PMN Fc-gamma receptor, leading to speedy phagocytosis and removal. Inflammation is resolved by PMN apoptosis, and subsequent efferocytosis of these dead cells by macrophages. When PMNs die by apoptosis, the cell membrane remains intact; however, necrosis or lysis of PMNs results in the release of cellular components that can increase inflammation and cause collateral damage to the host.[114] Similarly, when macrophages are absent, dying and fractured neutrophils will eventually release their proinflammatory contents.[115]

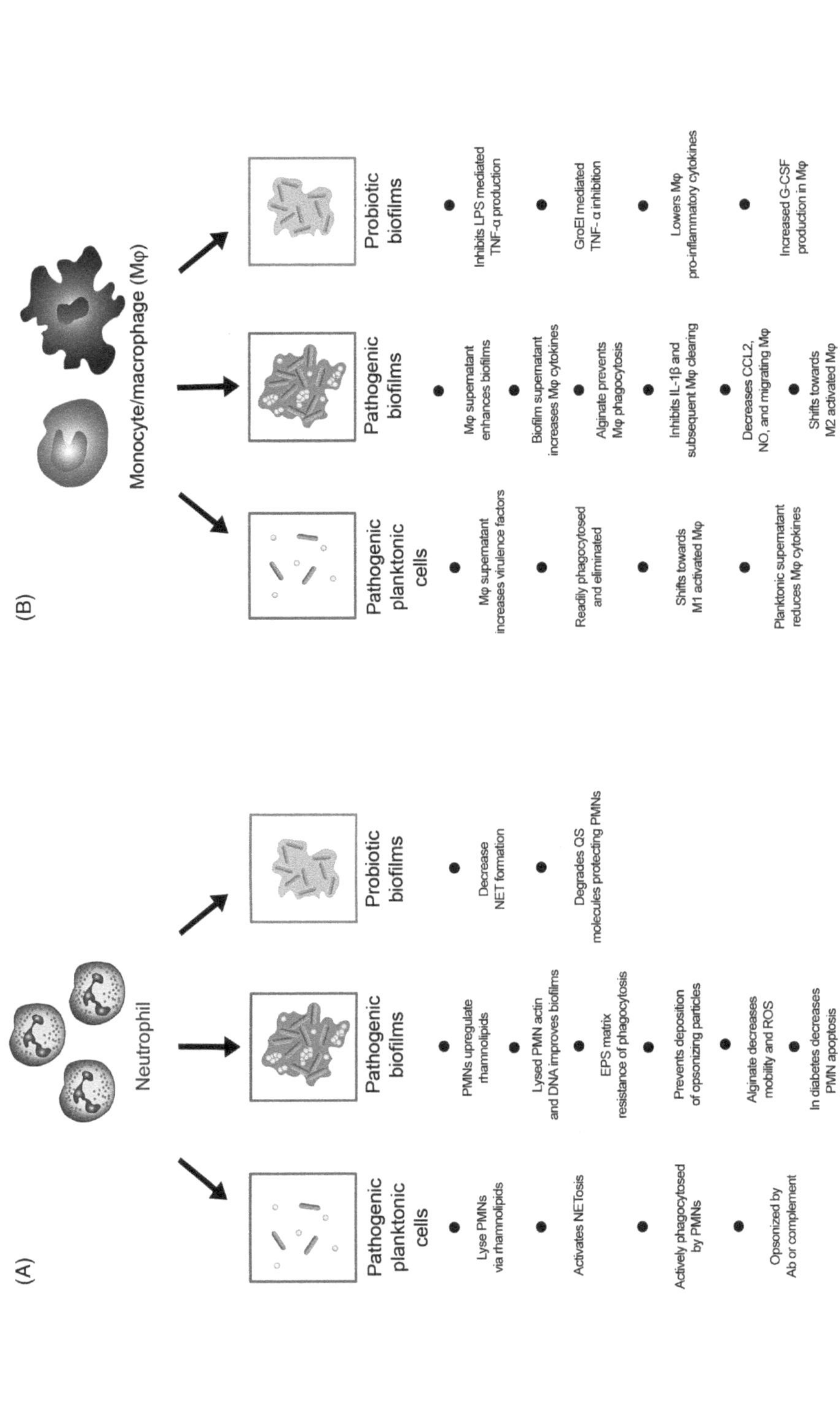
(A)
Neutrophil
Pathogenic planktonic cells
Lyse PMNs via rhamnolipids
Activates NETosis
Actively phagocytosed by PMNs
Opsonized by Ab or complement
Pathogenic biofilms
PMNs upregulate rhamnolipids
Lysed PMN actin and DNA improves biofilms
EPS matrix resistance of phagocytosis
Prevents deposition of opsonizing particles
Alginate decreases mobility and ROS
In diabetes decreases PMN apoptosis
Probiotic biofilms
Decrease NET formation
Degrades QS molecules protecting PMNs
(B)
Monocyte/macrophage (Mφ)
Pathogenic planktonic cells
Mφ supernatant increases virulence factors
Readily phagocytosed and eliminated
Shifts towards M1 activated Mφ
Planktonic supernatant reduces Mφ cytokines
Pathogenic biofilms
Mφ supernatant enhances biofilms
Biofilm supernatant increases Mφ cytokines
Alginate prevents Mφ phagocytosis
Inhibits IL-1β and subsequent Mφ clearing
Decreases CCL2, NO, and migrating Mφ
Shifts towards M2 activated Mφ
Probiotic biofilms
Inhibits LPS mediated TNF-α production
GroEI mediated TNF- α inhibition
Lowers Mφ pro-inflammatory cytokines
Increased G-CSF production in Mφ

As the frontline defenders of the body, PMNs are one of the first host cell populations to interact with both planktonic and biofilm bacterial cells. The majority of work dealing with PMNs and biofilms has been conducted in vitro with the well-studied biofilm formers, *Staphylococcus* and *P. aeruginosa*. Various microscopic techniques have revealed *S. aureus* biofilms of 2, 6, and 15 days to be susceptible to destruction via PMN mediated phagocytosis.[116] The authors observed direct migration toward and into the *S. aureus* biofilms, and found that PMN clearance was reduced with the age of the biofilm. This same group went on to show that 2 and 6 day old biofilms were again destroyed by PMNs via phagocytosis, degranulation, and NETosis.[117] The authors next reported that PMN attachment, phagocytosis, and degranulation of *S. aureus* biofilms was unaffected by complement opsonization mediated by C3bi.[118] A potential problem with these three prior studies is that biofilms were grown with constant shaking, which could have disrupted proper attachment, and led to more planktonic growth.[116–118] The ability of PMNs to clear *S. aureus* biofilms was shown to be IL-β dependent, as mice deficient in this cytokine displayed increased biofilm formation, and a 50% drop in neutrophil recruitment.[119]

Nguyen et al. found that *S. aureus* biofilms decreased the oxidative activity of PMNs in a murine diabetic wound infection model.[120] When compared to nondiabetic wounds, the diabetic wounds were found to have decreased levels of TLR2/4, TNF-α, and IL-β, leading to elevated bacterial loads and wound healing failure, but similar levels of neutrophils. This suggests that *S.*

◀ **Figure 2** Host immune cell interactions with pathogenic and probiotic bacteria. The host immune system under normal conditions readily clears planktonic pathogenic bacteria, though *Pseudomonas aeruginosa* can produce rhamnolipids which rapidly lyse PMNs. Pathogenic biofilms frustrate host immune cells through numerous mechanisms including reducing bacterial opsonization, decreasing the phagocytic potential of phagocytes, immobilizing PMNs, and inhibiting ROS production. In addition, bacteria have evolved to utilize components of both PMNs and macrophages to enhance biofilm formation, resulting in enhanced tolerance to host immunity and antimicrobial agents. The frustrated immune response and more robust biofilms contribute to chronic inflammation, causing collateral damage to host tissue. In contrast, probiotic biofilms seem overwhelmingly antiinflammatory, as the host doesn't respond with antiinflammatory cytokines, and phagocytic inflammation is reduced. Abbreviations: *PMN*, polymorphonuclear neutrophil; *NETosis*, release of neutrophil extracellular traps; *Ab*, antibody; *EPS*, extracellular polymeric substance; *ROS*, reactive oxygen species; *QS*, quorum sensing; *Mφ*, macrophage, *CCL2*, chemokine (C–C motif) ligand 2; *NO*, nitric oxide; *LPS*, lipopolysaccharide; *G-CSF*, granulocyte-colony stimulating factor.

aureus biofilm clearance is independent of neutrophil numbers alone but instead caused by the diminished diabetic immunity. Another study with diabetic mice infected with intraperitoneal *S. aureus* biofilms found that PMN apoptosis was decreased as compared to the wild type,[121] which led to decreased removal of PMNs by macrophages, and prolonged production of TNF-α. The authors heavily speculated that decreased apoptosis in PMNs correlated with the deficiency that diabetic patients have in clearing *S. aureus* biofilm associated infections. When *Staphylococcus epidermidis* biofilms were opsonized and incubated with human PMNs, they displayed increased resistance to PMN killing as compared to a biofilm EPS mutant strain and their planktonic counterparts.[122] As PMN mediated killing of planktonic *S. epidermidis* is complement mediated, they measured levels of complement activation and deposition. Surprisingly, *S. epidermidis* biofilms increased levels of C3a, a complement chemoattractant, while decreasing the amount of C3b that was attached to the biofilm matrix as compared to planktonic or mutant cells.[122] Similarly, in *Streptococcus pneumoniae* biofilms, C3b deposition was decreased, as was PMN phagocytosis.[123]

In their seminal study, conducted with fresh human PMNs and *P. aeruginosa* biofilms, Costerton et. al reported that the oxidative burst of PMNs was reduced to 25% by opsonized and nonopsonized biofilms, as compared to planktonic *P. aeruginosa*.[124] The authors linked this inhibition of PMN activity to alginate, which they measured in this nonmucoid strain, and found to be similar to mucoid *P. aeruginosa* strains. The authors went on to show that this effect on oxidative burst was returned to planktonic levels when the biofilm was mechanically dispersed.[125] Another study demonstrated that when human PMNs were incubated with *P. aeruginosa* biofilms they were immobilized and unpolarized, but retained phagocytic and partial degranulation activity.[126] Consequently, the biofilm cells responded to the PMNs by increasing oxygen intake, and activating an escape phenotype, resulting in a bolus release of planktonic cells, which surrounded the immobilized PMNs.

Another study with *P. aeruginosa* biofilms found that human PMNs rapidly phagocytized planktonic cells, but phagocytosis was averted with biofilms or stationary phase aggregates.[127] The authors proposed a mechanism whereby *P. aeruginosa* biofilms subverted PMN phagocytosis with rhamnolipid. Rhamnolipids are biosurfactant glycolipids which are produced during *P. aeruginosa* QS.[128] Rhamnolipids are controlled by the "Rhl" QS system, and are actively produced by planktonic cells and, while produced in biofilms, the amount is decreased as the transcriptional regulator AlgR represses rhamnolipid production during biofilm development.[129] However, rhamnolipids in the supernatants of both planktonic and biofilm cultures rapidly lysed human neutrophils and inhibited chemotaxis.[26] Incubation in vitro with 4 day old PAO1 biofilms, but not *lasR/rhlR* mutants, also resulted in the rapid lysis of PMNs. Additionally, PAO1 supernatants (biofilm vs planktonic undefined), but not *lasR/rhlR* mutants, lysed whole blood. The authors even showed that sputum from 9/13 cystic fibrosis patients infected with *P. aeruginosa* was able to lyse PMNs in vitro, while sputum samples from patients without *P. aeruginosa* infection had no effect.

While this group had previously shown that QS biofilm mutants were easily phagocytosed in comparison to wild type biofilms,[130] they further reported that *P. aeruginosa* biofilms actively upregulated production of QS virulence factors when in the presence of PMNs.[131] Rhamnolipids were one of the primary factors upregulated, and the authors proposed a "rhamnolipid shield model," in which rhamnolipid encompasses the biofilm and eliminates invading PMNs. In support of this, QS mutants were easily cleared in a murine lung infection model,[130] and bronchoalveolar lavage fluid from QS mutants contained more viable neutrophils than observed in wild type samples.[26] In addition, when a garlic-based rhamnolipid inhibitor, ajoene, was used in a murine pulmonary infection model, *P. aeruginosa* biofilms were quickly cleared.[132] In a murine silicone intraperitoneal infection model, *rhlA* mutants unable to produce rhamnolipids were phagocytosed by migrating PMNs, while wild type *P. aeruginosa* developed robust biofilms by day 1, which encased and lysed PMNs.[133]

In conjunction with rhamnolipid lysis, there is a growing body of evidence suggesting that PMNs themselves can enhance *P. aeruginosa* biofilm, resulting in increased protection from clearance and antibiotics. The seminal publication on this topic found that either lysed or intact PMNs enhanced biofilm biomass, and the EPS matrix, over 72 h.[134] PMNs were able to reduce planktonic cells (though they remained at $\sim 10^6$ CFUs), but had a negligible impact on viable biofilm cells. Fluorescent microscopy was used to show that *P. aeruginosa* incorporated actin and DNA from lysed PMNs to enhance biofilm thickness. Comparison to CF sputum revealed similar biofilm localization of DNA and actin, and *P. aeruginosa* binding was mediated specifically by F-actin. Treatment with DNase I readily dispersed the actin/DNA rich *P. aeruginosa* biofilms.[134]

Another study showed that incorporation of lysed PMNs into *P. aeruginosa* biofilms resulted in increased protection from aminoglycosides.[135] This was further explored in a murine diabetic wound infection model, where insulin treatment increased PMN migration, resulting in *P. aeruginosa* wound biofilms that displayed increased DNA content and tolerance to the aminoglycoside gentamicin.[136] Overall, these studies suggest that PMNs are effective at clearing most planktonic species, but biofilms are able to evade PMN killing mechanisms, and can even use PMN components to build better biofilms (Fig. 2). *P. aeruginosa* biofilms seem especially problematic for PMNs due to a variety of virulence factors including rhamnolipids, alginate, and the incorporation of DNA and actin into the EPS matrix.

2.2 Macrophages and Pathogenic Biofilms

Tissue macrophages are responsible for phagocytosing foreign material and microbes, and then stimulating an immune response with the recruitment of PMNs. Following PMN phagocytosis of pathogens and subsequent apoptosis, immature macrophages, known as monocytes, are recruited from the bloodstream.[137] Monocytes are circulating leukocytes, and inflammation or bacterial infection will recruit these cells to the infected area by the systemic release of recruiting chemokines CCL2 and CCL7, microbe-specific molecules such as *N*-formyl-methionyl peptides, and proinflammatory cytokines.[138] To be effective at the tissue site, macrophages must be activated either by the classical route (M1 via IFN-γ), or alternative activation (M2 via IL-4).[139,140]

Activated macrophages are involved in a wide array of immune activities, including increased phagocytosis, antigen presentation, production of cytokines TNF-α, IL-1, and IL-6, and efferocytosis.[141] Macrophages are the "professional" phagocytes, in their ability to phagocytose large materials quickly, recognize numerous receptors for uptake, engulf bacteria in blood, and persist in the tissue.[142] Once bacteria have been phagocytosed, macrophages like neutrophils produce ROS and nitric oxide (NO) to kill the engulfed bacteria. Though there are fewer studies describing the interactions between bacterial biofilms and macrophages, there have been some recent studies that provide great insight. A seminal study on macrophages and biofilms focused on *S. epidermidis* and separated out the "bacterial slime" into three fractions through column separation.[143] The authors found that, following a 20 h incubation with rabbit alveolar macrophages, 2/3 EPS fractions reduced the oxidative burst of the macrophages.

In another study, key components of *S. epidermidis* biofilms [polysaccharide intercellular adhesion (PIA), and intercellular adhesins Aap and Embp] were tested to determine their importance during immune evasion.[144]

Utilizing three different isogenic strains with PIA, Embp, or Aap as the biofilm component, exposure to murine macrophages revealed similar levels of protection from phagocytosis for all strains. The authors went on to show that *S. epidermidis* biofilms inhibited NF-κB activation, and subsequent production of IL-1β, as compared to biofilm deficient strains. Furthermore, when *S. epidermidis* biofilms were induced to produce increased nonculturable dormant cells and then incubated with macrophages, the levels of proinflammatory cytokines produced (TNF-α, IL-1, and IL-6) by these macrophages was reduced when compared to those incubated with biofilms harboring fewer dormant cells.[145] These data suggest that *S. epidermidis* biofilms display numerous mechanisms to evade macrophages, and prevent detection by the immune system.

P. aeruginosa biofilms that produced alginate were shown to be protected from killing after 4 h of incubation with human macrophages.[9] However, the biofilm of a particular *P. aeruginosa* alginate mutant was cleared by ~98% through an IFN-γ dependent mechanism. Unlike *S. epidermidis* biofilms, when six different strains of *P. aeruginosa* were exposed to murine macrophages, biofilm formation almost doubled the levels of different cytokines when compared to planktonic cells.[146] Interestingly, when biofilms were cultured in macrophage supernatant, alginate production doubled, as did other *P. aeruginosa* virulence factors. Even more surprising was that the macrophage supernatant stimulated the number of both planktonic and biofilm-associated viable cells. These data suggest that *P. aeruginosa* biofilms can be enhanced by macrophages, as with PMNs. A similar phenomenon was observed when *Candida albicans* biofilms were incubated with monocyte/macrophage supernatants.[147] The authors found that macrophages, and supernatant from macrophage cell cultures, enhanced *C. albicans* biofilm formation, but macrophage components from the cell surface did not.

When macrophages were cultured with supernatant from mature *S. aureus* biofilms, macrophage phagocytosis decreased and cytotoxicity increased.[148] This effect was found to be dependent on alpha-toxin (hla) and leukocidin AB (lukAB). The authors then tested a double *lukAB/hla S. aureus* mutant in an orthopedic infection model, and found that, without these virulence factors, bacterial load decreased and macrophage numbers expanded.[148] This group had previously examined the interactions of *S. aureus* biofilms and macrophages in a murine catheter infection model.[149–151] The authors first found that *S. aureus* biofilm infection decreased CCL2 levels, migrating macrophages, and NO production, while arginase-1 was increased.[149] The authors pointed out that NO and arginase-1 compete for arginine, and that by increasing arginase, fibrosis is favored over bacterial clearance. The authors next tested the importance of MyD88-signaling pathways (IL-1 receptor and TLRs) in containing *S. aureus* biofilms.[150] The authors found that, in a murine catheter model, bacterial load, fibrosis and numbers of migrating, alternatively activated, M2 macrophages were enhanced in MyD88 knockout mice, while proinflammatory cytokines were decreased. This group then wanted to determine if the shift toward M2 activated macrophages led to *S. aureus* biofilm protection.[151] In the catheter model, exogenously added M1 activated macrophages inhibited *S. aureus* biofilm formation, and enhanced proinflammatory cytokines. Addition of MyD88-mutant macrophages (M2 skewed) failed to reduce biofilm formation, highlighting the importance of classically activated M1 macrophages in *S. aureus* biofilm clearance.

Overall, it appears that the EPS components of biofilms greatly frustrate the phagocytic and ROS capacity of macrophages (Fig. 2). However, *Staphylococcus* biofilms tend to remain under the immune systems radar more effectively than other species. Curiously, macrophage supernatant is able to stimulate EPS production, and subsequent biofilm formation. While many of these studies were conducted in vivo, none looked at multispecies biofilms, which may provide unique immune profiles not observed with single-species studies.

2.3 Adaptive Immune Response and Pathogenic Biofilms

The adaptive immune response encompasses both the humoral- and cell-mediated arms of the immune system and works cooperatively with innate immunity to clear bacterial infections and induce a memory response. Humoral immunity is mediated by B cells, which can recognize antigens on bacterial surfaces, and produce antibodies that can bind to and neutralize bacterial toxins, and opsonize bacterial antigenic surfaces for subsequent phagocytosis.[152] Cell mediated immunity is dominated by populations of T cells, both CD4+ helper T cells, which mediate Th1, Th2, Treg, and Th17 responses, and CD8+ cytotoxic T cells. T cells recognize processed antigens presented by antigen presenting cells (APC's), such as epithelial cells, macrophages, and dendritic cells. Recognition of these antigens subsequently activate and mature T cells, which stimulate cytokine release and modulate the immune response appropriately.[152] Humoral immunity is classically Th1 mediated to tackle extracellular bacteria; while cell mediated immunity is

Th2 dependent, targeting intercellular pathogens.[153] T-helper-17 (Th17) cells are thought to be important for clearing extracellular bacteria, but recently they have been implicated in contributing to biofilm-associated chronic inflammation.[154] Development of Th17 cells is dependent on the cytokines TGFβ and IL-6, and these cells subsequently shift the immune response through the production of cytokines IL-17, IL-22, and IL-26.[155] This section will address the few studies that have investigated the interactions of bacterial biofilms and the adaptive immune response.

As mentioned previously, opsonization of bacteria by antibodies is critical to clearance via phagocytosis, and when *S. epidermidis* biofilms were incubated with IgG, there was decreased deposition of IgG, which resulted in delayed PMN killing.[118] However, IgG deposition and opsonization was not impeded in *S. aureus* biofilms, and was found to be unnecessary for PMN attachment and phagocytosis of the biofilm cells.[118] However, IgG mediated opsonization was shown to be critical for ROS production and biofilm destruction. The antibody response to *S. aureus* biofilms was characterized in a porcine osteomyelitis infection model, displaying increased antibody levels and systemic IL-6.[156] The authors suggested that since IL-6 is a proinflammatory cytokine linked to both Th1 and Th17 responses, this suggests that *S. aureus* biofilms can evade Th2 mediated killing as they are indeed intercellular pathogens. The antibody response was further elucidated in a murine *S. aureus* biofilm implant model revealing heightened levels of the Th1 associated antibody isotype, IgG2b, early on in the infection up to day 14.[157] As the infection persisted to day 28, increased numbers of IgG1 were observed, which is a Th2 antibody isotype. At day 7, local cytokine levels were shifted toward Th1 (IL-2, IL-12, and TNF-α), and Th17 (IL-17 and IL-17) cells were also considerably increased. To further test the role of Th1/Th17 in preventing biofilm clearance, this group next investigated mice who display skewed immunity in clearing *S. aureus* biofilm implant infections.[158] C57BL/6 mice have a Th1 skewed immune response and all mice were unable to clear *S. aureus* biofilms, while 75% of the Th2 skewed BALB/c mice cleared the infection. Th2 mediated clearance was associated with enhanced levels of relevant cytokines and T-regulatory cells.

In bacterial infections, myeloid-derived suppressor cells (MDSCs) have been reported to inhibit T cells, resulting in chronic infection. The Kielian group utilized a murine orthopedic infection model to examine the role of MDSCs in *S. aureus* biofilm infections.[153,159] In this in vivo model, MDSC activators (IL-10, Arg-1) were increased and local T-cells numbers were reduced.[159] After immunodepletion of MDSCs and PMNs, migrating macrophages were more effective at clearing *S. aureus* biofilms. This group then reported that adding exogenous MDSCs to p40 knockout mice (to knockout IL-12) increased bacterial loads during infection.[153] In addition, tissue samples taken from patients with orthopedic infections displayed similar levels of MDSCs with decreased T-cell infiltrates. These data highlight how *S. aureus* biofilms modulate the immune response to avoid clearance, and suggest a promising therapeutic avenue. The bulk of studies on biofilms and adaptive immunity have been done with *Staphylococcus* biofilms, revealing the importance of both humoral and cell mediated immunity, and highlighting ways to potentially improve patient outcomes. However, more studies are warranted involving the adaptive response in conjunction with diverse single and multispecies biofilms.

3. INTERACTIONS OF COMMENSAL BIOFILMS WITH THE IMMUNE SYSTEM

Another important component of immunity is the host's microbiome; a mixture of bacterial, viral, fungal, parasitic, and host cell constituents. This normal flora is essential for the host, both in the development of a competent immune system, and the continued protection against pathogens. One effective method to beneficially modulate the microbiome is via bacteriotherapy, which is the direct addition of live bacteria (mainly ingested, but sometimes applied topically) to improve patient health. Bacteriotherapy most commonly refers to probiotics, which are bacteria that reside in the gut, and thus prefer an acidic environment, growing best at pH 3–5. The gut, vagina, skin, and oral/pharyngeal cavities are often targeted with probiotic biofilms[160] or their supernatants.[161] The idea behind using probiotics is that they can decrease pH, produce antimicrobials, exclude pathogens from binding sites, and beneficially modulate the immune response.[162] Probiotics often incorporate into the host's microbial biofilm populations, and alter a variety of immune responses based on the bacteriotherapy strain, planktonic versus biofilm cell state, delivery route, and host disease state. This section will focus on the interactions of probiotic biofilms with the immune system, expounding first on probiotic biofilms, discussing beneficial interactions between biofilms and macrophages and neutrophils, with the remainder divided up by notable regions of the body where these bacterial and host immune cells interact.

3.1 Overview of Probiotic Biofilms

Probiotics are typically lactic acid producing bacteria of the genus *Lactobacillus* and *Bifidobacterium*. These bacteria are primarily facultative anaerobes or strict anaerobes and are thus incubated statically under anaerobic or microaerobic conditions[161,163–165] for extended time periods, resulting in the formation of probiotic biofilms, with cells adhered to each other, or to the surface that they are grown on. When probiotics are grown planktonically, they are typically cultured to log phase,[166] under aerobic[167] or shaking[168] conditions. The most employed probiotic strains based on efficacy and tradition include: *Lactobacillus casei, L. rhamnosus, L. plantarum, L. reuteri, L. acidophilus, Bifidobacterium bifdum, and B. breve*. These probiotics have been reported to form robust biofilms in vitro, and shown to attach to various host biofilm sites. For example, *L. rhamnosus strain GG* has been reported to attach to the human intestinal tract, and to persist for at least a week following oral ingestion in a healthy adult population.[169] Additionally when pregnant women ingested *L. rhamnosus GG*, their respective infants retained colonization for up to 24 months.[170] Although probiotics form similar biofilm modalities as pathogens, research and evaluation of these "good" biofilms has only occurred in the last 10 years.

One of the initial studies examined the impact of various host factors on biofilms formed by *L. rhamnosus GG* and found that low concentrations of bile and mucin enhanced formation, while pH of 4 and high osmolarity decreased formation.[163] The ability of *L. plantarum* biofilms to withstand environmental stressors was determined by exposing 24 h planktonic cultures or biofilms to up to 11% acetic acid and 40% ethanol for 30 and 60 min respectively. While all *L. plantarum* planktonic cells were killed off at these concentrations, ~10% of the biofilm cells survived acetic acid treatment, and 0.5% survived ethanol treatment. Further exposure of *L. plantarum* biofilms to other antimicrobial chemicals suggests that probiotic biofilms similarly display increased levels of tolerance to antimicrobials as their pathogenic counterparts.[164] *L. reuteri* biofilms were shown to produce more reuterin, a broad spectrum antimicrobial, than planktonic cells. Additionally, while planktonic cultures decrease as reuterin concentrations increase, *L. reuteri* biofilms remain 92% viable following maximal reuterin production.[161]

Probiotics can also outcompete other pathogenic and environmental microbes in the biofilm mode of growth. For instance, when *L. plantarum* strain PA21 was cultured in a multispecies biofilm for 6 days with the mupirocin producing *Pseudomonas fluorescens*, pathogenic *Salmonella enterica*, the opportunistic pathogens *Aeromonas hydrophila* and *Bacillus cereus*, all other strains were reduced to undetectable numbers except for *S. enterica*. Although *S. enterica* biofilm cell numbers were reduced, this pathogen was able to persist in the presence of *L. plantarum* planktonic and biofilm cells.[171] These studies point to the validity of probiotic biofilms and the subsequent sections will explore the curious interactions of the immune response and these potentially therapeutic biofilms.

3.2 Neutrophils, Monocytes/Macrophages, and Probiotic Biofilms

Regardless of how bacteria are classified as commensal, probiotic, or pathogenic, the initial immune response is "one size fits all" when it comes to the stimulation and migration of PMNs to the site of invading bacteria. Monocytes circulate through the blood stream, chemotaxing to chemokines like chemotactic protein-1, which draw this cell population to the bodily threat. In the tissue, these cells eventually differentiate into macrophages that phagocytose bacterial cells, and efferocytose apoptotic host cells and other cellular debris. This section will thus focus on how these three critical cell populations, neutrophils, monocytes, and macrophages individually interact with probiotic biofilms (Fig. 2).

3.2.1 Neutrophils

While PMNs represent some of the first responders to bacterial stimuli, very few studies have examined the biological dialog between PMNs and probiotic biofilms. What has mainly been studied is how milk, fermented with planktonic probiotics and imbibed by healthy subjects, can enhance the following qualities of blood derived PMNs: phagocytic capacity, respiratory burst, and expression of complement receptor 3 and other receptors critical to phagocytosis.[172–174] These studies suggest that intake of planktonic probiotics enables PMNs to better fight off bacteria. Mixing *L. plantarum* supernatant with that of *P. aeruginosa* was also reported to directly protect human PMNs from *P. aeruginosa*-mediated apoptosis. This protective mechanism was linked to the destruction of *P. aeruginosa* quorum sensing chemical molecules present in the supernatant.[175]

One of the mechanisms by which neutrophils combat bacterial cells is through NETosis. Although NETs have been shown to be useful against pathogenic microbes, they have also been linked with chronic inflammation that can be detrimental to the human host. Probiotics have been shown to have a curious impact on NETs;when PMNs were activated with phorbol

12-myristate 13-acetate (PMA) or *S. aureus*, incubation with planktonic *L. rhamnosus GG* decreased ROS production and phagocytosis, and adding exogenous H_2O_2 did not restore NET initiation.[166] Unlike *S. aureus*, *L. rhamnosus GG* failed to induce NET formation without stimuli, while the *E. coli* probiotic strain N.1917 could stimulate NETs. In another study, a pathogenic strain of *E. coli* was shown to activate NETs in vitro, which caused damage to the integrity of intestinal epithelium.[176]

Diabetes, and specifically hyperglycemia, has also been reported to prime PMNs for NET formation, increasing both NETosis and blood derived markers of NETosis.[177,178] However, treatment with DNase 1 enhanced wound healing in diabetic and nondiabetic mice, suggesting that the prevention of NETosis, or reduction of extracellular DNA present in wounds, could be helpful.[17] It is also worth noting that DNase 1 expression is reduced in patients suffering from inflammatory bowel disease,[179] which could also be related to NET formation. Still, much remains to be elucidated about PMN and probiotic interactions including strain differences, planktonic versus biofilm cell types, and the impact of these probiotics in disease states where NETs cause collateral damage to the host.

3.2.2 Monocytes

There have been several interesting studies documenting the impact of probiotic biofilms on monocytes. The supernatants from four strains of *L. reuteri*, grown up planktonically or as biofilms, were incubated with the monocyte cell line THP-1, and TNF-α production was measured by ELISA. Curiously, supernatants from two *L. reuteri* biofilm strains failed to induce THP-1 TNF-α production, while supernatants from the respective planktonic cells did. However, when THP-1 cells were first incubated with LPS and then the biofilm or planktonic supernatants, planktonic supernatants were more effective at dampening TNF-α production.[161] Overall, the different strain types of *L. reuteri* were more demonstrative of the antiinflammatory capabilities of the supernatants than the cultured phenotype (planktonic vs biofilms).[161] The impact on TNF-α production in THP-1 cells was similarly evaluated using supernatants from either biofilm or planktonic cultures of *L. plantarum* (two strains) and *L. fermentum* (three strains).[180] In this study, when THP-1 cells were incubated with LPS, all supernatants significantly decreased TNF-α production. Unlike the prior study; however, biofilm supernatants caused a more drastic decrease in TNF-α production (~25–35% of the control), as compared to planktonic cells (~65% of the control).[180] Another study investigated the impact of biofilm supernatants from *L. casei* on THP-1 cells, and similarly found an anti-TNF-α effect, while the planktonic supernatant had no effect.[160] This antiinflammatory effect was shown to be dose dependent, and inactivated by both heat and proteases. As the nuclear factor-κB (NF-κB) signaling pathway regulates TNF-α expression and the activation of NF-κB is dependent on the phosphorylation of IκB,[181] the impact of supernatant on this protein was tested. Supernatants taken from *L. casei* biofilms decreased the phosphorylation of IκB in THP-1 cells, as compared to LPS alone, or LPS in combination with planktonic supernatant. In attempts to find a mechanism of action, the role of probiotic associated proteins p40 and p75 were tested and found not to influence TNF-α.[160] These proteins are associated with the *Lactobacillus* cell wall from *L. casei* and *L. rhamnosus* strains and have been reported to protect human intestinal epithelial cells from apoptosis.[160] However, when GroEL, a heat shock surface protein, was immunoprecipitated out of the supernatant, the inhibitory activity of biofilm supernatants was abrogated. Additionally, biofilm supernatants were found to produce a larger amount of GroEL, as compared to planktonic supernatant. GroEL is a chaperone protein that assists with protein folding in the cytoplasm, is common to many types of enteric bacteria, and has been associated with host cell adhesion.[182] The authors finalized their study by showing that biofilm supernatants from multiple strains of *L. casei* displayed similar inhibitory effects on TNF-α. Biofilms from other species of *Lactobacillus* were also able to inhibit TNF-α production; specifically *L.fermentum* and *L. plantarum*, although *L. plantarum* was not as inhibitory.[160] These studies suggest that the biofilm mode of growth may enhance the antiinflammatory properties of products produced by probiotics (*L. plantarum*, *L. fermentum*,[160,183] and *L. casei*[160]), though it seems to be strain dependent (*L. reuteri* biofilms not as antiinflammatory[161]). This highlights the importance of evaluating multiple probiotic species and strains, for even among closely related *Lactobacilli* there can be profound differences in antiinflammatory and antimicrobial activity.

3.2.3 Macrophages

The few studies with macrophages and *Lactobacillus spp.* biofilms have yielded similar antiinflammatory trends as with monocytes, with some additional interesting results. When two strains (GG and GR-1) of 48-h-old *L. rhamnosus* cells were incubated individually with LPS or *E.coli*-treated, immortalized, murine macrophages, the level of TNF-α decreased, as compared to control. *L. rhamnosus* incubated alone with these mouse macrophages slightly increased TNF-α, while the level of granulocyte-colony stimulating factor (G-CSF) was increased by both *L. rhamnosus* strains when compared to pathogens *E. coli* and *Enterococcus faecalis*. The authors went on to show that the TNF-α inhibition was mediated through G-CSF, which specifically activated STAT3 and inhibited c-Jun-N-terminal kinases. G-CSF stimulates neutrophils in the bone marrow, but also quells production of proinflammatory cytokines, proving that, while the probiotic biofilms may decrease inflammation, this does not leave the host defenseless.[184] This same group went on to a more detailed characterization of the effect of *L. rhamnosus* GR-1 on murine macrophages and found similarly that the presence of these cells resulted in lower levels of TNF-α, as well as GM-CSF, IL-6, and IL-12 and greatly increased G-CSF production.[185]

3.3 Probiotic Biofilms and Gut Immunity

Over 100 years ago, the father of cell immunology, Ilya Metchnikoff hypothesized that ingesting "good" bacteria could result in health-promoting benefits.[186] As such, probiotics have predominantly been studied in regard to the gastrointestinal tract. These ingested probiotics can inhibit other bacteria directly, form a biofilm in the colon, thereby blocking the adhesion of pathogens, and shift the immune response to lower inflammation by various mechanisms. One striking example of the antipathogenic activity of probiotics was when two bacteriocin producing probiotic strains, *L. plantarum* and *Enterococcuscasseliflavus*, were grown up as biofilms for 7 days, and the attached cells were coincubated with *Listeria monocytogenes* for 10 more days under refrigeration, the 17 day old beneficial biofilms displayed the following antilisterial activity: *L. plantarum* reduced the number of planktonic *Listeria* cells by 5.4 logs and attached *Listeria* cells by 3.9 logs, while *E. casseliflavus* reduced planktonic *Listeria* cells by 4.8 logs and attached *Listeria* cells by 3.7 logs. This inhibition of *L. monocytogenes* was linked to the production of bacteriocins, rather than pH or other mechanisms.[187] It has been suggested that ingestion of probiotics leads to the integration of these bacteria into the microbiome; however, the longevity and effectiveness of attachment seems to depend on the current commensal population strain and species of the ingested probiotic.[169]

Species importance was highlighted in a study that examined the colonic cell binding ability of 14 strains isolated from a pulque, a fermented drink common to Mexico. *L. plantarum* (2/3 strains tested) and *L. paracase* (1/3 strains tested) displayed maximal adhesion to the mucous-producing human colorectal cell line HT-29-MTX, as compared to *L. casei*.[183] Similarly, fluorescent labeled *L. plantarum* displayed increased capacity to adhere in the gut of germ-free zebrafish larvae, as compared to *L. fermentum*, when they were exposed to *Lactobacillus* over 72 h.[188] Adherence of the two *L. plantarum* strains peaked at 6 and 24 h respectively, but by 72 h the fluorescence had diminished to very low levels, suggesting that ingested probiotics display temporary colonization. Another study showed that a GFP-labelled *L. casei* could adhere to the gut of germ free zebrafish larvae and persist for 4 days, as visualized by fluorescent and transmission electron microscopy.[160]

While the previously mentioned studies used cells or animals that were germ free, one study examined the ability of probiotics to colonize a variety of hosts with preestablished gut commensal floras.[189] Utilizing fluorescent in situ hybridization (FISH), preexisting and exogenously applied *Lactobacillus* gastrointestinal biofilms were visualized in murine, avian, and human hosts. In mice, FISH revealed robust commensal biofilms in the forestomach, and a *Lactobacillus* probe revealed that the bulk of these biofilms were composed of *Lactobacillus* cells. Murine colon samples revealed that the commensals formed thin biofilms. The crop of chickens also displayed strong, *Lactobacillus spp.* rich biofilms, while imaging of illeal biopsy samples from patients suffering from pouchitis revealed dispersed, *Lactobacilli*-depleted, cell populations in the mucus layer. Addition of exogenous *L. rhamnosus GG* in humans (4 weeks), and *Lactobacillus brevis* (3 days) in the animal populations failed to form biofilms in already colonized hosts.[189] These data bring into question the mechanism of action of ingested probiotics, and whether biofilm formation by ingested *Lactobacilli* is necessary for their therapeutic effects. However, this study does build a case for already established *Lactobacillus* biofilms in the gut acting as innate immune protectors of the underlying epithelial cells, excluders of pathogens, and maintainers of a low pH.

To evaluate the antiinflammatory properties in zebrafish larvae, they were first treated with 2,4,6-trinitrobenzenesulfonic acid (TNBS), which enhances gut inflammation by specifically increasing the production of NO, TNF-α, and IL-1β.[160] When zebrafish harboring 1-day old *L. casei* gut biofilms were treated with TNBS, RT-PCR revealed significantly lower levels of TNF-α and IL-1β mRNA than those that did not harbor biofilms. In addition, *L. casei*-treated zebrafish larvae displayed increased IL-10 transcripts, as compared to germ-free larvae. Flow cytometry revealed that while global macrophage recruitment was enhanced in TNBS-treated larvae, TNBS treatment in the larvae with 1-day *L. casei* biofilms resulted in an abrogation of recruited macrophages. The reduction in macrophages in *L. casei* zebrafish was also observed 3 days after TNBS treatment, in comparison to controls.[160] These data suggest that the presence of *L. casei* lowers inflammation in vivo.

Another study examined the antiinflammatory properties of 14 *Lactobacillus* strains isolated from pulque. They began by testing whether biofilm cells alone could modulate IL-8 production in HT-29 colon cells activated with TNF-α.[183] Following 6 h of incubation, *L. plantarum* (3/4 strains), and *Lactobacillus composti* (1/1 strains) significantly suppressed the production of IL-8 in HT-29 cells that had been stimulated with 5 or 10 ng of TNF-α. When these four antiinflammatory strains were inactivated with UV treatment and incubated with TNF-α stimulated HT-29 cells, IL-8 was not as strongly inhibited. However, the dead *Lactobacilli* still caused up to a 30% decrease in IL-8 production, suggesting the mechanism of action is linked a cell surface protein.[183] Three antiinflammatory strains (*L. plantarum*, *L. composti*, and *L. sanfranciscensis*) were then tested, at a concentration of 10^9 CFU, in a chronic colitis mouse model, after which weight loss, gut permeability, and cytokine production was monitored. Overall, mice given the *Lactobacillus* exhibited more weight loss, decreased gut permeability, and lower levels of several cytokines commonly associated with patients suffering from inflammatory bowel disease (namely IFN-γ, IL-22, IL-17A, and IL-5), in comparison to the untreated mice. Although the overall outcome was similar in mice given the three different *Lactobacillus* treatments, the *L. composti* and *L. sanfranciscensis* treatments led to a broader and more significant drop in proinflammatory cytokines.[183] Taken together, these studies provide support that Lactobacilli are an important component of the normal innate gut flora, and that the absence of these antiinflammatory biofilms can contribute to IBD and other related diseases.

3.4 Probiotic Biofilms and Vaginal Immunity

The Lactobacilli biofilms on vaginal epithelium highlights the importance of these beneficial biofilms in protecting the host tissue and serving as a component of the innate immune response. The vaginal microbiome of healthy females is dominated by *Lactobacillus crispatus*, *L. jensenii*, *L. gasseri*, and *L. iners*.[190] Under *Lactobacilli*-rich conditions, these biofilms can contribute to protection from sexually transmitted microbes and urinary tract infections, in addition to imparting better outcomes in pregnancy and delivery. These biofilms protect the vaginal epithelium by creating an acidic environment (~pH 4–4.5), competitive exclusion, and the release of antimicrobial compounds such as H_2O_2, bacteriocins, and other byproducts. However, when the normal vaginal commensals are displaced by anaerobic bacteria and fungal pathogens, this results in an increased pH (above 4.5), resulting in vaginosis. This compromised state of the vagina has been associated with increased incidences of preterm delivery, contraction of the human immunodeficiency virus, and pelvic inflammatory disease.[191]

Vaginosis can be caused by bacterial or fungal pathogens. The ability of several different pathogens to displace protective biofilms was examined by growing *L. crispatus* biofilms on glass slides and then incubating the slides with 10^9 CFU of five different causative bacterial agents of vaginosis. Each pathogen was able to attach to the slides, but did not readily displace the *L. crispatus* biofilms after a 30-min incubation. The anaerobic pathogen *Mobiluncus mulieris*, caused 23% of the *L. crispatus* biofilms to detach while the four other pathogens caused ~10% detachment respectively.[192] Repeating these studies in cervical cells showed a reverse correlation, for when *L. crispatus* biofilms were incubated with HeLa cells and then exposed to a pathogenic and commensal strain of *Gardonella vaginalis*, there were drastic reductions in the attached biofilm cells. Conversely, *L. crispatus* effectively prevented both *G. vaginalis* strains from attaching to the HeLa cells, resulting in only ~7% attachment when low inoculums of *G. vaginalis* were incubated with *L. crispatus* biofilms. Interestingly, the presence of *L. iners* enhanced attachment of the pathogenic *G. vaginalis*, and was less susceptible to displacement by both strains, as compared to *L. crispatus*.[193] This result could help explain why women who are colonized with *L. iners* are often more susceptible to bacterial vaginosis. Following treatment of bacterial vaginosis in patients, *L. jensenii* biofilms were isolated with vaginal epithelial cells and imaged with wet mount microscopy, displaying the clinical presence of these biofilms.[190]

In another study, *Lactobacillus* isolates from healthy versus fungal vaginosis patients were compared for their production of antimicrobial substances, and their ability to inhibit various *Candida* species. Overall the authors found that *Lactobacillus* isolates from either diseased or healthy patients produced similar amounts of lactic acid and hydrogen peroxide, while the isolates from diseased patients produced increased levels of diacetyl. Curiously, though, the healthy patient isolates were able to inhibit a maximum of 68% of the 35 different *Candida spp.* tested, while the isolates from the diseased patients only inhibited a maximum of 5.8% of the *Candida spp.*[194] These data suggest that the *Lactobacillus* isolates from healthy patients are producing more inhibitory compounds, which are most likely bacteriocin, or pH, related. They also suggest that *Lactobacilli* don't necessarily adapt by producing more antimicrobial compounds, when previously exposed to the pathogen. It would be interesting to test this hypothesis by continuously exposing a *Lactobacillus* biofilm to a bacterial or fungal pathogen to see if one could naturally select for more effective beneficial commensals that have been "immunized" against various pathogens then reintroduced to the host.

With the presence of *Lactobacillus* biofilms in the vagina, it also begs the question how do these beneficial biofilms interact with sperm, eggs, fertilization, and the development of the fetus? The investigation into these questions has been primarily addressed in regard to in vitro fertilization (IVF), with the overarching hypothesis that microbes adversely impact embryo transfer and ultimately a successful pregnancy.[195] In a study that determined the microbiome in the follicular fluid from 263 women, some curious correlations with the presence of *Lactobacillus* biofilms and reproduction were found. The follicular fluid is the liquid found in ovaries, which lubricates the eggs. Surprisingly, the authors noted that this fluid was often contaminated with *Lactobacillus*; the left ovary was dominated with *L. crispatus* and *L. gasseri*, while the right ovary housed *L. iners*. The presence of *Lactobacillus* in the left and right follicular fluids was associated with improved embryo transfer rates, and ultimately pregnancy outcomes. Similarly, a prior study in 91 women found that the presence of hydrogen peroxide-producing *Lactobacillus* in the vagina or embryo transfer catheter tip resulted in an increase in successful births.[196]

The placenta is proposed to protect the fetus against numerous assaults, including microbial invasion. However, pathogenic biofilms have even been identified in the placenta,[197] with an absence of *Lactobacillus*, and this was correlated with negative pregnancy outcomes. An analysis of 34 human placental samples, evaluated for the presence of *Lactobacillus spp.* or *Bifidobacterium spp.*, returned only nonculturable organisms. However, PCR revealed the presence of *Bifidobacteria* and *L. rhamnosus* DNA in over 90% of placental samples. This suggests that the mother transfers beneficial biofilm immunogens through the placenta that may contribute to the development of the fetal immunity.[198] Further culturing of a larger number of placentas from premature births (23–27 weeks gestation) revealed interesting correlations with *Lactobacillus* biofilms and newborn immune responses.[199] From the 527 placentas cultured, only 214 came back culture positive and fetal blood taken at days 1–3 was then analyzed for 25 proinflammatory markers and correlated with colonization of the detected microorganisms.

Lactobacillus biofilms isolated alone made up only 3.7% of the culture-positive placentas, and were correlated with undetectable levels of 13/25 proinflammatory cytokines found with bacterial vaginosis pathogens. *Lactobacillus* biofilms were also found in combination with other microbes making up 6% of the samples, and were able to lower the odds ratio for most proinflammatory markers compared to vaginal pathogens.[199]

To examine the impact of conditioned media from *L. rhamnosus* GR-1 biofilms on fetal inflammation, tissue was extracted from the placenta of healthy women undergoing term cesarean sections and the cells were separated and digested to establish placental trophoblast cell cultures. These cells were then stimulated with LPS, and one group was pretreated with *L. rhamnosus* GR-1 biofilm supernatant at a 1:20 dilution. Evaluation with ELISA revealed that the conditioned media inhibited TNF-α production, while upregulating the antiinflammatory cytokine 1L-10. Bacterial vaginosis increases prostaglandin expression, which can lead to early contractions and ultimately preterm birth[200]; therefore, sex specific prostaglandin enzymes were examined. Western blot analysis showed that *L. rhamnosus* supernatant led to decrease in PTGS2 (prostaglandin producing enzyme) by 30% during pregnancies in which the fetus was male. Interestingly, when the fetus was female, these supernatants caused an increase in PGDH, which metabolizes prostaglandin. So, for both sexes of fetus, the *Lactobacillus* biofilm conditioned media reduced, directly or indirectly, prostaglandin production, which theoretically should lead to better delivery outcomes in women suffering from bacterial vaginosis.[201] Similarly, the authors also tested the impact of *L. rhamnosus* supernatant on colony stimulating factor 3 (CSF3), a cytokine that primarily regulates the production of neutrophils, which are key in pathogenic bacteria clearance. Overall they found that the supernatant alone, or in combination with LPS, significantly enhanced CSF3 in female, but not male births. This finding suggests that there may be sex-specific responses to bacterial infection.[202]

L. rhamnosus supernatant was also evaluated in a pregnant mouse model to test if the supernatant could reduce preterm birth and modulate the immune response. Surprisingly, the supernatant did not impact LPS induced preterm birth, but did reduce proinflammatory cytokines maximally in the endometrium and maternal plasma.[203] Lactobacilli biofilms are well known to be important components of the vaginal innate immunity, but more studies are needed to elucidate specific interactions of these biofilms with the epithelium and other cell types. These and other recent advances in microbiome research have provided exciting results in terms of *Lactobacilli* and reproduction, and we are just beginning to understand the immune dynamics between fetal tissue and these beneficial biofilms.[191]

3.5 Probiotic Biofilms and Skin Immunity

While *Lactobacillus* biofilms do form on vaginal epithelium, in other regions of the skin these particular biofilms are absent. However, the overall pH of the skin ranges from 4 to 6 and recent findings have suggested that *Lactobacillus* cells could serve as an effective therapeutic in wound infections through antimicrobial and antiinflammatory mechanisms. When 10% *Bifidobacterium longum* lysates were applied to patients suffering from reactive skin, sensitivity decreased, as did transepidermal water loss.[204] This contrasts with the effect of pathogenic biofilms in a porcine wound infection model, which increased transepidermal water loss. These findings suggest that pathogenic biofilms may weaken the skin while *Bifidobacterium* lysate can potentially strengthen this barrier.[205] *L. rhamnosus* has been shown to adhere to human keratin in vitro; however, it could not prevent *S. aureus* from attaching to keratin. This is possibly because a mature biofilm had not formed yet on the keratin as the *L. rhamnosus* was only incubated for 1 h with the keratin, and thus was unable to effectively block *S. aureus* attachment.[206]

Utilizing primary human keratinocytes and various strains of *Lactobacillus* biofilms led to protection of human cells. Specifically, when *L. reuteri* and *L. rhamnosus* were applied before, or concurrently with, *S. aureus* they provided significant protection from pathogen induced cell death. However, when *S. aureus* cells were incubated with keratinocytes, *Lactobacillus* strains were unable to protect the skin cells. Further investigation into *L. reuteri*'s protective mechanism revealed that these cells did not inhibit *S. aureus* growth, but seemed to competitively exclude *S. aureus* via blocking the $\alpha 5\beta 1$ integrin.[207] Further investigation with *L. rhamnosus* GG in this model showed that whole cell, supernatants, and lysates applied prior to or simultaneously with *S. aureus* resulted in significant protection from *S. aureus*. Additionally, the lysate or live cells could dislodge *S. aureus* from keratinocytes.[208] This group went on to show that probiotic biofilm lysates enhanced tight junctions between human primary keratinocytes following 24 h incubation. This proved to be true for *B. longum*, *L. plantarum*, *L. reuteri*, and *L. rhamnosus* GG, while *L. fermentum* lysates decreased the viability of the cells.[209] In a keratinocyte monolayer scratch assay, application of lysates from *L. reuteri* and *L. rhamnosus* GG displayed enhanced rates of reepithelization. *L. rhamnosus* GG lysate displayed maximal rates of epithelization and subsequent microarray analysis from scratches where lysate had been applied revealed increased levels of CXCL2 expression. This chemokine stimulates keratinocyte growth and migration, and is critical to normal wound healing.[210]

The impact of *L. brevis* and *L. plantarum* biofilms on wound healing and the expression of inflammatory markers was also explored in an uninfected rat wound model.[211] The control groups were treated either with eucerin, which is a skin ointment often sold under the product name Aquaphor, or nothing at all. The authors specifically picked these two *Lactobacilli* for their increased ability to produce EPS, thus the final ointment preparations contained biofilms with EPS and viable *Lactobacilli*. Wound closure and the number of neutrophils, macrophages, and fibroblasts were enumerated at days 1, 3, 7, 14, and 21 postinjury and treatment. At all time points but day 1, wound closure was improved by both *Lactobacilli*, as compared to either control. At day 3–21, the numbers of neutrophils were significantly lower in the *Lactobacillus*-treated group, while macrophages were increased at day 3, but then lower through the course of the experiment. Fibroblasts were observed in a similar pattern, with high numbers at days 3 and 7, but numbers lower than the negative controls thereafter.[211] While *Lactobacillus* and Bifidobacterium are not common members of the normal skin flora (except vaginal epithelium), their ability to strengthen skin (decreasing water loss and enhancing tight junctions), protect and stimulate keratinocytes, and decrease inflammation provide strong support for their use as candidate wound infection therapeutics.

ACKNOWLEDGMENTS

We thank Katie Farris for contributing to the technical illustrations. Biofilm work in the KPR laboratory is supported in part by grant AI105763 from the National Institute of Allergy and Infectious Diseases and grant 62507-LS from the US Army Research Office. Probiotic and pathogenic biofilm research in the CMW laboratory is supported in part by US Navy Research and Development Command work unit number A1601. The views expressed in this article are those of the authors and do not necessarily reflect the official policy of the Department of the Navy, Department of Defense, or the US Government. The authors CMW and DKB are military service members of the US Government. This work was prepared as part of their official duties. Title 17 USC §105 provides that "Copyright protection under this title is not available for any work of the US Government." Title 17 USC §101 defines a US Government work as a work prepared by a military service member or employee of the US Government as part of that person's official duties.

REFERENCES

1. Flemming HC, Wingender J. The biofilm matrix. *Nat Rev Microbiol.* 2010;8(9):623–633.
2. Wingender J, Strathmann M, Rode A, Leis A, Flemming HC. Isolation and biochemical characterization of extracellular polymeric substances from *Pseudomonas aeruginosa. Methods Enzymol.* 2001;336:302–314.
3. Bales PM, Renke EM, May SL, Shen Y, Nelson DC. Purification and characterization of biofilm-associated EPS exopolysaccharides from ESKAPE organisms and other pathogens. *PLoS One.* 2013;8(6):e67950.
4. Limoli DH, Jones CJ, Wozniak DJ. Bacterial extracellular polysaccharides in biofilm formation and function. *Microbiol Spectr.* 2015;3(3).
5. Murofushi Y, Villena J, Morie K, et al. The toll-like receptor family protein RP105/MD1 complex is involved in the immunoregulatory effect of exopolysaccharides from *Lactobacillus plantarum* N14. *Mol Immunol.* 2015;64(1):63–75.
6. Raffatellu M, Chessa D, Wilson RP, Dusold R, Rubino S, Baumler AJ. The Vi capsular antigen of *Salmonella enterica* serotype Typhi reduces Toll-like receptor-dependent interleukin-8 expression in the intestinal mucosa. *Infect Immun.* 2005;73(6):3367–3374.
7. Bylund J, Burgess LA, Cescutti P, Ernst RK, Speert DP. Exopolysaccharides from *Burkholderia cenocepacia* inhibit neutrophil chemotaxis and scavenge reactive oxygen species. *J Biol Chem.* 2006;281(5):2526–2532.
8. Pier GB, Coleman F, Grout M, Franklin M, Ohman DE. Role of alginate O acetylation in resistance of mucoid *Pseudomonas aeruginosa* to opsonic phagocytosis. *Infect Immun.* 2001;69(3):1895–1901.
9. Leid JG, Willson CJ, Shirtliff ME, Hassett DJ, Parsek MR, Jeffers AK. The exopolysaccharide alginate protects *Pseudomonas aeruginosa* biofilm bacteria from IFN-gamma-mediated macrophage killing. *J Immunol.* 2005;175(11):7512–7518.
10. Aslam SN, Newman MA, Erbs G, et al. Bacterial polysaccharides suppress induced innate immunity by calcium chelation. *Curr Biol.* 2008;18(14):1078–1083.
11. Lin MH, Yang YL, Chen YP, et al. A novel exopolysaccharide from the biofilm of Thermus aquaticus YT-1 induces the immune response through Toll-like receptor 2. *J Biol Chem.* 2011;286(20):17736–17745.
12. Graveline R, Segura M, Radzioch D, Gottschalk M. TLR2-dependent recognition of *Streptococcus suis* is modulated by the presence of capsular polysaccharide which modifies macrophage responsiveness. *Int Immunol.* 2007;19(4):375–389.
13. Zhao G, Usui ML, Lippman SI, et al. Biofilms and inflammation in chronic wounds. *Adv Wound Care.* 2013;2(7):389–399.
14. Attinger C, Wolcott R. Clinically Addressing Biofilm in Chronic Wounds. *Adv Wound Care.* 2012;1(3):127–132.
15. Lovewell RR, Patankar YR, Berwin B. Mechanisms of phagocytosis and host clearance of *Pseudomonas aeruginosa. Am J Physiol Lung Cell Mol Physiol.* 2014;306(7):L591–L603.
16. Mishra M, Byrd MS, Sergeant S, et al. *Pseudomonas aeruginosa* Psl polysaccharide reduces neutrophil phagocytosis and the oxidative response by limiting complement-mediated opsonization. *Cellular Microbiol.* 2012;14(1):95–106.
17. Okshevsky M, Meyer RL. The role of extracellular DNA in the establishment, maintenance and perpetuation of bacterial biofilms. *Crit Rev Microbiol.* 2015;41(3):341–352.
18. Das T, Sehar S, Manefield M. The roles of extracellular DNA in the structural integrity of extracellular polymeric substance and bacterial biofilm development. *Environ Microbiol Rep.* 2013;5(6):778–786.
19. Jakubovics NS, Shields RC, Rajarajan N, Burgess JG. Life after death: the critical role of extracellular DNA in microbial biofilms. *Lett Appl Microbiol.* 2013;57(6):467–475.
20. Whitchurch CB, Tolker-Nielsen T, Ragas PC, Mattick JS. Extracellular DNA required for bacterial biofilm formation. *Science.* 2002;295(5559):1487.

21. Alhede M, Bjarnsholt T, Givskov M, Alhede M. *Pseudomonas aeruginosa* biofilms: mechanisms of immune evasion. *Adv Appl Microbiol.* 2014;86:1–40.
22. Krieg AM. CpG motifs in bacterial DNA and their immune effects. *Annu Rev Immunol.* 2002;20:709–760.
23. Krieg AM, Wagner H. Causing a commotion in the blood: immunotherapy progresses from bacteria to bacterial DNA. *Immunol Today.* 2000;21(10):521–526.
24. Krieg AM, Yi AK, Matson S, et al. CpG motifs in bacterial DNA trigger direct B-cell activation. *Nature.* 1995;374(6522):546–549.
25. Trevani AS, Chorny A, Salamone G, et al. Bacterial DNA activates human neutrophils by a CpG-independent pathway. *Eu J Immunol.* 2003;33(11):3164–3174.
26. Jensen PO, Bjarnsholt T, Phipps R, et al. Rapid necrotic killing of polymorphonuclear leukocytes is caused by quorum-sensing-controlled production of rhamnolipid by *Pseudomonas aeruginosa. Microbiology.* 2007;153(Pt 5):1329–1338.
27. Bjarnsholt T, Jensen PO, Fiandaca MJ, et al. *Pseudomonas aeruginosa* biofilms in the respiratory tract of cystic fibrosis patients. *Pediatr Pulmonol.* 2009;44(6):547–558.
28. Fuxman Bass JI, Russo DM, Gabelloni ML, et al. Extracellular DNA: a major proinflammatory component of *Pseudomonas aeruginosa* biofilms. *J Immunol.* 2010;184 (11):6386–6395.
29. Mulcahy H, Charron-Mazenod L, Lewenza S. Extracellular DNA chelates cations and induces antibiotic resistance in *Pseudomonas aeruginosa* biofilms. *PLoS Pathog.* 2008;4 (11):e1000213.
30. Brinkmann V, Reichard U, Goosmann C, et al. Neutrophil extracellular traps kill bacteria. *Science.* 2004;303(5663):1532–1535.
31. Jiao Y, Cody GD, Harding AK, et al. Characterization of extracellular polymeric substances from acidophilic microbial biofilms. *Appl Environ Microbiol.* 2010;76 (9):2916–2922.
32. Lasa I, Penades JR. Bap: a family of surface proteins involved in biofilm formation. *Res Microbiol.* 2006;157(2):99–107.
33. Speziale P, Pietrocola G, Foster TJ, Geoghegan JA. Protein-based biofilm matrices in Staphylococci. *Front Cell Infect Microbiol.* 2014;4:171.
34. Muthukrishnan G, Quinn GA, Lamers RP, et al. Exoproteome of *Staphylococcus aureus* reveals putative determinants of nasal carriage. *J Proteome Res.* 2011;10(4):2064–2078.
35. Zhang X, Bishop PL. Biodegradability of biofilm extracellular polymeric substances. *Chemosphere.* 2003;50(1):63–69.
36. Kaplan JB. Biofilm dispersal: mechanisms, clinical implications, and potential therapeutic uses. *J Dent Res.* 2010;89(3):205–218.
37. Goodman SD, Obergfell KP, Jurcisek JA, et al. Biofilms can be dispersed by focusing the immune system on a common family of bacterial nucleoid-associated proteins. *Mucosal Immunol.* 2011;4(6):625–637.
38. Borlee BR, Goldman AD, Murakami K, Samudrala R, Wozniak DJ, Parsek MR. *Pseudomonas aeruginosa* uses a cyclic-di-GMP-regulated adhesin to reinforce the biofilm extracellular matrix. *Mol Microbiol.* 2010;75(4):827–842.
39. Passmore IJ, Nishikawa K, Lilley KS, Bowden SD, Chung JC, Welch M. Mep72, a metzincin protease that is preferentially secreted by biofilms of *Pseudomonas aeruginosa. J Bacteriol.* 2015;197(4):762–773.
40. Lourenco A, de Las Heras A, Scortti M, Vazquez-Boland J, Frank JF, Brito L. Comparison of *Listeria monocytogenes* Exoproteomes from biofilm and planktonic state: Lmo2504, a protein associated with biofilms. *Appl Environ Microbiol.* 2013;79 (19):6075–6082.
41. Dharmaprakash A, Mutt E, Jaleel A, Ramanathan S, Thomas S. Proteome profile of a pandemic *Vibrio parahaemolyticus* SC192 strain in the planktonic and biofilm condition. *Biofouling.* 2014;30(6):729–739.

42. Carvalhais V, Cerveira F, Vilanova M, Cerca N, Vitorino R. An immunoproteomic approach for characterization of dormancy within *Staphylococcus epidermidis* biofilms. *Mol Immunol.* 2015;65(2):429–435.
43. Yi L, Wang Y, Ma Z, et al. Biofilm formation of *Streptococcusequi* ssp. zooepidemicus and comparative proteomic analysis of biofilm and planktonic cells. *Curr Microbiol.* 2014;69 (3):227–233.
44. Zijnge V, Kieselbach T, Oscarsson J. Proteomics of protein secretion by *Aggregatibacter actinomycetemcomitans*. *PLoS One.* 2012;7(7):e41662.
45. Panaro MA, Acquafredda A, Sisto M, Lisi S, Maffione AB, Mitolo V. Biological role of the N-formyl peptide receptors. *Immunopharmacol Immunotoxicol.* 2006;28(1):103–127.
46. He HQ, Troksa EL, Caltabiano G, Pardo L, Ye RD. Structural determinants for the interaction of formyl peptide receptor 2 with peptide ligands. *J Biol Chem.* 2014;289 (4):2295–2306.
47. Schiffmann E, Corcoran BA, Wahl SM. N-formylmethionyl peptides as chemoattractants for leucocytes. *Proc Natl Acad Sci USA.* 1975;72(3):1059–1062.
48. Nauseef WM. How human neutrophils kill and degrade microbes: an integrated view. *Immunol Rev.* 2007;219:88–102.
49. Zhang Q, Raoof M, Chen Y, et al. Circulating mitochondrial DAMPs cause inflammatory responses to injury. *Nature.* 2010;464(7285):104–107.
50. Takeda K, Akira S. Toll-like receptors in innate immunity. *Int Immunol.* 2005;17 (1):1–14.
51. Brady RA, Leid JG, Camper AK, Costerton JW, Shirtliff ME. Identification of *Staphylococcus aureus* proteins recognized by the antibody-mediated immune response to a biofilm infection. *Infect Immun.* 2006;74(6):3415–3426.
52. Enany S, Yoshida Y, Magdeldin S, et al. Two dimensional electrophoresis of the exoproteome produced from community acquired methicillin resistant *Staphylococcusaureus* belonging to clonal complex 80. *Microbiol Res.* 2013;168(8):504–511.
53. Gil C, Solano C, Burgui S, et al. Biofilm matrix exoproteins induce a protective immune response against *Staphylococcus aureus* biofilm infection. *Infect Immun.* 2014;82 (3):1017–1029.
54. Sadowska B, Wieckowska-Szakiel M, Paszkiewicz M, Rozalska B. The immunomodulatory activity of *Staphylococcus aureus* products derived from biofilm and planktonic cultures. *Arch Immunol Ther Exp.* 2013;61(5):413–420.
55. Sanui T, Gregory RL. Analysis of Streptococcus mutans biofilm proteins recognized by salivary immunoglobulin A. *Oral Microbiol Immunol.* 2009;24(5):361–368.
56. Purschke FG, Hiller E, Trick I, Rupp S. Flexible survival strategies of *Pseudomonas aeruginosa* in biofilms result in increased fitness compared with *Candida albicans*. *Mol Cell Proteomics.* 2012;11(12):1652–1669.
57. Bostanci N, Bao K, Wahlander A, Grossmann J, Thurnheer T, Belibasakis GN. Secretome of gingival epithelium in response to subgingival biofilms. *Mol Oral Microbiol.* 2015;30(4):323–335.
58. Schooling SR, Beveridge TJ. Membrane vesicles: an overlooked component of the matrices of biofilms. *J Bacteriol.* 2006;188(16):5945–5957.
59. Couto N, Schooling SR, Dutcher JR, Barber J. Proteome profiles of outer membrane vesicles and extracellular matrix of *Pseudomonas aeruginosa* biofilms. *J Proteome Res.* 2015;14(10):4207–4222.
60. Ellis TN, Leiman SA, Kuehn MJ. Naturally produced outer membrane vesicles from *Pseudomonas aeruginosa* elicit a potent innate immune response via combined sensing of both lipopolysaccharide and protein components. *Infect Immun.* 2010;78 (9):3822–3831.
61. Kulp A, Kuehn MJ. Biological functions and biogenesis of secreted bacterial outer membrane vesicles. *Annu Rev Microbiol.* 2010;64:163–184.

62. Toyofuku M, Roschitzki B, Riedel K, Eberl L. Identification of proteins associated with the *Pseudomonas aeruginosa* biofilm extracellular matrix. *J Proteome Res.* 2012;11 (10):4906–4915.
63. Ng WL, Bassler BL. Bacterial quorum-sensing network architectures. *Annu Rev Genet.* 2009;43:197–222.
64. Everett JA, Rumbaugh KP. *Biofilms, Quorum Sensing and Crosstalk in Medically Important Microbes.* In: Elsevier; 2015:235–247. Molecular Medical Microbiology; vol. 1
65. Rumbaugh KP, Kaufmann GF. Exploitation of host signaling pathways by microbial quorum sensing signals. *Curr Opin Microbiol.* 2012;15(2):162–168.
66. Lowery CA, Dickerson TJ, Janda KD. Interspecies and interkingdom communication mediated by bacterial quorum sensing. *Chem Soc Rev.* 2008;37(7):1337–1346.
67. Massion PP, Inoue H, Richman-Eisenstat J, et al. Novel Pseudomonas product stimulates interleukin-8 production in airway epithelial cells in vitro. *J Clin Invest.* 1994;93(1):26–32.
68. Smith RS, Fedyk ER, Springer TA, Mukaida N, Iglewski BH, Phipps RP. IL-8 production in human lung fibroblasts and epithelial cells activated by the *Pseudomonas autoinducer* N-3-oxododecanoyl homoserine lactone is transcriptionally regulated by NF-kappa B and activator protein-2. *J Immunol.* 2001;167(1):366–374.
69. Shiner EK, Terentyev D, Bryan A, et al. *Pseudomonas aeruginosa* autoinducer modulates host cell responses through calcium signalling. *Cell Microbiol.* 2006;8(10):1601–1610.
70. Tateda K, Ishii Y, Horikawa M, et al. The *Pseudomonas aeruginosa* autoinducer N-3-oxododecanoyl homoserine lactone accelerates apoptosis in macrophages and neutrophils. *Infect Immun.* 2003;71(10):5785–5793.
71. Schwarzer C, Wong S, Shi J, et al. *Pseudomonas aeruginosa* Homoserine lactone activates store-operated cAMP and cystic fibrosis transmembrane regulator-dependent Cl- secretion by human airway epithelia. *J Biol Chem.* 2010;285(45):34850–34863.
72. Holban AM, Bleotu C, Chifiriuc MC, Bezirtzoglou E, Lazar V. Role of *Pseudomonas aeruginosa* quorum sensing (QS) molecules on the viability and cytokine profile of human mesenchymal stem cells. *Virulence.* 2014;5(2):303–310.
73. Li H, Wang L, Ye L, et al. Influence of *Pseudomonas aeruginosa* quorum sensing signal molecule N-(3-oxododecanoyl) homoserine lactone on mast cells. *Med Microbiol Immunol.* 2009;198(2):113–121.
74. Jacobi CA, Schiffner F, Henkel M, et al. Effects of bacterial N-acyl homoserine lactones on human Jurkat T lymphocytes-OdDHL induces apoptosis via the mitochondrial pathway. *Int J Med Microbiol.* 2009;299(7):509–519.
75. Kravchenko VV, Kaufmann GF, Mathison JC, et al. N-(3-Oxo-acyl)homoserine lactones signal cell activation through a mechanism distinct from the canonical pathogen-associated molecular pattern recognition receptor pathways. *J Biol Chem.* 2006;281 (39):28822–28830.
76. Sperandio V, Torres AG, Jarvis B, Nataro JP, Kaper JB. Bacteria-host communication: the language of hormones. *Proc Natl Acad Sci USA.* 2003;100(15):8951–8956.
77. Lyte M. Microbial endocrinology and infectious disease in the 21st century. *Trends Microbiol.* 2004;12(1):14–20.
78. Shiner EK, Rumbaugh KP, Williams SC. Interkingdom signaling: deciphering the language of acyl homoserine lactones. *FEMS Microbiol Rev.* 2005;29(5):935–947.
79. Rumbaugh KP. Convergence of hormones and autoinducers at the host/pathogen interface. *Anal Bioanal Chem.* 2007;387(2):425–435.
80. Jahoor A, Patel R, Bryan A, et al. Peroxisome proliferator-activated receptors mediate host cell proinflammatory responses to *Pseudomonas aeruginosa* autoinducer. *J Bacteriol.* 2008;190(13):4408–4415.
81. Cooley MA, Whittall C, Rolph MS. Pseudomonas signal molecule 3-oxo-C12-homoserine lactone interferes with binding of rosiglitazone to human PPARgamma. *Microbes Infect.* 2010;12(3):231–237.

82. Lee RJ, Xiong G, Kofonow JM, et al. T2R38 taste receptor polymorphisms underlie susceptibility to upper respiratory infection. *J Clin Invest.* 2012;122(11):4145–4159.
83. Karlsson T, Musse F, Magnusson KE, Vikstrom E. N-Acylhomoserine lactones are potent neutrophil chemoattractants that act via calcium mobilization and actin remodeling. *J Leukoc Biol.* 2012;91(1):15–26.
84. Romling U, Galperin MY, Gomelsky M. Cyclic di-GMP: the first 25 years of a universal bacterial second messenger. *Microbiol Mol Biol Rev.* 2013;77(1):1–52.
85. Jenal U, Malone J. Mechanisms of cyclic-di-GMP signaling in bacteria. *Annu Rev Genet.* 2006;40:385–407.
86. Romling U. Cyclic di-GMP, an established secondary messenger still speeding up. *Environ Microbiol.* 2012;14(8):1817–1829.
87. Commichau FM, Dickmanns A, Gundlach J, Ficner R, Stulke J. A jack of all trades: the multiple roles of the unique essential second messenger cyclic di-AMP. *Mol Microbiol.* 2015;97(2):189–204.
88. Valle J, Solano C, Garcia B, Toledo-Arana A, Lasa I. Biofilm switch and immune response determinants at early stages of infection. *Trends Microbiol.* 2013;21(8):364–371.
89. McWhirter SM, Barbalat R, Monroe KM, et al. A host type I interferon response is induced by cytosolic sensing of the bacterial second messenger cyclic-di-GMP. *J Exp Med.* 2009;206(9):1899–1911.
90. Woodward JJ, Iavarone AT, Portnoy DA. c-di-AMP secreted by intracellular *Listeria monocytogenes* activates a host type I interferon response. *Science.* 2010;328(5986):1703–1705.
91. Jayaseelan S, Ramaswamy D, Dharmaraj S. Pyocyanin: production, applications, challenges and new insights. *World J Microbiol Biotechnol.* 2014;30(4):1159–1168.
92. Manago A, Becker KA, Carpinteiro A, et al. *Pseudomonas aeruginosa* pyocyanin induces neutrophil death via mitochondrial reactive oxygen species and mitochondrial acid sphingomyelinase. *Antioxid Redox Signal.* 2015;22(13):1097–1110.
93. Das T, Manefield M. Pyocyanin promotes extracellular DNA release in *Pseudomonas aeruginosa. PLoS One.* 2012;7(10):e46718.
94. Das T, Kutty SK, Kumar N, Manefield M. Pyocyanin facilitates extracellular DNA binding to *Pseudomonas aeruginosa* influencing cell surface properties and aggregation. *PLoS One.* 2013;8(3):e58299.
95. Rada B, Jendrysik MA, Pang L, et al. Pyocyanin-enhanced neutrophil extracellular trap formation requires the NADPH oxidase. *PLoS One.* 2013;8(1):e54205.
96. Logters T, Margraf S, Altrichter J, et al. The clinical value of neutrophil extracellular traps. *Med Microbiol Immunol.* 2009;198(4):211–219.
97. Garcia-Romo GS, Caielli S, Vega B, et al. Netting neutrophils are major inducers of type I IFN production in pediatric systemic lupus erythematosus. *Sci Transl Med.* 2011;3(73):73ra20.
98. Manzenreiter R, Kienberger F, Marcos V, et al. Ultrastructural characterization of cystic fibrosis sputum using atomic force and scanning electron microscopy. *J Cyst Fibros.* 2012;11(2):84–92.
99. Fothergill JL, Panagea S, Hart CA, Walshaw MJ, Pitt TL, Winstanley C. Widespread pyocyanin over-production among isolates of a cystic fibrosis epidemic strain. *BMC Microbiol.* 2007;7:45.
100. Lau GW, Hassett DJ, Ran H, Kong F. The role of pyocyanin in *Pseudomonas aeruginosa* infection. *Trends Mol Med.* 2004;10(12):599–606.
101. Dowd SE, Wolcott RD, Sun Y, McKeehan T, Smith E, Rhoads D. Polymicrobial nature of chronic diabetic foot ulcer biofilm infections determined using bacterial tag encoded FLX amplicon pyrosequencing (bTEFAP). *PLoS One.* 2008;3(10):e3326.
102. Wolcott RD, Hanson JD, Rees EJ, et al. Analysis of the chronic wound microbiota of 2,963 patients by 16S rDNA pyrosequencing. *Wound Repair Regen.* 2015.

103. Rumbaugh KP, Diggle SP, Watters CM, Ross-Gillespie A, Griffin AS, West SA. Quorum sensing and the social evolution of bacterial virulence. *Curr Biol.* 2009;19(4):341–345.
104. Gjodsbol K, Christensen JJ, Karlsmark T, Jorgensen B, Klein BM, Krogfelt KA. Multiple bacterial species reside in chronic wounds: a longitudinal study. *Int Wound J.* 2006;3(3):225–231.
105. Sun Y, Dowd SE, Smith E, Rhoads DD, Wolcott RD. In vitro multispecies Lubbock chronic wound biofilm model. *Wound Repair Regen.* 2008;16(6):805–813.
106. DeLeon S, Clinton A, Fowler H, Everett J, Horswill AR, Rumbaugh KP. Synergistic interactions of *Pseudomonas aeruginosa* and *Staphylococcus aureus* in an in vitro wound model. *Infect Immun.* 2014;82(11):4718–4728.
107. Freudenberg MA, Tchaptchet S, Keck S, et al. Lipopolysaccharide sensing an important factor in the innate immune response to Gram-negative bacterial infections: benefits and hazards of LPS hypersensitivity. *Immunobiology.* 2008;213(3–4):193–203.
108. Summers C, Rankin SM, Condliffe AM, Singh N, Peters AM, Chilvers ER. Neutrophil kinetics in health and disease. *Trends Immunol.* 2010;31(8):318–324.
109. De Filippo K, Dudeck A, Hasenberg M, et al. Mast cell and macrophage chemokines CXCL1/CXCL2 control the early stage of neutrophil recruitment during tissue inflammation. *Blood.* 2013;121(24):4930–4937.
110. Miyamoto M, Prause O, Sjostrand M, Laan M, Lotvall J, Linden A. Endogenous IL-17 as a mediator of neutrophil recruitment caused by endotoxin exposure in mouse airways. *J Immunol.* 2003;170(9):4665–4672.
111. Witko-Sarsat V, Rieu P, Descamps-Latscha B, Lesavre P, Halbwachs-Mecarelli L. Neutrophils: molecules, functions and pathophysiological aspects. *Lab Invest.* 2000;80(5):617–653.
112. Amulic B, Cazalet C, Hayes GL, Metzler KD, Zychlinsky A. Neutrophil function: from mechanisms to disease. *Annu Rev Immunol.* 2012;30:459–489.
113. van Kessel KP, Bestebroer J, van Strijp JA. Neutrophil-mediated phagocytosis of *Staphylococcus aureus*. *Front Immunol.* 2014;5:467.
114. Berger M. Inflammation in the lung in cystic fibrosis. a vicious cycle that does more harm than good? *Clin Rev Allergy.* 1991;9(1–2):119–142.
115. Cox G, Crossley J, Xing Z. Macrophage engulfment of apoptotic neutrophils contributes to the resolution of acute pulmonary inflammation in vivo. *Am J Resp Cell Mol Biol.* 1995;12(2):232–237.
116. Gunther F, Wabnitz GH, Stroh P, et al. Host defence against *Staphylococcus aureus* biofilms infection: phagocytosis of biofilms by polymorphonuclear neutrophils (PMN). *Mol Immunol.* 2009;46(8–9):1805–1813.
117. Meyle E, Stroh P, Gunther F, Hoppy-Tichy T, Wagner C, Hansch GM. Destruction of bacterial biofilms by polymorphonuclear neutrophils: relative contribution of phagocytosis, DNA release, and degranulation. *The International journal of artificial organs.* 2010;33(9):608–620.
118. Stroh P, Gunther F, Meyle E, Prior B, Wagner C, Hansch GM. Host defence against *Staphylococcus aureus* biofilms by polymorphonuclear neutrophils: oxygen radical production but not phagocytosis depends on opsonisation with immunoglobulin G. *Immunobiology.* 2011;216(3):351–357.
119. Bernthal NM, Pribaz JR, Stavrakis AI, et al. Protective role of IL-1beta against post-arthroplasty *Staphylococcus aureus* infection. *J Orthop Res.* 2011;29(10):1621–1626.
120. Nguyen KT, Seth AK, Hong SJ, et al. Deficient cytokine expression and neutrophil oxidative burst contribute to impaired cutaneous wound healing in diabetic, biofilm-containing chronic wounds. Wound repair and regeneration: official publication of the Wound Healing Society [and] the European Tissue Repair Society; 2013;21(6):833–841

121. Hanses F, Park S, Rich J, Lee JC. Reduced neutrophil apoptosis in diabetic mice during staphylococcal infection leads to prolonged Tnfalpha production and reduced neutrophil clearance. *PloS One*. 2011;6(8):e23633.
122. Kristian SA, Birkenstock TA, Sauder U, Mack D, Gotz F, Landmann R. Biofilm formation induces C3a release and protects *Staphylococcus* epidermidis from IgG and complement deposition and from neutrophil-dependent killing. *J Infect Dis*. 2008;197(7):1028–1035.
123. Domenech M, Ramos-Sevillano E, Garcia E, Moscoso M, Yuste J. Biofilm formation avoids complement immunity and phagocytosis of Streptococcus pneumoniae. *Infect Immun*. 2013;81(7):2606–2615.
124. Jensen ET, Kharazmi A, Lam K, Costerton JW, Hoiby N. Human polymorphonuclear leukocyte response to *Pseudomonas aeruginosa* grown in biofilms. *Infect Immun*. 1990;58(7):2383–2385.
125. Jensen ET, Kharazmi A, Hoiby N, Costerton JW. Some bacterial parameters influencing the neutrophil oxidative burst response to *Pseudomonas aeruginosa* biofilms. *APMIS*. 1992;100(8):727–733.
126. Jesaitis AJ, Franklin MJ, Berglund D, et al. Compromised host defense on *Pseudomonas aeruginosa* biofilms: characterization of neutrophil and biofilm interactions. *J Immunol*. 2003;171(8):4329–4339.
127. Alhede M, Kragh KN, Qvortrup K, et al. Phenotypes of non-attached *Pseudomonas aeruginosa* aggregates resemble surface attached biofilm. *PLoS One*. 2011;6(11):e27943.
128. Dusane DH, Zinjarde SS, Venugopalan VP, McLean RJ, Weber MM, Rahman PK. Quorum sensing: implications on rhamnolipid biosurfactant production. *Biotechnol Genet Eng Rev*. 2010;27:159–184.
129. Morici LA, Carterson AJ, Wagner VE, et al. *Pseudomonas aeruginosa* AlgR represses the Rhl quorum-sensing system in a biofilm-specific manner. *J Bacteriol*. 2007;189(21):7752–7764.
130. Bjarnsholt T, Jensen PO, Burmolle M, et al. *Pseudomonas aeruginosa* tolerance to tobramycin, hydrogen peroxide and polymorphonuclear leukocytes is quorum-sensing dependent. *Microbiology*. 2005;151(Pt 2):373–383.
131. Alhede M, Bjarnsholt T, Jensen PO, et al. *Pseudomonas aeruginosa* recognizes and responds aggressively to the presence of polymorphonuclear leukocytes. *Microbiology*. 2009;155(Pt 11):3500–3508.
132. Jakobsen TH, van Gennip M, Phipps RK, et al. Ajoene, a sulfur-rich molecule from garlic, inhibits genes controlled by quorum sensing. *Antimicrob Agents Chemother*. 2012;56(5):2314–2325.
133. van Gennip M, Christensen LD, Alhede M, et al. Interactions between polymorphonuclear leukocytes and *Pseudomonas aeruginosa* biofilms on silicone implants in vivo. *Infect Immun*. 2012;80(8):2601–2607.
134. Walker TS, Tomlin KL, Worthen GS, et al. Enhanced *Pseudomonas aeruginosa* biofilm development mediated by human neutrophils. *Infect Immun*. 2005;73(6):3693–3701.
135. Chiang WC, Nilsson M, Jensen PO, et al. Extracellular DNA shields against aminoglycosides in *Pseudomonas aeruginosa* biofilms. *Antimicrob Agents Chemother*. 2013;57(5):2352–2361.
136. Watters C, Everett JA, Haley C, Clinton A, Rumbaugh KP. Insulin treatment modulates the host immune system to enhance *Pseudomonas aeruginosa* wound biofilms. *Infect Immun*. 2014;82(1):92–100.
137. Dale DC, Boxer L, Liles WC. The phagocytes: neutrophils and monocytes. *Blood*. 2008;112(4):935–945.
138. Shi C, Pamer EG. Monocyte recruitment during infection and inflammation. *Nat Rev Immunol*. 2011;11(11):762–774.

139. Schroder K, Hertzog PJ, Ravasi T, Hume DA. Interferon-gamma: an overview of signals, mechanisms and functions. *J Leukoc Biol.* 2004;75(2):163–189.
140. Gordon S. Alternative activation of macrophages. *Nat Rev Immunol.* 2003;3(1):23–35.
141. Mosser DM, Edwards JP. Exploring the full spectrum of macrophage activation. *Nat Rev Immunol.* 2008;8(12):958–969.
142. Colucci-Guyon E, Tinevez JY, Renshaw SA, Herbomel P. Strategies of professional phagocytes in vivo: unlike macrophages, neutrophils engulf only surface-associated microbes. *J Cell Sci.* 2011;124(Pt 18):3053–3059.
143. Myrvik QN, Wagner W, Barth E, Wood P, Gristina AG. Effects of extracellular slime produced by *Staphylococcus* epidermidis on oxidative responses of rabbit alveolar macrophages. *J Invest Surg.* 1989;2(4):381–389.
144. Schommer NN, Christner M, Hentschke M, Ruckdeschel K, Aepfelbacher M, Rohde H. *Staphylococcus* epidermidis uses distinct mechanisms of biofilm formation to interfere with phagocytosis and activation of mouse macrophage-like cells 774A.1. *Infect Immun.* 2011;79(6):2267–2276.
145. Cerca F, Andrade F, Franca A, et al. *Staphylococcus* epidermidis biofilms with higher proportions of dormant bacteria induce a lower activation of murine macrophages. *J Med Microbiol.* 2011;60(Pt 12):1717–1724.
146. Mittal R, Sharma S, Chhibber S, Harjai K. Effect of macrophage secretory products on elaboration of virulence factors by planktonic and biofilm cells of *Pseudomonas aeruginosa. Comp Immunol Microbiol Infect Dis.* 2006;29(1):12–26.
147. Chandra J, McCormick TS, Imamura Y, Mukherjee PK, Ghannoum MA. Interaction of *Candida albicans* with adherent human peripheral blood mononuclear cells increases *C. albicans* biofilm formation and results in differential expression of pro- and anti-inflammatory cytokines. *Infect Immun.* 2007;75(5):2612–2620.
148. Scherr TD, Hanke ML, Huang O, et al. *Staphylococcus aureus* biofilms induce macrophage dysfunction through leukocidin AB and alpha-toxin. *mBio.* 2015;6(4).
149. Thurlow LR, Hanke ML, Fritz T, et al. *Staphylococcus aureus* biofilms prevent macrophage phagocytosis and attenuate inflammation in vivo. *J Immunol.* 2011;186 (11):6585–6596.
150. Hanke ML, Angle A, Kielian T. MyD88-dependent signaling influences fibrosis and alternative macrophage activation during *Staphylococcus aureus* biofilm infection. *PloS One.* 2012;7(8):e42476.
151. Hanke ML, Heim CE, Angle A, Sanderson SD, Kielian T. Targeting macrophage activation for the prevention and treatment of *Staphylococcus aureus* biofilm infections. *J Immunol.* 2013;190(5):2159–2168.
152. Stephen TL, Groneck L, Kalka-Moll WM. The modulation of adaptive immune responses by bacterial zwitterionic polysaccharides. *Int J Microbiol.* 2010;2010:917075.
153. Heim CE, Vidlak D, Scherr TD, Hartman CW, Garvin KL, Kielian T. IL-12 promotes myeloid-derived suppressor cell recruitment and bacterial persistence during *Staphylococcus aureus* orthopedic implant infection. *J Immunol.* 2015;194(8):3861–3872.
154. Peck A, Mellins ED. Precarious balance: Th17 cells in host defense. *Infect Immun.* 2010;78(1):32–38.
155. Aujla SJ, Dubin PJ, Kolls JK. Th17 cells and mucosal host defense. *Semin Immunol.* 2007;19(6):377–382.
156. Jensen LK, Jensen HE, Koch J, Bjarnsholt T, Eickhardt S, Shirtliff M. Specific antibodies to *Staphylococcus aureus* biofilm are present in serum from pigs with osteomyelitis. *In Vivo.* 2015;29(5):555–560.
157. Prabhakara R, Harro JM, Leid JG, Harris M, Shirtliff ME. Murine immune response to a chronic *Staphylococcus aureus* biofilm infection. *Infect Immun.* 2011;79(4):1789–1796.
158. Prabhakara R, Harro JM, Leid JG, Keegan AD, Prior ML, Shirtliff ME. Suppression of the inflammatory immune response prevents the development of chronic biofilm

infection due to methicillin-resistant *Staphylococcus aureus*. *Infect Immun.* 2011;79 (12):5010–5018.
159. Heim CE, Vidlak D, Scherr TD, et al. Myeloid-derived suppressor cells contribute to *Staphylococcus aureus* orthopedic biofilm infection. *J Immunol.* 2014;192(8):3778–3792.
160. Rieu A, Aoudia N, Jego G, et al. The biofilm mode of life boosts the anti-inflammatory properties of *Lactobacillus*. *Cell Microbiol.* 2014;16(12):1836–1853.
161. Jones SE, Versalovic J. Probiotic *Lactobacillus reuteri* biofilms produce antimicrobial and anti-inflammatory factors. *BMC Microbiol.* 2009;9:35.
162. Sonal Sekhar M, Unnikrishnan MK, Vijayanarayana K, Rodrigues GS, Mukhopadhyay C. Topical application/formulation of probiotics: will it be a novel treatment approach for diabetic foot ulcer? *Med Hypotheses.* 2014;82(1):86–88.
163. Lebeer S, Verhoeven TL, Perea Velez M, Vanderleyden J, De Keersmaecker SC. Impact of environmental and genetic factors on biofilm formation by the probiotic strain *Lactobacillus rhamnosus* GG. *Appl Environ Microbiol.* 2007;73(21):6768–6775.
164. Kubota H, Senda S, Nomura N, Tokuda H, Uchiyama H. Biofilm formation by lactic acid bacteria and resistance to environmental stress. *J Biosci Bioeng.* 2008;106(4): 381–386.
165. Bove P, Capozzi V, Garofalo C, Rieu A, Spano G, Fiocco D. Inactivation of the ftsH gene of *Lactobacillus plantarum* WCFS1: effects on growth, stress tolerance, cell surface properties and biofilm formation. *Microbiol Res.* 2012;167(4):187–193.
166. Vong L, Lorentz RJ, Assa A, Glogauer M, Sherman PM. Probiotic *Lactobacillus rhamnosus* inhibits the formation of neutrophil extracellular traps. *J Immunol.* 2014;192 (4):1870–1877.
167. Kubota H, Senda S, Tokuda H, Uchiyama H, Nomura N. Stress resistance of biofilm and planktonic *Lactobacillus plantarum* subsp. plantarum JCM 1149. *Food Microbiol.* 2009;26(6):592–597.
168. Dong H, Rowland I, Yaqoob P. Comparative effects of six probiotic strains on immune function in vitro. *Br J Nutr.* 2012;108(3):459–470.
169. Alander M, Satokari R, Korpela R, et al. Persistence of colonization of human colonic mucosa by a probiotic strain, *Lactobacillus rhamnosus* GG, after oral consumption. *Appl Environ Microbiol.* 1999;65(1):351–354.
170. Schultz M, Gottl C, Young RJ, Iwen P, Vanderhoof JA. Administration of oral probiotic bacteria to pregnant women causes temporary infantile colonization. *J Pediatr Gastroenterol Nutr.* 2004;38(3):293–297.
171. Jalilsood T, Baradaran A, Song AA, et al. Inhibition of pathogenic and spoilage bacteria by a novel biofilm-forming *Lactobacillus* isolate: a potential host for the expression of heterologous proteins. *Microb Cell Fact.* 2015;14:96.
172. Donnet-Hughes A, Rochat F, Serrant P, Aeschlimann JM, Schiffrin EJ. Modulation of nonspecific mechanisms of defense by lactic acid bacteria: effective dose. *J Dairy Sci.* 1999;82(5):863–869.
173. Baiu DC, Prechl J, Tchorbanov A, et al. Modulation of the humoral immune response by antibody-mediated antigen targeting to complement receptors and Fc receptors. *J Immunol.* 1999;162(6):3125–3130.
174. Arunachalam K, Gill HS, Chandra RK. Enhancement of natural immune function by dietary consumption of *Bifidobacterium lactis* (HN019). *Eur J Clin Nutr.* 2000;54 (3):263–267.
175. Ramos AN, Gobbato N, Rachid M, Gonzalez L, Yantorno O, Valdez JC. Effect of *Lactobacillus plantarum* and *Pseudomonas aeruginosa* culture supernatants on polymorphonuclear damage and inflammatory response. *Int Immunopharmacol.* 2010;10(2):247–251.
176. Marin-Esteban V, Turbica I, Dufour G, et al. Afa/Dr diffusely adhering *Escherichia coli* strain C1845 induces neutrophil extracellular traps that kill bacteria and damage human enterocyte-like cells. *Infect Immun.* 2012;80(5):1891–1899.

177. Wong SL, Demers M, Martinod K, et al. Diabetes primes neutrophils to undergo NETosis, which impairs wound healing. *Nat Med.* 2015;21(7):815–819.
178. Menegazzo L, Ciciliot S, Poncina N, et al. NETosis is induced by high glucose and associated with type 2 diabetes. *Acta Diabetol.* 2015;52(3):497–503.
179. Malickova K, Duricova D, Bortlik M, et al. Impaired deoxyribonuclease I activity in patients with inflammatory bowel diseases. *Autoimmune Dis.* 2011;2011:945861.
180. Aoudia N, Rieu A, Briandet R, et al. Biofilms of *Lactobacillus plantarum* and *Lactobacillus fermentum*: Effect on stress responses, antagonistic effects on pathogen growth and immunomodulatory properties. *Food Microbiol.* 2016;53(Pt A):51–59.
181. Kretz-Remy C, Mehlen P, Mirault ME, Arrigo AP. Inhibition of I kappa B-alpha phosphorylation and degradation and subsequent NF-kappa B activation by glutathione peroxidase overexpression. *J Cell Biol.* 1996;133(5):1083–1093.
182. Bergonzelli GE, Granato D, Pridmore RD, Marvin-Guy LF, Donnicola D, Corthesy-Theulaz IE. GroEL of *Lactobacillus johnsonii* La1 (NCC 533) is cell surface associated: potential role in interactions with the host and the gastric pathogen *Helicobacter pylori*. *Infect Immun.* 2006;74(1):425–434.
183. Torres-Maravilla E, Lenoir M, Mayorga-Reyes L, et al. Identification of novel anti-inflammatory probiotic strains isolated from pulque. *Appl Microbiol Biotechnol.* 2015;100(1):385–396.
184. Kim SO, Sheikh HI, Ha SD, Martins A, Reid G. G-CSF-mediated inhibition of JNK is a key mechanism for *Lactobacillus rhamnosus*-induced suppression of TNF production in macrophages. *Cell Microbiol.* 2006;8(12):1958–1971.
185. Martins AJ, Spanton S, Sheikh HI, Kim SO. The anti-inflammatory role of granulocyte colony-stimulating factor in macrophage-dendritic cell crosstalk after *Lactobacillus rhamnosus* GR-1 exposure. *J Leukoc Biol.* 2011;89(6):907–915.
186. Mackowiak PA. Recycling metchnikoff: probiotics, the intestinal microbiome and the quest for long life. *Front Public Health.* 2013;1:52.
187. Elisa Guerrieri SdN, Patrizia Messi, Carla Sabia, Ramona Iseppi, Immacolata Anacarso, Moreno Bondi. Use of lactic acid bacteria (LAB) biofilms for the control of *Listeria monocytogenes*. *Food Control.* 2009;20(9):861–865
188. Russo P, Iturria I, Mohedano ML, et al. Zebrafish gut colonization by mCherry-labelled lactic acid bacteria. *Appl Microbiol Biotechnol.* 2015;99(8):3479–3490.
189. Lebeer S, Verhoeven TL, Claes IJ, et al. FISH analysis of *Lactobacillus* biofilms in the gastrointestinal tract of different hosts. *Lett Appl Microbiol.* 2011;52(3):220–226.
190. Ventolini G. Vaginal *Lactobacillus*: biofilm formation in vivo—clinical implications. *Int J Womens Health.* 2015;7:243–247.
191. Marrazzo JM. Vaginal biofilms and bacterial vaginosis: of mice and women. *J Infect Dis.* 2013;207(10):1481–1483.
192. Machado A, Jefferson KK, Cerca N. Interactions between *Lactobacillus crispatus* and bacterial vaginosis (BV)-associated bacterial species in initial attachment and biofilm formation. *Int J Mol Sci.* 2013;14(6):12004–12012.
193. Castro J, Henriques A, Machado A, Henriques M, Jefferson KK, Cerca N. Reciprocal interference between *Lactobacillus* spp. and *Gardnerella vaginalis* on initial adherence to epithelial cells. *Int J Med Sci.* 2013;10(9):1193–1198.
194. Ogunshe AA, Omotoso MA, Bello VB. The in vitro antimicrobial activities of metabolites from *Lactobacillus* strains on Candida species implicated in *Candida vaginitis*. *MJMS.* 2011;18(4):13–25.
195. Pelzer ES, Allan JA, Waterhouse MA, Ross T, Beagley KW, Knox CL. Microorganisms within human follicular fluid: effects on IVF. *PloS One.* 2013;8(3):e59062.
196. Moore DE, Soules MR, Klein NA, Fujimoto VY, Agnew KJ, Eschenbach DA. Bacteria in the transfer catheter tip influence the live-birth rate after in vitro fertilization. *Fertil Steril.* 2000;74(6):1118–1124.

197. Romero R, Schaudinn C, Kusanovic JP, et al. Detection of a microbial biofilm in intraamniotic infection. *Am J Obstet Gynecol.* 2008;198(1). 135 e131-135.
198. Satokari R, Gronroos T, Laitinen K, Salminen S, Isolauri E. Bifidobacterium and *Lactobacillus* DNA in the human placenta. *Lett Appl Microbiol.* 2009;48(1):8–12.
199. Fichorova RN, Onderdonk AB, Yamamoto H, et al. Maternal microbe-specific modulation of inflammatory response in extremely low-gestational-age newborns. *mBio.* 2011;2(1):e00280–e00310.
200. Romero R, Mazor M, Wu YK, Avila C, Oyarzun E, Mitchell MD. Bacterial endotoxin and tumor necrosis factor stimulate prostaglandin production by human decidua. *Prostaglandins Leukot Essent Fatty Acids.* 1989;37(3):183–186.
201. Yeganegi M, Watson CS, Martins A, et al. Effect of *Lactobacillus rhamnosus* GR-1 supernatant and fetal sex on lipopolysaccharide-induced cytokine and prostaglandin-regulating enzymes in human placental trophoblast cells: implications for treatment of bacterial vaginosis and prevention of preterm labor. *Am J Obstet Gynecol.* 2009;200(5). 532 e531-538.
202. Yeganegi M, Leung CG, Martins A, et al. *Lactobacillus rhamnosus* GR-1 stimulates colony-stimulating factor 3 (granulocyte) (CSF3) output in placental trophoblast cells in a fetal sex-dependent manner. *Biol Reprod.* 2011;84(1):18–25.
203. Yang S, Li W, Challis JR, Reid G, Kim SO, Bocking AD. Probiotic *Lactobacillus rhamnosus* GR-1 supernatant prevents lipopolysaccharide-induced preterm birth and reduces inflammation in pregnant CD-1 mice. *Am J Obstet Gynecol.* 2014;211(1). 44 e41–44 e12.
204. Gueniche A, Bastien P, Ovigne JM, et al. Bifidobacterium longum lysate, a new ingredient for reactive skin. *Exp Dermatol.* 2010;19(8):e1–e8.
205. Roy S, Elgharably H, Sinha M, et al. Mixed-species biofilm compromises wound healing by disrupting epidermal barrier function. *J Pathol.* 2014;233(4):331–343.
206. Ouwehand AC, Batsman A, Salminen S. Probiotics for the skin: a new area of potential application? *Lett Appl Microbiol.* 2003;36(5):327–331.
207. Prince T, McBain AJ, O'Neill CA. *Lactobacillus reuteri* protects epidermal keratinocytes from *Staphylococcus aureus*-induced cell death by competitive exclusion. *Appl Environ Microbiol.* 2012;78(15):5119–5126.
208. Mohammedsaeed W, McBain AJ, Cruickshank SM, O'Neill CA. *Lactobacillus rhamnosus* GG inhibits the toxic effects of *Staphylococcus aureus* on epidermal keratinocytes. *Appl Environ Microbiol.* 2014;80(18):5773–5781.
209. Sultana R, McBain AJ, O'Neill CA. Strain-dependent augmentation of tight-junction barrier function in human primary epidermal keratinocytes by *Lactobacillus* and *Bifidobacterium* lysates. *Appl Environ Microbiol.* 2013;79(16):4887–4894.
210. Mohammedsaeed W, Cruickshank S, McBain AJ, O'Neill CA. *Lactobacillus rhamnosus* GG lysate increases re-epithelialization of keratinocyte scratch assays by promoting migration. *Sci Rep.* 2015;5:16147.
211. Zahedi F, Heydari Nasrabadi M, Tajabadi Ebrahimi M, Aboutalebi H. Comparison of the effects of *Lactobacillus brevis* and *Lactobacillus plantarum* on cutaneous wound healing in rats. *Afr J Microbiol Res.* 2011;5(24):4226–4233.

CHAPTER EIGHT

Viral interactions with components of the splicing machinery

F. Meyer[1]
Department of Biochemistry & Molecular Biology, Entomology & Plant Pathology, Mississippi State University, Starkville, MS, USA
[1] Corresponding author. E-mail address: florencia.meyer@msstate.edu

Contents

Abstract

Eukaryotic genes are often interrupted by stretches of sequence with no protein coding potential or obvious function. After transcription, these interrupting sequences must be removed to give rise to the mature messenger RNA. This fundamental process

Progress in Molecular Biology and Translational Science, Volume 142
ISSN 1877-1173
http://dx.doi.org/10.1016/bs.pmbts.2016.05.008

is called RNA splicing and is achieved by complicated machinery made of protein and RNA that assembles around the RNA to be edited. Viruses also use RNA splicing to maximize their coding potential and economize on genetic space, and use clever strategies to manipulate the splicing machinery to their advantage. This article gives an overview of the splicing process and provides examples of viral strategies that make use of various components of the splicing system to promote their replicative cycle. Representative virus families have been selected to illustrate the interaction with various regulatory proteins and ribonucleoproteins. The unifying theme is fine regulation through protein–protein and protein–RNA interactions with the spliceosome components and associated factors to promote or prevent spliceosome assembly on given splice sites, in addition to a strong influence from *cis*-regulatory sequences on viral transcripts. Because there is an intimate coupling of splicing with the processes that direct mRNA biogenesis, a description of how these viruses couple the regulation of splicing with the retention or stability of mRNAs is also included. It seems that a unique balance of suppression and activation of splicing and nuclear export works optimally for each family of viruses.

1. INTRODUCTION

Splicing was first observed in Adenovirus 2 infected cells using DNA–RNA hybridization followed by electron microscope imaging.[1] Those studies elucidated the now well-established concept that in eukaryotic organisms, primary transcripts transcribed from genomic DNA don't direct the synthesis of proteins, but must be edited first into mature messenger RNA (mRNA). With advances in molecular biology methods we now know that most eukaryotic genes carry interrupting sequences that are not part of the final gene product. The transcribed noncoding sequences (introns), must be precisely removed from primary transcripts in a process called splicing. The remaining sequences (exons) are joined together to give rise to the mature protein-coding mRNA which is exported to the cytosol to direct protein synthesis.[2,3]

Splicing is a core step in efficient gene expression. In its absence or dysregulation, functional proteins cannot be synthesized.[4] Specific cellular machinery exists to precisely excise introns from primary transcripts (or pre-mRNAs). An elaborate association of RNA, splicing factors and other ribonucleoproteins (RNPs) assemble as a large complex termed the spliceosome[2,5] to process the primary transcripts into mature mRNA. The complexity of the spliceosome composition, its dynamic assembly and the splicing reaction itself offer opportunity for extensive regulation.[6,7]

2. THE SPLICEOSOME

The spliceosome is a large structure composed of several RNAs and approximately 100 other associated proteins.[8] The assembly depends on both complementary base pairing between the small nuclear RNAs and the intron and exon substrates, and on extensive protein–RNA and protein–protein interactions. These interactions are fundamental for proper spliceosomal function. The addition of splicing factors on pre-mRNAs occurs in detailed and timely coordination.[3]

At the core of the spliceosome are five small nuclear RNAs (snRNAs) termed U1, U2, U4, U5, and U6. U1-U5 snRNAs are produced by RNA polymerase II transcription and are 5′- capped with 7-methylguanosine, while U6 is transcribed by RNA Pol III and has a different cap structure. Each of the snRNAs is associated with a set of 8 Sm proteins, B/B′, D3, D2, D1, E, F, and G.[9] The Sm proteins bind to each other and to a highly conserved sequence on U1, U2, U4, and U5 snRNAs.[10] The association of Sm proteins and snRNAs occurs in the cytoplasm in a controlled and precise order, and the 5′cap structure of snRNAs must be hypermethylated before the small nuclear ribonucleoprotein (snRNP) complex is imported back into the nucleus.[9,11] U6 snRNA diverges from this assembly pathway and associates to Sm-like proteins LSm2, LSm3, LSm4, LSm5, LSm6, LSm7, and LSm8 in the nucleus.[9] Once in the nucleus, U6 and U4 snRNPs form a stable association by extensive base pairing of the snRNAs and with aid from LSm proteins.[12] The U4/U6 snRNP then binds the U5 snRNP to form the tri-snRNP particle.[2]

While the vast majority of introns are spliced by the U1, U2, U4, U5, and U6 snRNPs (U2 type introns), a minor class of introns that do not have the canonical sequence landmarks of most introns[13] are spliced by a 'novel spliceosome' made of the much less abundant snRNPs U11 and U12 snRNPs, but still use U5 snRNP.[14,15] These are called U12 introns and have been observed in a small number of spliced genes. Interestingly, this minor spliceosome seems to localize to the cytoplasm.[16]

Accessory proteins are equally essential components of the spliceosome. Two particularly important proteins are the U2AF dimer and PSF. The U2 auxiliary factor subunit 35 ($U2AF_{35}$) is central to the spliceosome in the recognition and binding of the canonical AG dinucleotide at the 3′ splice site (intron–exon junction) (Fig. 1A). It is also essential for recruiting and stably binding to $U2AF_{65}$ to form a heterodimer.[17] $U2AF_{65}$ recognizes a stretch of pyrimidines that exists upstream of the 3′ splice junction. This heterodimer is

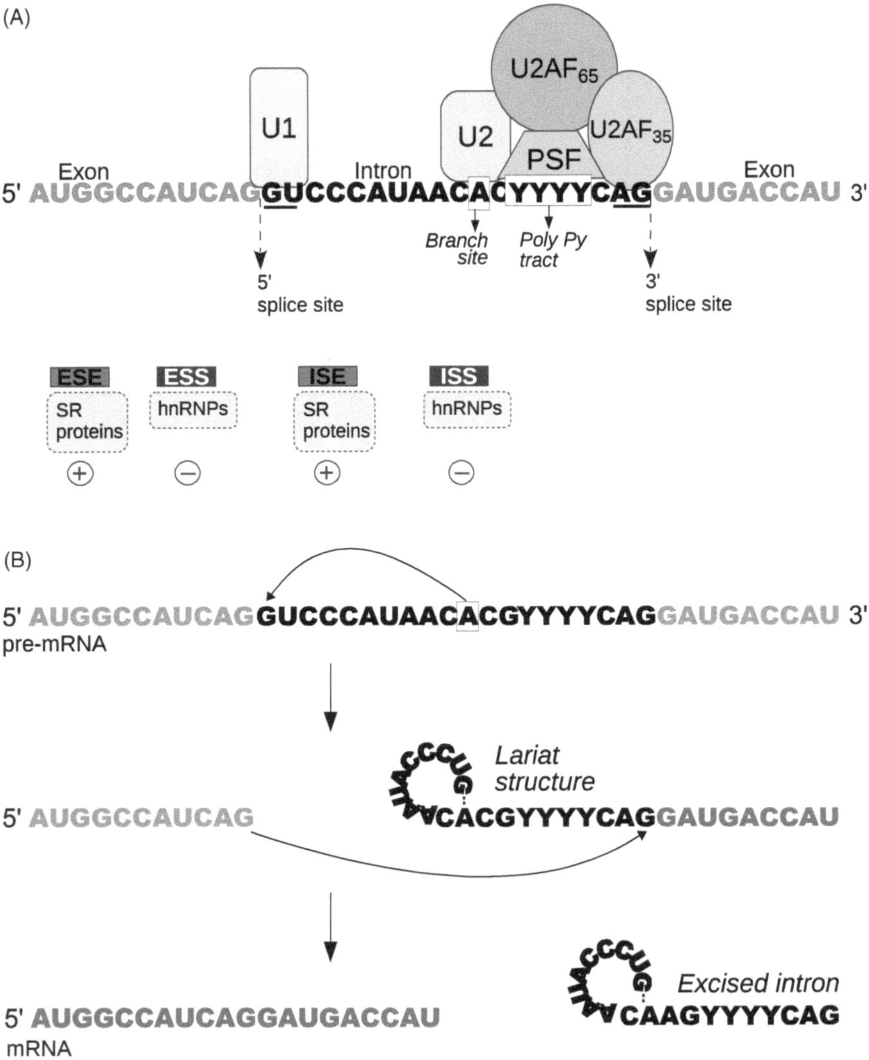

Figure 1 (A) Diagram of important factors required for initial spliceosome assembly. An intron (represented with black ribonucleotides) is flanked by two exons (gray). The canonical dinucleotides at the intron edges (GU-AG) are underlined. U1 snRNP (U1) and U2 snRNP (U2) bind to the 5′ splice site and to the branch site (boxed), respectively. The small subunit of U2AF interacts with the 3' splice site and PSF is responsible for binding the poly pyrimidine (Poly Py, boxed) tract, as well as $U2AF_{65}$. Below, the boxes represent exonic splicing enhancers (ESE) or silencers (ESS), and intronic splicing enhancers (ISE) or silencers (ISS) that bind regulatory proteins such as SR proteins or hnRNPs to modulate splicing. PSF, poly-pyrimidine tract splicing factor; U2AF, U2 accessory factor. (B) Diagram representing the two catalytic steps of the splicing reaction (described in the text).

necessary for, and precedes, the binding of U2 to the branch site.[18,19] The polypyrimidine tract splicing factor (PSF, also known as SFPQ) binds strongly to the polypyrimidine tract and together with the other factors aids in the recognition of the 3′ intron–exon junction by U2 snRNP.[20,21] Many other proteins termed spliceosome-associated or accessory proteins (SAPs) further bind to the assembling structure to provide stability and regulatory checkpoints. For example, splicing factors 3A (SF3A) and SF3B are both components of the U2 snRNP subunit of the spliceosome.[22,23] SF3A and SF3B are multimeric proteins associated with the U2 snRNP, and are essential for initiating the splicing reaction.[24] For complete lists of spliceosomal components see previous studies.[5,8,25]

SR proteins comprise a family of eight proteins with a modular organization and high conservation among metazoans.[26–29] The amino terminus contains one or more RNA binding domains which bind to divergent RNA sequences.[27] A C-terminal domain enriched in serine and arginine dimers (SR) gives the proteins their name. This domain is mainly responsible for protein–protein interactions, recruiting additional splicing factors to the pre-mRNA complex. SR proteins are essential to the splicing reaction.[28,29] However the functions of SR proteins are not restricted to splicing; an increasing body of evidence suggests that SR proteins are key factors in the integration of several other processes related to RNA metabolism.[30] Proteins outside of this conserved family can also contain SR domains which engage in similar protein–protein and protein–RNA interactions. For example, the SR protein ASF/SF2 interacts with snRNA U1-70 KDa protein[31] as well as with the small subunit of U2AF,[32] while SR proteins ASF/SF2 (also called SRSF1), SC35 and SRp55 were shown to interact with the tri-snRNP particle-associated proteins to direct the U4, U5, and U6 tri-snRNP to the assembling spliceosome complex.[33,34] Similar interactions amongst SR domain-containing proteins play comparable roles in bringing the components together and stabilizing the spliceosome. The timely regulation of SR domain phosphorylation plays an important role in promoting or repressing the interaction between SR-containing proteins with each other and with RNA.[32,35]

Additional pieces of the spliceosome constitute the heterogeneous nuclear ribonucleoproteins (hnRNPs). Heterogeneous nuclear RNA (hnRNA) refers to the bulk of transcribed RNA that has not been processed. This RNA is bound by proteins cotranscriptionally, forming complexes termed hnRNPs.[36] About 30 proteins are known to bind to the various hnRNAs, giving each hnRNP complex its name (eg, hnRNP H, hnRNP K, etc.).[37,38] Collectively hnRNPs are involved in various functions related to mRNA biogenesis such as nucleocytoplasmic trafficking, alternative splicing, pre-mRNA 3′-end processing, mRNA translation and turnover.[37]

The interplay between hnRNPs and SR proteins during splicing can be extremely complex, and often times hnRNPs and SR proteins have opposing functions regarding the assembly stages or during catalysis.[39,40] HnRNPs typically bind to exonic splicing silencers (ESS) while SR proteins tend to bind to exonic splicing enhancers (ESE)[29,41] but this may not always be the case. For example, hnRNPA1 is an antagonist of the SR protein SF2/ASF[42] and it may act by displacing an SR protein, SC35, from an ESE rather than binding to an ESS, with the overall downregulation of splicing.[39] Exceptions to the rule of hnRNPs being repressive exist; hnRNP H is required for bridging the interaction between snRNP U1 and the exonic 5′ splice site, stabilizing spliceosome assembly.[43] Other functions of hnRNPs may be to ensure proper binding between spliceosomal components. For instance hnRNPA1 exerts a proofreading activity of the 3′ splice sites recognized by $U2AF_{35}$.[44] These and other functions have been reviewed in Ref. [36]

The spliceosome is believed to assemble in a precise stepwise manner around pre-mRNA substrates as needed. The assembly of the snRNPs and other splicing factors must occur before the catalytic steps, and a separate spliceosome forms on each individual intron to be excised.[2] The assembly is dynamic and reversible. However, a functional spliceosomal complex composed of the five main snRNPs and dozens of other pre-mRNA splicing factors and associated proteins was isolated in human cells[8] and in yeast,[25] suggesting that this large structure, or at least part of its components, may be a ready-made complex. In addition to the expected spliceosome content of the human spliceosome revealed by mass spectrometry, an extensive list of "novel" proteins was also discovered. These novel proteins were not previously known to associate with the spliceosome.[8] These include proteins related to RNA processing, transport, or catabolism, as well as DNA transcription, translation, and several other functions (Tables 2 and 3 in Ref. [8]). These findings point to the notion of a deep interconnection between cellular processes that we are only beginning to unveil.

3. THE SPLICING REACTION

The pre-mRNA is the substrate for assembly of the spliceosome and the information needed for the recruitment of factors is contained in the intronic sequences themselves. Essential sequence components are (1) the conserved intron/exon junctions both at the 5′ and 3′ splice sites; (2) the branch site; (3) the polypyrimidine tract (or poly Py) (Fig. 1A). The strong conservation of intron/exon boundaries across metazoan intron-containing genes show that the large majority of introns contain the dinucleotides /GT and AG/ at the 5′ and 3′ splice sites, respectively (/represents the intron boundary). These have been termed GT-AG introns. The 5′ dinucleotide GU on the pre-mRNA is recognized by U1 snRNP and the AG dinucleotide present at the 3′ splice site is recognized by $U2AF_{35}$. A poly-pyrimidine tract situated between the branch site and the 3′ end of the intron is important for the binding of PSF factor as well as $U2AF_{65}$, which is stabilized as it forms a heterodimer with $U2AF_{35}$. The factors described so far facilitate the recruitment and the stable interaction of U2 snRNP with the branch site, a single adenosine residue positioned near the 3′ end of the intron (Fig. 1A). The branch site provides the 2′ OH nucleophile for the first transesterification reaction.[2,3]

The preassembled tri-snRNP particle (U4, U5, and U6 snRNPs) later joins the partially assembled factors on the pre-mRNA. The splicing factor PSF is responsible for binding the 3′ splice site and facilitates the recruitment of U4-U6-U5 snRNPs, and is essential to the catalytic step II.[20] U6 snRNP makes contact with U2 snRNP and a structural transition weakens the association with U4 snRNP, which then separates from the complex and does not participate in the catalytic events.[45] U1 snRNP also seems to destabilize its union and departs from the complex. The new connection between U6 and U2 snRNPs juxtaposes the branch site (adenosine) to the 5′ splice site. In the first catalytic step, the 2′ hydroxyl of the branch site performs a nucleophilic attack on the 5′ phosphate of the first intronic nucleotide (usually a guanosine), breaking the exon/intron bond and forming a new one with the branch site (Fig. 1B). In the second transesterification, the newly generated free 3′ OH group now performs a nucleophilic attack on the last intronic nucleotide (a guanosine), thereby ligating the exons together (Fig. 1B).[2,3,6]

Intron excision is achieved by an extensive network of protein–protein interaction and RNA–RNA interactions between the components of (snRNAs and Sm proteins) snRNPs. The accessory proteins of the spliceosome play fundamental roles in the splicing reaction by correctly positioning the splice sites and bridging interactions between the various components. The energy for the splicing process appears to derive, in part, from the hydrolysis of GTP, which is facilitated by snu114, an essential associated protein to snRNP U5 that has a functional GTPase domain.[46,47]

4. VIRUSES AND ALTERNATIVE SPLICING

Exonic splicing is fundamental to proper gene expression. However, genes frequently contain several exons and not all of them are incorporated into the mRNA. In other words, some multiexon genes exhibit selectivity when editing their transcripts. This selective usage of exons, termed alternative splicing, may be influenced by nutritional status or developmental stage, and the opportunity to rearrange gene fragments increases the coding capacity of the otherwise static genetic instructions on the genome.[7,48] The use of alternative splicing from a single gene can produce one or more variant proteins that can impart profound changes in the regulation of cellular processes, by changing the sequence specificity to binding partners, the enzymatic properties or the cellular localization of each new variant.[7] Three quarters of human multiexon genes are alternatively spliced.[49]

Viruses have acquired exceptional capabilities in alternatively splicing due in part to lack of abundant genetic space. Viruses encode overlapping genes and promoters on the same or the antisense strand, and employ translational frameshifting and stop-codon read through strategies.[50] Thus, alternative splicing is vital to the replicative success of most viruses. What determines whether a gene undergoes alternative splicing is the combinatorial effect of host factors (SR proteins, hnRNPs, the U snRNPs) and viral elements (viral RNA transcripts and viral proteins) that must interact with each other. Herpesviruses are the least dependent on splicing since only about 5% of their genes are alternatively spliced. These include regulatory proteins,[51] proteins related to the establishment and maintenance of latency[52] and important virulence factors.[53,54] Other viruses such as retroviruses are completely dependent on alternative splicing. The ~10Kbp genome of HIV can only generate three proteins from translating the unspliced genomic-length transcript. Through alternative splicing of the full transcript, HIV generates a group of about 40 transcripts that code for 7 additional regulatory small proteins.[55–57]

The sections that follow explore in more detail some of the strategies used by Retroviruses, Herpesviruses, and Influenza viruses with respect to enhancing or blocking the splicing of cellular and viral transcripts. In addition, the splicing reaction and the immediate or delayed nuclear export of transcripts seems to be intimately connected and directly affects splicing efficiency. The highlight strategies are summarized in Table 1.

Table 1 Summary of Viral Strategies Used to Regulate Splicing and Nuclear Export.

Virus Family (exemplars)	Features	Effect on Splicing	Effect on RNA Export

Retroviridae
Human immunodeficiency virus
	• (+) RNA genome, 7–11 Kbp
	• Complex alternative splicing essential for viral cycle
		• Suboptimal viral splicing
		  • Imperfect splice sites
		  • Enhancer and silencer sequences
			• Exploit CRM1 export system to export unspliced viral mRNA
			  • Genome-bound Rev recruits CRM1
			• Suppress NXT1 export system

Herpesviridae Herpes simplex virus 1 Epstein Barr virus Kaposi's sarcoma-associated virus	• dsDNA genome, 120–240 Kbp • Mostly intronless genes • A few regulatory genes are alternatively spliced • Viral nuclease degrades cellular mRNA	• ICP27 and homologs inhibit splicing • Interference with U2 and U1 snRNPs and SR protein • Promote splicing of viral transcripts	• Expedite export of mRNA • Exploit NXF1 and CRM1 export systems
Orthomyxoviridae Influenza A virus	• Segmented (-) RNA genome, 13.5 Kbp • 2 fragments must undergo alternative splicing	• NS1 promotes splicing of viral mRNAs • Inhibit splicing • Binds U6	• Prevent cellular mRNA export • Exploit NFX1 export system

5. RETROVIRUSES

The retroviral genome is a homo dimer of single-stranded linear RNA molecules with positive polarity that encodes 4-9 overlapping genes (depending on the retrovirus). The RNA is copied into double stranded DNA and integrated into the host genome as a provirus.[58] Human retroviruses such as human immunodeficiency virus (HIV) or human T-cell leukemia virus (HTLV) present a complex transcriptional and translational landscape in which a single genome-length RNA molecule serves as both the viral RNA genome, as well as the template to generate alternatively spliced mRNAs. Collectively the transcripts encode three structural proteins or enzymes *Gag, Pol*, and *Env*, two essential regulatory proteins *Tat* and *Rev*, and four accessory proteins *Vpu*, *Vif*, *Vpr*, and *Nef*.[57,59,60] For this purpose, there are four different 5′ splice sites and eight 3′ splice sites on the genome[56] that facilitate the generation of approximately 40 mRNAs of different lengths. Both alternatively spliced and unspliced transcripts must be exported for efficient protein expression and viral packaging.[59] Therefore, a careful balance between the splicing rates and nuclear export of the different types of viral RNA must be maintained for efficient replication. This is brought about both by *cis*-acting regulatory sequences present on viral RNA as well as by proteins of the splicing machinery such as the SR family members and snRNPs.

5.1 Suboptimal Splicing

In order to achieve a balanced splicing of its transcripts, the splice sites on the HIV genome differ in their binding capacity for various components of the spliceosome machinery. While the 5′ sites are said to be 'strong', the 3′ splice sites are 'weak' and are defined by sub-optimal intron-exon junctions, poly-pyrimidine tracts or unusual branch points (eg, uridine instead of adenine)[61–63] reducing the splicing activity by as much as 60% on those sites.[64] In addition, the presence of ESE or ESS sequences that often mediate their effect at a distance also regulates the efficiency of splicing.[55] ESS and ESE sequences typically bind hnRNPs and SR proteins, respectively, to promote or block one or more steps in the formation of the spliceosome,[65–67] and often act in concert with neighboring 5′ or 3′ splice sites.[41,68,69] A variety of studies support the concept that hnRNP proteins strongly bind to ESS sequences to negatively regulate splicing in HIV, in particular family members hnNRNP A/B and hnRNPA1.[70–72]

The viral protein *Tat* is a regulatory protein with a traditionally prominent role in transcription.[57] However it has recently also been assigned a role in regulation of viral splicing activity. When Tat becomes acetylated it is able to bind, among other proteins, to a cofactor of an important SR protein ASF/SF2, p32,[73] as well as to the cyclin-dependent kinase 13 (CDK13), a kinase that phosphorylates ASF/SF2.[74] The phosphorylation of ASF/SF2 appears necessary for facilitating protein-protein interactions while preventing non-specific binding to RNA.[34] The apparent sequestration of p32 and CDK13 by acetylated *Tat* serves both to deprive SF2/ASF from its cofactor and to prevent its phosphorylation by CDK13, leading to a decrease in splicing activity and favoring the generation of longer unspliced transcripts. Increasing evidence suggests that ASF/SF2 is a prominent cellular regulatory factor for HIV transcription[75] and splicing.[76]

Several other SR proteins play fundamental roles in the regulation of viral RNA splicing aside from ASF/SF2. SRp55 and SRp75 selectively promote splicing from some splice sites while inhibiting others to produce the *vpr proein* mRNA, while SRp40 utilizes a ESE sequences to induce the splicing of *tat protein.*[77] Other studies show that at the same time SRp40 and SRp55 enhance protein translation from the unspliced HIV RNA by binding to the full-length genomic RNA, expediting its export before splicing occurs.[78] SR proteins like ASF/SF2, SC35 or 9G8 also have differential effects in splicing when acting on the various splice sites.[79] This study and others highlight that the strong effect that ASF/SF2 has on the activation of viral splicing could be detrimental for viral replication because of significant reduction in the production of some structural proteins that are produced from the long (unspliced) mRNAs, which are needed late during infection for the assembly of new infectious particles.[75,79] Other retroviruses such as Rous sarcoma virus encode negative regulatory sequences that function as splice site decoys,[80,81] redirecting vital splicing machinery such as SR proteins thereby decreasing overall splicing efficiency.[82,83] The binding of hnRNP H to this complex has a stimulatory role in polyadenylation.[84] The authors postulate that a complex of SR proteins assembles on the negative regulatory sequence that prevents splicing, while hnRNP K also binds and stimulates polyadenylation, with the overall goal of producing unspliced genome-length mRNAs.[85]

A fine example of the detail of splicing regulation is provided by the observation that some HIV isolates naturally carry a point mutation in the *env* coding region. A single U-to-C mutation is enough to activate a pair of 5′ and 3′ splice sites that define an otherwise inactive exon. As a result, this exon is incorporated into viral mRNAs and the result is a dramatic reduction in viral replication.[86] That mutation disrupts an exonic splicing silencer while at the same time promoting the binding of SR proteins SC35 and ASF/SF2, in addition to recruiting hnRNP H to an adjacent and partially overlapped sequence on the same exon. This single mutation, therefore, can be said to transform the silencer into a splicing enhancer.[43] The juxtaposition of ESSs and ESEs with varied binding capacities for hnRNPs, SR proteins and snRNPs constitutes a fine regulatory tool in both host and viral genomes.[87,88]

Together these studies highlight the importance of splicing regulation in retroviruses. This is a fundamental process for the retroviral life cycle and consequently offers the opportunity for the development of antiviral drugs that target important splicing factors.[75,89]

5.2 Nuclear Export

The viral protein *Rev* promotes nuclear export of unspliced RNA by binding to Rev-responsive elements (RREs) on *gag-pol* and *env* mRNAs, stabilizing the mRNAs and steering them away from splicing,[90] and promoting their nuclear export by exploiting the cellular CRM1 export system.[91,92] The CRM1 export system shuttles all ribosomal RNAs and the U snRNAs to the cytoplasm. However, cellular checkpoints exist to ensure that mRNAs that are not completely spliced do not exit the nucleus through this pathway.[93] Hence, the CRM1 protein and its associated Ran-GTP is mostly responsible for export of RNA other than mRNA. Only a subset of mRNAs that contains AU-rich elements use this pathway for nuclear export.[93] The viral protein *Rev* binds to RRE motifs on viral transcripts and recruits the CRM1 protein in the form of a dimer to the viral RNAs[94,95] to promote their export. The lack of RREs on the genome of simpler retroviruses such as Mason-Pfizer monkey virus is compensated by structural stem and loop-like motif that can recruit export proteins to intron-containing transcripts.[96]

At the same time that the virus exploits the interaction with CRM1 export system for unspliced transcripts, it suppresses the NXF1 export system.[97] This nuclear export system is used for bulk mRNA export and comprises a multi-protein complex, including proteins such as Aly and the Tap-Nxt export factor, to interact with components of the nuclear pore complex.[92,98] The Rev protein inhibits this export pathway by interacting with the cap-binding protein on mRNAs and blocking access to Aly/REF binding, a central adaptor protein of the NXF1 export system.[97] However, the suppression of NXT1-dependent mRNA export may not be complete. HIV exploits the host hnRNP C1/C2 splicing factor, a tetramer complex which binds to nascent transcripts longer than 200–300 nucleotides. The binding of hnRNP C1/C2 to long transcripts favors their nuclear export via the NXT1 system by keeping them away from mRNA processing proteins and at the same time recruiting the Aly/REF adaptor.[99]

Over the years we have progressively learned that the processes of transcription, splicing, and mRNA editing may be more interconnected than previously recognized[100] and it appears that SR proteins are major actors in the coupling of transcription to mRNA splicing and export.[30] These processes are connected or coupled and proteins or RNPs participating in one process escort the RNA to the next stage, blurring at times the classifications regarding protein function. In conclusion, promoting the export of singly- or unspliced transcripts is a common strategy among retroviruses to strike the right balance of the variously sized transcripts available for translation or for genome packaging.[101]

6. HERPESVIRUSES

Herpesviruses are large double-stranded DNA viruses of 120–240 Kbp in length that typically encode 70–100 genes, most of which are intron-less. Therefore, the vast majority (~95%) of viral transcripts originate in the unspliced form. However, there are important genes that require careful splicing for proper function. These include genes that encode pivotal regulatory proteins[102,103] and factors important for the establishment and maintenance of latency.[52] In some cases, the presence of pathogenic traits has been shown to depend on the generation of splicing variants.[54]

A central common feature in the *Herpesviridae* family of viruses is the host shutoff mechanism by which host protein synthesis is abruptly diminished after infection.[104] This is important because splicing depends on the availability of pre-mRNA, and during herpesviral infection the majority of mRNA is degraded by a viral protein called vhs (*viral host shutoff*) that destabilizes and degrades both viral and host mRNA.[105,106] Some studies suggest that vhs discriminates between viral and host mRNA in the context of the infected cell and selectively degrades host mRNA.[107] Similar vhs activity exists in most herpesviruses.[104,108].

6.1 Splicing Inhibition

In herpes simplex type 1 (HSV-1), ICP27 is a nuclear phosphoprotein that performs a variety of functions related to mRNA biogenesis and transport.[109] Among its many roles is the obstruction of splicing. Homologs of ICP27 in other herpesviruses are Mta/SM in Epstein-Barr virus (EBV),[110] the UL69 in human cytomegalovirus (HCMV),[111] ORF4 in varicella-zoster virus (VZV),[112] and ORF57 in Kaposi's sarcoma associated herpesvirus (KSHV).[113] The overall function of this protein seems to be conserved in the various herpesviruses.

The ICP27 protein is known to inhibit splicing at many levels: assembly, catalytic activity, and regulatory activity. The initial observations showed that ICP27 induced a marked redistribution of snRNPs and various splicing factors in the nucleus of infected cells.[114,115] It was later shown that ICP27 physically interacts with core components of the spliceosome. ICP27 interacts with U1 70 KDa associated protein via its C-terminal region,[116] to which the splicing repressor domain had been previously mapped.[115] In the case of U2 snRNP, the interaction with ICP27 is not direct; ICP27 interacts with U2 snRNP-associated protein 145[117] blocking its interaction with SAP49, a protein which binds to pre-mRNA upstream of the branch site. The blockade caused by ICP27 contributes to the halting of spliceosome assembly at the early stage.[118] Finally, ICP27 associates with SR protein SRp20 as well as with the SR protein kinase 1,[119] a kinase that phosphorylates SR proteins.[120] SR proteins play such a fundamental role in the interconnection of spliceosomal components that modification of SR protein phosphorylation levels can have a profound influence on splicing.[121,122] In this regard, the control of phosphorylation has also been observed in the context of Adenoviral infections.[123,124]

6.2 Nuclear Export

The described interaction between ICP27 and the SR protein SRp20 may have a dual purpose in HSV-1 infection because SRp20 and 9G8 are SR proteins that also participate in nuclear export of mRNA.[125] ICP27 colocalizes with these splicing factors in infected cells, bringing HSV-1 mRNA to the proximity of components of the nuclear export machinery such as NXF1[126], a factor responsible for the export of most mRNAs.[93] EBV ORF57 also exploits this export pathway by interacting with the protein complex containing Aly/REF and Tap proteins, and promoting export of mRNA.[127] An alternate strategy for exporting viral transcripts by herpesviruses is through the CRM1 export machinery, which is mainly responsible for the export of rRNA and snRNAs.[93] EBV protein SM binds to intronless mRNAs and facilitates its export by binding to CRM1 RNA export machinery to expedite nuclear export and therefore gene expression.[128]

In its central role of regulating splicing and mRNA export, ICP27 also interacts with hnRNP K. HnRNP K is a multifunctional protein involved in a large variety of cellular processes.[129] hnRNP K exhibits a modular organization with both DNA and RNA binding domains and is phosphorylated by several kinases. HnRNP K acts like a central docking and regulatory protein.[129] ICP27 and its KSHV homolog ORF57 have been shown to interact with hnRNP K,[130–132] which is reported to enhance splicing in influenza virus-infected cells.[133] HnRNP K is also an important factor for the replicative cycle of other viruses such as Dengue virus,[134] vesicular stomatitis virus[135] or hepatitis C.[136] Because of hnRNP K's ability for bidirectional nucleocytoplasmic shuttling,[137] it is possible that these viruses exploit the interaction for mRNA transport.

Collectively, the activities of ICP27 and homolog proteins in other herpesviruses contribute to the shutoff of host protein synthesis by down-modulation of splicing in various ways. Because unspliced transcripts are not exported in the context of normal cellular function, the number of cellular transcripts available for translation is limited. In addition, ICP27 was found to physically interact with the vhs protein in HSV-1,[138] reinforcing the role of ICP27 in ensuring that cellular mRNA is either not spliced or degraded by vhs nuclease activity.

6.3 Splicing Enhancement of Viral RNA

Although most of the herpesviral genes are intronless, the viruses do contain genes that must be spliced for proper function. Moreover, some of these genes encode important regulatory proteins that orchestrate gene expression and immune evasion, among other functions.[139] Therefore, herpesviruses must also find ways to enhance splicing of their own genes. The KSHV protein ORF57 acts as a splicing factor, by binding SR protein SC35 and the snRNAs, and recruiting them to viral pre-mRNAs. ORF57 was observed to colocalize in specific nuclear regions (speckles) where splicing occurs.[140] In line with the enhancement of splicing, ORF57 also counteracts the repressive actions of SRSF3 on the splicing of KSHV genes.[141] The homolog protein in EBV, SM, also functions as an alternative splicing factor[142] by interacting with SR proteins even if it does not contain an SR domain.[143] Additionally this protein increases the processing of DNA polymerase transcript due in part to its interaction with splicing factors hnRNPC1/C2 and with hnRNP A1/A2.[144]

Also supporting the notion that herpesviruses must enhance their own splicing is the study linking viral replication to a cellular protein called host cell factor 1 (HCF-1). HCF-1 is known in the context of HSV-1 infection as a protein that, in conjunction with viral protein VP16 and cellular transcription factor Oct-1, aids in the efficient transcription of viral immediate-early genes.[145,146] HCF-1 was subsequently found to interact with U2 and U5 snRNPs of the spliceosome and to prepare the newly generated transcripts for splicing.[147] The virus could exploit this interaction to promote its own splicing in addition to transcription enhancement.

7. INFLUENZA VIRUSES

Influenza viruses belong to the *Orthomyxoviridae* family. The influenza A and B virus genomes are made of eight single-stranded negative-sense RNA segments. With the exception of the NS and the M fragment, each of the other fragments encodes a single protein, giving rise to three components of the viral RNA-dependent RNA polymerase (PB1, PB2, and PA), hemaglutinin (HA), neuraminidase (NA) and the nucleoprotein (NP). Influenza virus replicates in the nucleus of infected cells, and the viral polymerase copies the genome into two positive sense mRNAs: one of these is exported to the cytoplasm to serve as template for viral proteins, and the other is retained in the nucleus to serve as the template for genomic RNA.[148] The NS segment encodes the non-structural proteins NS1 and NS2 by means of alternative splicing; the same is true for the M fragment, which codes for the matrix M1 and M2 proteins.[148,149] In 2012 an Influenza A clinical isolate was found to produce a third protein, NS3, from the NS fragment.[150] The M fragment can theoretically also yield three alternatively spliced transcripts, yet only two proteins have been detected,[151] with a third transcript found in a few influenza isolates.[152] In both NS and M fragments, the splicing events shift the reading frame, thereby diversifying the viral coding potential.[151,153]

7.1 Splicing Enhancement

Perhaps because the life cycle of influenza viruses is much faster than most viruses, the virus must quickly complete the replicative cycle before it can activate any cellular responses.[149] Infectious particles are released from infected cells as early as 6 h after infection.[154] Therefore the viruses use the splicing machinery to efficiently splice segments NS and M in a timely fashion to provide all the structural components for virion assembly and release. Influenza is thought to efficiently mimic the host splice sites. The dinucleotides (GU and AG) as well as the surrounding sequences that flank the 5′ and 3′ splice sites are very similar to consensus cellular splice sites, more so in the case of the NS transcripts than in the M transcripts, and are able to recruit the needed spliceosome factors to execute efficient splicing.[149] There is an obvious reorganization of nuclear structures and splicing factors soon after infection,[155,156] suggesting that viral infection affects splicing. The newly made viral proteins can also bind to spliceosome components such as hnRNPs M, H1 and A1, PSF/, SFR3, and U6 snRNA[157,158] to recruit and further direct them to viral splice sites. However, the splicing is not completely efficient due to the presence of weak 3′ splice sites on the genome.[159,160] The delay in splicing leads to the accumulation of unspliced transcripts that delay protein expression and virion assembly. The existence of weak or strong splice sites could perhaps be explained by the formation of RNA secondary structures that temporarily mask the required splicing sequences.[149,161,162]

NS1 is a multifunctional regulatory protein involved in blocking interferon signaling, among many other functions.[163,164] In addition, it enhances the splicing of its own segment (NS) and of the M segment by interacting with a host protein called NS1 binding protein (NS1-BP),[165] which promotes splicing of the M segment to produce the M1 mRNA for producing a functional M1 protein. The splicing enhancement is also dependent on the integrity of cellular protein hnRNP K.[133]

The control of splicing of the M segment is very important as well. The viral M2 gene encodes an ion channel protein that forms part of the assembled virion.[148] The mRNA for M2 protein is generated by splicing the M1 mRNA. There is a pair of 5′ splice sites on the M1 transcript that are important for this regulation. The distal 5′ site is the stronger of the two because it matches more closely the consensus 5′ splice sequences while the proximal site is less conserved and as a consequence less efficient.[166] As transcription occurs, the nascent RNA transcripts are bound by splicing factors, polyadenylation and 5′-cap molecular machinery.[100] When the stronger distal 5′ splice site is partially blocked by the viral polymerase complex, the weaker or proximal site remains free to recruit and assemble snRNP U1.[166] The splicing needed to generate M2 is also dependent on the SR protein SF2/ASF, which specifically binds to an ESE sequence downstream of the 3′ splice site on the M1 mRNA (on the second exon). There is a strong correlation between the amount of M2 protein and mRNA and the levels of SF2/ASF protein in various types of infected cells.[167]

7.2 Splicing Downregulation

In addition to splicing enhancement the NS1 protein can also engage spliceosomal players in an opposite way. NS1 can bind to U2 and U6 snRNPs to block normal interactions within the spliceosome during the assembly stages. The binding of NS1 to U6 snRNA prevents the interaction between U4 and U6, which is extensive and necessary for spliceosome assembly.[2,3] In addition, NS1 perturbs the interaction between U2 and U6 during the splicing reaction.[168] NS1 can also bind U6atac, an snRNP that participates in the assembly of a different type of spliceosome that splices introns with AT-AC splicing junctions.[14,169]

The downregulation of splicing also occurs due to inefficient splicing. The usage of a suboptimal splice site on the NS fragment causes an imbalance in the ratio of spliced (which produces the nuclear export protein (NEP)) versus unspliced (which produces NS1) mRNAs.[160] The slow accumulation of NEP disturbs the coordinated export of viral RNAs and affects the timing of infection by incorrectly signaling the switch from transcription to replication.[170]

7.3 Nuclear Export

To avoid recognition by the host immune system most viruses have negative regulatory mechanisms to modulate their own gene expression. Influenza viruses use the multifunctional NS1, which is involved in hindering mRNA processing and nuclear export. NS1 binds to the cleavage and polyadenylation specificity factor (CPSF), a protein needed for proper 3′ end cleavage and polyadenylation of cellular transcripts. NS1's interaction with CPSF blocks CPSF normal function. NS1 interaction blocks contact with the transcripts, which are not completely processed and therefore not exported.[171] The interaction forms a blockade at the nuclear pore and mRNA export is inhibited.[172] Because viral mRNA is not polyadenylated by this method,[173] the binding of NS1 to CPSF prevents cellular, but not viral, transcripts from being exported to the cytoplasm. In addition to the nuclear retention of cellular transcripts which favors the viral infection, NS1 interacts with several proteins of the NXF1/Tap complex that is used for normal export of mRNAs.[98]

8. CONCLUDING REMARKS

The spliceosome is an incredibly complex molecular motor made of RNA and protein. The extensive protein-protein, RNA-protein and RNA-RNA interactions that take place are precise and regulated in time and space. SR proteins and SR domain-containing proteins establish a vast network of interactions and appear to be at the core of splicing regulation. Aside from the highly specific interactions that take place between SR proteins, hnRNPs and pre-mRNA, the relative concentrations of individual factors adds another regulatory aspect which relates to the efficiency of each factor's gene expression. When all is considered, the regulatory network that metazoans have in place for the efficient and well-coordinated functioning of the spliceosome is stunning, and offers an equally impressive opportunity for additional regulation by other cellular processes. For instance, a slight disturbance in the activity of kinases that phosphorylate SR proteins can modify the SR protein binding partners, which in turn influences the efficiency of splicing. Both retroviruses and herpesviruses seem to use this strategy of hypophosphorylating SR proteins to suppress splicing.[74,119] Adenoviruses also depend on SR protein phosphorylation for regulation of splicing.[123,124]

The manipulation of nuclear export systems is also a common strategy used by these viral families. Perhaps because splicing is closely coupled to other aspects of mRNA biogenesis,[30,100] the same proteins or complexes of proteins that participate in the splicing are often the ones that escort the mRNA to the nuclear pore for export. The CRM1 export system has been adopted by herpesviruses and retroviruses to transport incompletely spliced transcripts, while all three families of viruses described here exploit the NXF1 mRNA export system. In addition to transporting viral mRNAs, the binding of one or more proteins on NXF1 or CRM1 transport systems means that the machinery is no longer available (or less available) for cellular RNA export, overall benefiting viral gene expression. When this is coupled with the degradation of cellular mRNAs, accomplished by the vhs protein in herpesviruses, the balance tips in favor of viral success.

It is interesting how the inefficiency of certain splice sites can themselves constitute a regulatory mechanism that affects viral replication. The presence of *cis*-regulatory sequences plays a fundamental role in determining the strength of a splice site, by determining the RS proteins or hnRNPs that they bind or exclude. The continuous adjustment of the strength of individual splice sites on a viral genome may very well be one of the driving forces behind virus adaptation by constantly refining their fitness. This is a provoking thought already proposed by others,[174] and the recent discovery of a strain of influenza virus that produces a new protein due to the [apparently random] creation of a splice site[150] highlights the importance of splicing and splicing regulation in viral success.

ACKNOWLEDGMENTS

I would like to thank Dr Yuhua Farnell and Dr Jonas King for critical reading and suggestions.

REFERENCES

1. Berget SM, Moore C, Sharp PA. Spliced segments at the 5′ terminus of adenovirus 2 late mRNA. *Proc Natl Acad Sci USA*. 1977;74(8):3171–3175.
2. Lamond AI. The spliceosome. *Bioessays*. 1993;15(9):595–603.
3. Will CL, Lührmann R. Spliceosome structure and function. *Cold Spring Harb Perspect Biol*. 2011;3(7). a003707.
4. Smith CW, Patton JG, Nadal-Ginard B. Alternative splicing in the control of gene expression. *Annu Rev Genet*. 1989;23:527–577.
5. Jurica MS, Moore MJ. Pre-mRNA splicing: awash in a sea of proteins. *Mol Cell*. 2003;12(1):5–14.
6. House AE, Lynch KW. Regulation of alternative splicing: more than just the ABCs. *J Biol Chem*. 2008;283(3):1217–1221.
7. Kelemen O, Convertini P, Zhang Z, et al. Function of alternative splicing. *Gene*. 2013;514(1):1–30.
8. Rappsilber J, Ryder U, Lamond AI, Mann M. Large-scale proteomic analysis of the human spliceosome. *Genome Res*. 2002;12(8):1231–1245.
9. Will CL, Lührmann R. Spliceosomal UsnRNP biogenesis, structure and function. *Curr Opin Cell Biol*. 2001;13(3):290–301.
10. Branlant C, Krol A, Ebel JP, Lazar E, Haendler B, Jacob M. U2 RNA shares a structural domain with U1, U4, and U5 RNAs. *EMBO J*. 1982;1(10):1259–1265.
11. Raker VA, Plessel G, Lührmann R. The snRNP core assembly pathway: identification of stable core protein heteromeric complexes and an snRNP subcore particle in vitro. *EMBO J*. 1996;15(9):2256–2269.
12. Achsel T, Brahms H, Kastner B, Bachi A, Wilm M, Lührmann R. A doughnut-shaped heteromer of human Sm-like proteins binds to the 3′-end of U6 snRNA, thereby facilitating U4/U6 duplex formation in vitro. *EMBO J*. 1999;18(20):5789–5802.
13. Hall SL, Padgett RA. Conserved sequences in a class of rare eukaryotic nuclear introns with non-consensus splice sites. *J Mol Biol*. 1994;239(3):357–365.
14. Tarn WY, Steitz JA. A novel spliceosome containing U11, U12, and U5 snRNPs excises a minor class (AT-AC) intron in vitro. *Cell*. 1996;84:801–811.
15. Hall SL, Padgett RA. Requirement of U12 snRNA for in vivo splicing of a minor class of eukaryotic nuclear pre-mRNA introns. *Science*. 1996;271(5256): 1716–1718.
16. König H, Matter N, Bader R, Thiele W, Müller F. Splicing segregation: the minor spliceosome acts outside the nucleus and controls cell proliferation. *Cell*. 2007;131 (4):718–729.
17. Guth S, Tange TØ. Kellenberger E, Valcárcel J. Dual function for U2AF(35) in AG-dependent pre-mRNA splicing. *Mol Cell Biol*. 2001;21(22):7673–7681.

18. Ruskin B, Zamore PD, Green MR. A factor, U2AF, is required for U2 snRNP binding and splicing complex assembly. *Cell.* 1988;52(2):207–219.
19. Graveley BR, Hertel KJ, Maniatis T. The role of U2AF35 and U2AF65 in enhancer-dependent splicing. *RNA.* 2001;7(6):806–818.
20. Gozani O, Patton JG, Reed R. A novel set of spliceosome-associated proteins and the essential splicing factor PSF bind stably to pre-mRNA prior to catalytic step II of the splicing reaction. *EMBO J.* 1994;13(14):3356–3367.
21. Shav-Tal Y, Zipori D. PSF and p54nrb/NonO—multi-functional nuclear proteins. *FEBS Lett.* 2002;531(2):109–114.
22. Lin P-C, Xu R-M. Structure and assembly of the SF3a splicing factor complex of U2 snRNP. *EMBO J.* 2012;31(6):1579–1590.
23. Brosi R, Groning K, Behrens S, Luhrmann R, Kramer A. Interaction of mammalian splicing factor SF3a with U2 snRNP and relation of its 60-kD subunit to yeast PRP9. *Science (80-).* 1993;262(5130):102–105.
24. Huang C-J, Ferfoglia F, Raleff F, Krämer A. Interaction domains and nuclear targeting signals in subunits of the U2 small nuclear ribonucleoprotein particle-associated splicing factor SF3a. *J Biol Chem.* 2011;286(15):13106–13114.
25. Stevens SW, Ryan DE, Ge HY, et al. Composition and functional characterization of the yeast spliceosomal penta-snRNP. *Mol Cell.* 2002;9(1):31–44.
26. Manley JL, Tacke R. SR proteins and splicing control. *Genes Dev.* 1996;10 (13):1569–1579.
27. Graveley BR. Sorting out the complexity of SR protein functions. *RNA.* 2000;6 (9):1197–1211.
28. Huang Y, Steitz JA. SRprises along a messenger's journey. *Mol Cell.* 2005;17 (5):613–615.
29. Long JC, Caceres JF. The SR protein family of splicing factors: master regulators of gene expression. *Biochem J.* 2009;417(1):15.
30. Zhong X-Y, Wang P, Han J, Rosenfeld MG, Fu X-D. SR proteins in vertical integration of gene expression from transcription to RNA processing to translation. *Mol Cell.* 2009;35(1):1–10.
31. Kohtz JD, Jamison SF, Will CL, et al. Protein-protein interactions and 5′-splice-site recognition in mammalian mRNA precursors. *Nature.* 1994;368(6467):119–124.
32. Xiao SH, Manley JL. Phosphorylation-dephosphorylation differentially affects activities of splicing factor ASF/SF2. *EMBO J.* 1998;17(21):6359–6367.
33. Roscigno RF, Garcia-Blanco MA. SR proteins escort the U4/U6.U5 tri-snRNP to the spliceosome. *RNA.* 1995;1(7):692–706.
34. Makarova OV, Makarov EM, Lührmann R. The 65 and 110 kDa SR-related proteins of the U4/U6.U5 tri-snRNP are essential for the assembly of mature spliceosomes. *EMBO J.* 2001;20(10):2553–2563.
35. Cao W, Jamison SF, Garcia-Blanco MA. Both phosphorylation and dephosphorylation of ASF/SF2 are required for pre-mRNA splicing in vitro. *RNA.* 1997;3(12):1456–1467.
36. Martinez-Contreras R, Cloutier P, Shkreta L, Fisette J-F, Revil T, Chabot B. hnRNP proteins and splicing control. *Adv Exp Med Biol.* 2007;623:123–147.
37. Krecic AM, Swanson MS. hnRNP complexes: composition, structure, and function. *Curr Opin Cell Biol.* 1999;11(3):363–371.
38. Dreyfuss G, Matunis MJ, Piñol-Roma S, Burd CG. hnRNP proteins and the biogenesis of mRNA. *Annu Rev Biochem.* 1993;62:289–321.
39. Zhu J, Mayeda A, Krainer AR. Exon identity established through differential antagonism between exonic splicing silencer-bound hnRNP A1 and enhancer-bound SR proteins. *Mol Cell.* 2001;8(6):1351–1361.
40. Wu JY, Maniatis T. Specific interactions between proteins implicated in splice site selection and regulated alternative splicing. *Cell.* 1993;75(6):1061–1070.

41. Wang J, Smith PJ, Krainer AR, Zhang MQ. Distribution of SR protein exonic splicing enhancer motifs in human protein-coding genes. *Nucleic Acids Res.* 2005;33 (16):5053–5062.
42. Mayeda A, Krainer AR. Regulation of alternative pre-mRNA splicing by hnRNP A1 and splicing factor SF2. *Cell.* 1992;68(2):365–375.
43. Caputi M, Zahler AM. SR proteins and hnRNP H regulate the splicing of the HIV-1 tev-specific exon 6D. *EMBO J.* 2002;21(4):845–855.
44. Tavanez JP, Madl T, Kooshapur H, Sattler M. Valcárcel J. hnRNP A1 proofreads 3′ splice site recognition by U2AF. *Mol Cell.* 2012;45(3):314–329.
45. Yean SL, Lin RJ. U4 small nuclear RNA dissociates from a yeast spliceosome and does not participate in the subsequent splicing reaction. *Mol Cell Biol.* 1991;11(11): 5571–5577.
46. Brenner TJ, Guthrie C. Assembly of Snu114 into U5 snRNP requires Prp8 and a functional GTPase domain. *RNA.* 2006;12(5):862–871.
47. Fabrizio P, Laggerbauer B, Lauber J, Lane WS, Lührmann R. An evolutionarily conserved U5 snRNP-specific protein is a GTP-binding factor closely related to the ribosomal translocase EF-2. *EMBO J.* 1997;16(13):4092–4106.
48. Tazi J, Bakkour N, Stamm S. Alternative splicing and disease. *Biochim Biophys Acta.* 2009;1792(1):14–26.
49. Johnson JM, Castle J, Garrett-Engele P, et al. Genome-wide survey of human alternative pre-mRNA splicing with exon junction microarrays. *Science.* 2003;302(5653): 2141–2144.
50. Hernandez-Lopez HR, Graham SV. Alternative splicing in human tumour viruses: a therapeutic target? *Biochem J.* 2012;445(2):145–156.
51. Roizmann B, Knipe D, Whitley R. Herpes simplex viruses. In: Knipe DM, Howley PM, eds. In: *Fields Virology 6th ed. Philadelphia: Lippincott Williams & Wilkins; 2013:.*
52. Wagner EK, Flanagan WM, Devi-Rao G, et al. The herpes simplex virus latency-associated transcript is spliced during the latent phase of infection. *J Virol.* 1988;62 (12):4577–4585.
53. Korom M, Davis KL, Morrison LA. Up to four distinct polypeptides are produced from the γ34.5 open reading frame of herpes simplex virus 2. *J Virol.* 2014;88(19): 11284–11296.
54. Tang S, Guo N, Patel A, Krause PR. Herpes simplex virus 2 expresses a novel form of ICP34.5, a major viral neurovirulence factor, through regulated alternative splicing. *J Virol.* 2013;87(10):5820–5830.
55. Stoltzfus CM, Madsen JM. Role of viral splicing elements and cellular RNA binding proteins in regulation of HIV-1 alternative RNA splicing. *Curr HIV Res.* 2006;4 (1):43–55.
56. Karn J, Stoltzfus CM. Transcriptional and posttranscriptional regulation of HIV-1 gene expression. *Cold Spring Harb Perspect Med.* 2012;2(2):a006916.
57. Freed EO, Martin MA. Human immunodeficiency viruses: replication. In: Knipe DM, Howley PM, eds. In: *Fields Virology 6th ed. Philadelphia: Lippnicott Williams & Wilkins; 2013:.*
58. Goff SP. Retroviridae. In: Knipe DM, Howley PM, eds. In: *Fields Virology 6th ed. Philadelphia: Lippnicott Williams & Wilkins; 2013:.*
59. Purcell DF, Martin MA. Alternative splicing of human immunodeficiency virus type 1 mRNA modulates viral protein expression, replication, and infectivity. *J Virol.* 1993;67 (11):6365–6378.
60. Cavallari I, Rende F, apos D, Agostino DM, Ciminale V. Converging strategies in expression of human complex retroviruses. *Viruses.* 2011;3(8):1395–1414.
61. O'Reilly MM, McNally MT, Beemon KL. Two strong 5′ splice sites and competing, suboptimal 3′ splice sites involved in alternative splicing of human immunodeficiency virus type 1 RNA. *Virology.* 1995;213(2):373–385.

62. Dyhr-Mikkelsen H, Kjems J. Inefficient spliceosome assembly and abnormal branch site selection in splicing of an HIV-1 transcript in vitro. *J Biol Chem.* 1995;270 (41):24060–24066.
63. Si Z, Amendt BA, Stoltzfus CM. Splicing efficiency of human immunodeficiency virus type 1 tat RNA is determined by both a suboptimal 3′ splice site and a 10 nucleotide exon splicing silencer element located within tat exon 2. *Nucleic Acids Res.* 1997;25 (4):861–867.
64. Madsen JM, Stoltzfus CM. A suboptimal 5′ splice site downstream of HIV-1 splice site A1 is required for unspliced viral mRNA accumulation and efficient virus replication. *Retrovirology.* 2006;3:10.
65. Staffa A, Cochrane A. Identification of positive and negative splicing regulatory elements within the terminal tat-rev exon of human immunodeficiency virus type 1. *Mol Cell Biol.* 1995;15(8):4597–4605.
66. Amendt BA, Hesslein D, Chang LJ, Stoltzfus CM. Presence of negative and positive cis-acting RNA splicing elements within and flanking the first tat coding exon of human immunodeficiency virus type 1. *Mol Cell Biol.* 1994;14(6):3960–3970.
67. Amendt BA, Si ZH, Stoltzfus CM. Presence of exon splicing silencers within human immunodeficiency virus type 1 tat exon 2 and tat-rev exon 3: evidence for inhibition mediated by cellular factors. *Mol Cell Biol.* 1995;15(8):4606–4615.
68. Si ZH, Rauch D, Stoltzfus CM. The exon splicing silencer in human immunodeficiency virus type 1 Tat exon 3 is bipartite and acts early in spliceosome assembly. *Mol Cell Biol.* 1998;18(9):5404–5413.
69. Exline CM, Feng Z, Stoltzfus CM. Negative and positive mRNA splicing elements act competitively to regulate human immunodeficiency virus type 1 vif gene expression. *J Virol.* 2008;82(8):3921–3931.
70. Bilodeau PS, Domsic JK, Mayeda A, Krainer AR, Stoltzfus CM. RNA splicing at human immunodeficiency virus type 1 3′ splice site A2 is regulated by binding of hnRNP A/B proteins to an exonic splicing silencer element. *J Virol.* 2001;75 (18):8487–8497.
71. Caputi M, Mayeda A, Krainer AR. Zahler AM. hnRNP A/B proteins are required for inhibition of HIV-1 pre-mRNA splicing. *EMBO J.* 1999;18(14):4060–4067.
72. Del Gatto-Konczak F, Olive M, Gesnel MC, Breathnach R. hnRNP A1 recruited to an exon in vivo can function as an exon splicing silencer. *Mol Cell Biol.* 1999;19 (1):251–260.
73. Berro R, Kehn K, de la Fuente C, et al. Acetylated Tat regulates human immunodeficiency virus type 1 splicing through its interaction with the splicing regulator p32. *J Virol.* 2006;80(7):3189–3204.
74. Berro R, Pedati C, Kehn-Hall K, et al. CDK13, a new potential human immunodeficiency virus type 1 inhibitory factor regulating viral mRNA splicing. *J Virol.* 2008;82 (14):7155–7166.
75. Paz S, Lu ML, Takata H, Trautmann L, Caputi M. SRSF1 RNA recognition motifs are strong inhibitors of HIV-1 replication. *J Virol.* 2015;89(12):6275–6286.
76. Paz S, Krainer AR, Caputi M. HIV-1 transcription is regulated by splicing factor SRSF1. *Nucleic Acids Res.* 2014;42(22):13812–13823.
77. Tranell A, Fenyö EM, Schwartz S. Serine- and arginine-rich proteins 55 and 75 (SRp55 and SRp75) induce production of HIV-1 vpr mRNA by inhibiting the 5′-splice site of exon 3. *J Biol Chem.* 2010;285(41):31537–31547.
78. Swanson CM, Sherer NM, Malim MH. SRp40 and SRp55 promote the translation of unspliced human immunodeficiency virus type 1 RNA. *J Virol.* 2010;84(13):6748–6759.
79. Jacquenet S, Decimo D, Muriaux D, Darlix J-L. Dual effect of the SR proteins ASF/SF2, SC35 and 9G8 on HIV-1 RNA splicing and virion production. *Retrovirology.* 2005;2:33.

80. Stoltzfus CM, Fogarty SJ. Multiple regions in the Rous sarcoma virus src gene intron act in cis to affect the accumulation of unspliced RNA. *J Virol.* 1989;63(4):1669–1676.
81. Arrigo S, Beemon K. Regulation of Rous sarcoma virus RNA splicing and stability. *Mol Cell Biol.* 1988;8(11):4858–4867.
82. Gontarek RR, McNally MT, Beemon K. Mutation of an RSV intronic element abolishes both U11/U12 snRNP binding and negative regulation of splicing. *Genes Dev.* 1993;7(10):1926–1936.
83. McNally LM, McNally MT. SR protein splicing factors interact with the Rous sarcoma virus negative regulator of splicing element. *J Virol.* 1996;70(2):1163–1172.
84. Fogel BL, McNally MT. A cellular protein, hnRNP H, binds to the negative regulator of splicing element from Rous sarcoma virus. *J Biol Chem.* 2000;275(41):32371–32378.
85. Fogel BL, McNally LM, McNally MT. Efficient polyadenylation of Rous sarcoma virus RNA requires the negative regulator of splicing element. *Nucleic Acids Res.* 2002;30 (3):810–817.
86. Wentz MP, Moore BE, Cloyd MW, Berget SM, Donehower LA. A naturally arising mutation of a potential silencer of exon splicing in human immunodeficiency virus type 1 induces dominant aberrant splicing and arrests virus production. *J Virol.* 1997;71 (11):8542–8551.
87. Zahler AM, Damgaard CK, Kjems J, Caputi M. SC35 and heterogeneous nuclear ribonucleoprotein A/B proteins bind to a juxtaposed exonic splicing enhancer/exonic splicing silencer element to regulate HIV-1 tat exon 2 splicing. *J Biol Chem.* 2004;279 (11):10077–10084.
88. Kan JL, Green MR. Pre-mRNA splicing of IgM exons M1 and M2 is directed by a juxtaposed splicing enhancer and inhibitor. *Genes Dev.* 1999;13(4):462–471.
89. Tazi J, Bakkour N, Marchand V, Ayadi L, Aboufirassi A, Branlant C. Alternative splicing: regulation of HIV-1 multiplication as a target for therapeutic action. *FEBS J.* 2010;277(4):867–876.
90. Felber BK, Hadzopoulou-Cladaras M, Cladaras C, Copeland T, Pavlakis GN. rev protein of human immunodeficiency virus type 1 affects the stability and transport of the viral mRNA. *Proc Natl Acad Sci USA.* 1989;86(5):1495–1499.
91. Booth DS, Cheng Y, Frankel AD. The export receptor Crm1 forms a dimer to promote nuclear export of HIV RNA. *Elife.* 2014;3:e04121.
92. Cullen BR. Nuclear mRNA export: insights from virology. *Trends Biochem Sci.* 2003;28 (8):419–424.
93. Cullen BR. Nuclear RNA export. *J Cell Sci.* 2003;116(4):587–597.
94. Fischer U, Huber J, Boelens WC, Mattajt LW, Lührmann R. The HIV-1 Rev Activation Domain is a nuclear export signal that accesses an export pathway used by specific cellular RNAs. *Cell.* 1995;82(3):475–483.
95. Neville M, Stutz F, Lee L, Davis LI, Rosbash M. The importin-beta family member Crm1p bridges the interaction between Rev and the nuclear pore complex during nuclear export. *Curr Biol.* 1997;7(10):767–775.
96. Ernst RK, Bray M, Rekosh D, Hammarskjöld ML. A structured retroviral RNA element that mediates nucleocytoplasmic export of intron-containing RNA. *Mol Cell Biol.* 1997;17(1):135–144.
97. Taniguchi I, Mabuchi N, Ohno M. HIV-1 Rev protein specifies the viral RNA export pathway by suppressing TAP/NXF1 recruitment. *Nucleic Acids Res.* 2014;42 (10):6645–6658.
98. Stutz F. The interplay of nuclear mRNP assembly, mRNA surveillance and export. *Trends Cell Biol.* 2003;13(6):319–327.
99. McCloskey A, Taniguchi I, Shinmyozu K. Ohno M. hnRNP C tetramer measures RNA length to classify RNA polymerase II transcripts for export. *Science.* 2012;335 (6076):1643–1646.

100. Rosonina E, Blencowe BJ, Gene Expression:. The close coupling of transcription and splicing. *CurrBiol.* 2002;12(9):R319–R321.
101. Rojas-Araya B, Ohlmann T, Soto-Rifo R. Translational CONTROL of the HIV unspliced genomic RNA. *Viruses.* 2015;7(8):4326–4351.
102. Carter KL, Roizman B. Alternatively spliced mRNAs predicted to yield frame-shift proteins and stable intron 1 RNAs of the herpes simplex virus 1 regulatory gene alpha 0 accumulate in the cytoplasm of infected cells. *Proc Natl Acad Sci USA.* 1996;93 (22):12535–12540.
103. Awasthi S, Isler JA, Alwine JC. Analysis of splice variants of the immediate-early 1 region of human cytomegalovirus. *J Virol.* 2004;78(15):8191–8200.
104. Read GS. Virus-encoded endonucleases: expected and novel functions. *Wiley Interdiscip Rev RNA.* 2013;4(6). n/a - n/a.
105. Strom T, Frenkel N. Effects of herpes simplex virus on mRNA stability. *J Virol.* 1987;61 (7):2198–2207.
106. Everly DN, Feng P, Mian IS, Read GS. mRNA degradation by the virion host shutoff (Vhs) protein of herpes simplex virus: genetic and biochemical evidence that Vhs is a nuclease. *J Virol.* 2002;76(17):8560–8571.
107. Taddeo B, Zhang W, Roizman B. The herpes simplex virus host shutoff RNase degrades cellular and viral mRNAs made before infection but not viral mRNA made after infection. *J Virol.* 2013;87(8):4516–4522.
108. Berthomme H, Jacquemont B, Epstein A. The pseudorabies virus host-shutoff homolog gone: nucleotide sequence and comparison with alphaherpesvirus protein counterparts. *Virology.* 1993;193(2):1028–1032.
109. Sandri-Goldin RM. The many roles of the regulatory protein ICP27 during herpes simplex virus infection. *Front Biosci.* 2008;13:5241–5256.
110. Semmes OJ, Chen L, Sarisky RT, Gao Z, Zhong L, Hayward SD. Mta has properties of an RNA export protein and increases cytoplasmic accumulation of Epstein-Barr virus replication gene mRNA. *J Virol.* 1998;72(12):9526–9534.
111. Toth Z, Stamminger T. The human cytomegalovirus regulatory protein UL69 and its effect on mRNA export. *Front Biosci.* 2008;13:2939–2949.
112. Ote I, Lebrun M, Vandevenne P, et al. Varicella-zoster virus IE4 protein interacts with SR proteins and exports mRNAs through the TAP/NXF1 pathway. *PLoS One.* 2009;4 (11):e7882.
113. Schumann S, Jackson BR, Baquero-Perez B, Whitehouse A. Kaposi's sarcoma-associated herpesvirus ORF57 protein: exploiting all stages of viral mRNA processing. *Viruses.* 2013;5(8):1901–1923.
114. Phelan A, Carmo-Fonseca M, McLaughlan J, Lamond AI, Clements JB. A herpes simplex virus type 1 immediate-early gene product, IE63, regulates small nuclear ribonucleoprotein distribution. *Proc Natl Acad Sci USA.* 1993;90(19):9056–9060.
115. Sandri-Goldin RM, Hibbard MK, Hardwicke MA. The C-terminal repressor region of herpes simplex virus type 1 ICP27 is required for the redistribution of small nuclear ribonucleoprotein particles and splicing factor SC35; however, these alterations are not sufficient to inhibit host cell splicing. *J Virol.* 1995;69(10):6063–6076.
116. Sandri-Goldin RM, Hibbard MK. The herpes simplex virus type 1 regulatory protein ICP27 coimmunoprecipitates with anti-Sm antiserum, and the C terminus appears to be required for this interaction. *J Virol.* 1996;70(1):108–118.
117. Bryant HE, Wadd SE, Lamond AI, Silverstein SJ, Clements JB. Herpes simplex virus IE63 (ICP27) protein interacts with spliceosome-associated protein 145 and inhibits splicing prior to the first catalytic step. *J Virol.* 2001;75(9):4376–4385.
118. Champion-Arnaud P, Reed R. The prespliceosome components SAP 49 and SAP 145 interact in a complex implicated in tethering U2 snRNP to the branch site. *Genes Dev.* 1994;8(16):1974–1983.

119. Sciabica KS, Dai QJ, Sandri-Goldin RM. ICP27 interacts with SRPK1 to mediate HSV splicing inhibition by altering SR protein phosphorylation. *EMBO J.* 2003;22 (7):1608–1619.
120. Gui JF, Lane WS, Fu XD. A serine kinase regulates intracellular localization of splicing factors in the cell cycle. *Nature.* 1994;369(6482):678–682.
121. Sanford JR, Bruzik JP. Developmental regulation of SR protein phosphorylation and activity. *Genes Dev.* 1999;13(12):1513–1518.
122. Mermoud JE, Cohen PT, Lamond AI. Regulation of mammalian spliceosome assembly by a protein phosphorylation mechanism. *EMBO J.* 1994;13(23):5679–5688.
123. Kanopka A, Mühlemann O, Petersen-Mahrt S, Estmer C, Ohrmalm C, Akusjärvi G. Regulation of adenovirus alternative RNA splicing by dephosphorylation of SR proteins. *Nature.* 1998;393(6681):185–187.
124. Kanopka A, Mühlemann O, Akusjärvi G. Inhibition by SR proteins of splicing of a regulated adenovirus pre-mRNA. *Nature.* 1996;381(6582):535–538.
125. Huang Y, Steitz JA. Splicing factors SRp20 and 9G8 promote the nucleocytoplasmic export of mRNA. *Mol Cell.* 2001;7(4):899–905.
126. Escudero-Paunetto L, Li L, Hernandez FP, Sandri-Goldin RM. SR proteins SRp20 and 9G8 contribute to efficient export of herpes simplex virus 1 mRNAs. *Virology.* 2010;401 (2):155–164.
127. Malik P, Blackbourn DJ, Clements JB. The evolutionarily conserved Kaposi's sarcoma-associated herpesvirus ORF57 protein interacts with REF protein and acts as an RNA export factor. *J Biol Chem.* 2004;279(31):33001–33011.
128. Boyle SM, Ruvolo V, Gupta AK, Swaminathan S. Association with the cellular export receptor CRM 1 mediates function and intracellular localization of Epstein-Barr virus SM protein, a regulator of gene expression. *J Virol.* 1999;73(8):6872–6881.
129. Bomsztyk K, Denisenko O, Ostrowski J. hnRNP K: one protein multiple processes. *Bioessays.* 2004;26(6):629–638.
130. Malik P, Clements JB. Protein kinase CK2 phosphorylation regulates the interaction of Kaposi's sarcoma-associated herpesvirus regulatory protein ORF57 with its multifunctional partner hnRNP K. *Nucleic Acids Res.* 2004;32(18):5553–5569.
131. Wadd S, Bryant H, Filhol O, et al. The multifunctional herpes simplex virus IE63 protein interacts with heterogeneous ribonucleoprotein K and with casein kinase 2. *J Biol Chem.* 1999;274(41):28991–28998.
132. Koffa MD, Kean J, Zachos G, Rice SA, Clements JB. CK2 protein kinase is stimulated and redistributed by functional herpes simplex virus ICP27 protein. *J Virol.* 2003;77 (7):4315–4325.
133. Tsai P-L, Chiou N-T, Kuss S, Garcia-Sastre A, Lynch KW, Fontoura BMA. Cellular RNA binding proteins NS1-BP and hnRNP K regulate influenza A virus RNA splicing. *PLoS Pathog.* 2013;9(6):e1003460.
134. Dechtawewat T, Songprakhon P, Limjindaporn T, et al. Role of human heterogeneous nuclear ribonucleoprotein C1/C2 in dengue virus replication. *Virol J.* 2015;12:14.
135. Dinh PX, Das A, Franco R, Pattnaik AK. Heterogeneous nuclear ribonucleoprotein K supports vesicular stomatitis virus replication by regulating cell survival and cellular gene expression. *J Virol.* 2013;87(18):10059–10069.
136. Fan B, Lu K-Y, Reymond Sutandy FX, et al. A human proteome microarray identifies that the heterogeneous nuclear ribonucleoprotein K (hnRNP K) recognizes the 5′ terminal sequence of the hepatitis C virus RNA. *Mol Cell Proteomics.* 2014;13(1):84–92.
137. Ostareck-lederer A, Ostareck DH, Hentze MW. Cytoplasmic regulatory functions of the KH-domain proteins hnRNPs K and E1/E2. *Trends Biochem Sci.* 1998;23 (11):409–411.
138. Taddeo B, Zhang W, Roizman B. Role of herpes simplex virus ICP27 in the degradation of mRNA by virion host shutoff RNase. *J Virol.* 2010;84(19):10182–10190.

139. Pellett PE, Roizman B. Herpesviridae. In: Knipe DM, Howley PM, eds. In: *Fields Virology 6th ed. Philadelphia: Lippincott Williams & Wilkins; 2013:.*
140. Majerciak V, Yamanegi K, Allemand E, Kruhlak M, Krainer AR, Zheng Z-M. Kaposi's sarcoma-associated herpesvirus ORF57 functions as a viral splicing factor and promotes expression of intron-containing viral lytic genes in spliceosome-mediated RNA splicing. *J Virol.* 2008;82(6):2792–2801.
141. Majerciak V, Lu M, Li X, Zheng Z-M. Attenuation of the suppressive activity of cellular splicing factor SRSF3 by Kaposi sarcoma-associated herpesvirus ORF57 protein is required for RNA splicing. *RNA.* 2014;20(11):1747–1758.
142. Verma D, Swaminathan S. Epstein-Barr virus SM protein functions as an alternative splicing factor. *J Virol.* 2008;82(14):7180–7188.
143. Verma D, Bais S, Gaillard M, Swaminathan S. Epstein-barr virus SM protein utilizes cellular splicing factor SRp20 to mediate alternative splicing. *J Virol.* 2010;84 (22):11781–11789.
144. Key SCS, Yoshizaki T, Pagano JS. The Epstein-barr virus (EBV) SM protein enhances pre-mRNA processing of the EBV DNA polymerase transcript. *J Virol.* 1998;72 (11):8485–8492.
145. Spector D, Purves F, Roizman B. Role of alpha-transinducing factor (VP16) in the induction of alpha genes within the context of viral genomes. *J Virol.* 1991;65 (7):3504–3513.
146. Wilson AC, LaMarco K, Peterson MG, Herr W. The VP16 accessory protein HCF is a family of polypeptides processed from a large precursor protein. *Cell.* 1993;74 (1):115–125.
147. Ajuh P, Chusainow J, Ryder U, Lamond AI. A novel function for human factor C1 (HCF-1), a host protein required for herpes simplex virus infection, in pre-mRNA splicing. *EMBO J.* 2002;21(23):6590–6602.
148. Shaw M, Palese P. Orthomyxoviridae. In: Knipe DM, Howley PM, eds. In: *Fields Virology 6th ed. Philadelphia: Lippincott Williams & Wilkins; 2013:.*
149. Dubois J, Terrier O, Rosa-Calatrava M. Influenza viruses and mRNA splicing: doing more with less. *MBio.* 2014;5(3):e00070–e000714.
150. Selman M, Dankar SK, Forbes NE, Jia J-J, Brown EG. Adaptive mutation in influenza A virus non-structural gene is linked to host switching and induces a novel protein by alternative splicing. *Emerg Microbes Infect.* 2012;1(11):e42.
151. Lamb RA, Lai CJ, Choppin PW. Sequences of mRNAs derived from genome RNA segment 7 of influenza virus: colinear and interrupted mRNAs code for overlapping proteins. *Proc Natl Acad Sci USA.* 1981;78(7):4170–4174.
152. Shih SR, Suen PC, Chen YS, Chang SC. A novel spliced transcript of influenza A/WSN/33 virus. *Virus Genes.* 1998;17(2):179–183.
153. Lamb RA, Lai CJ. Sequence of interrupted and uninterrupted mRNAs and cloned DNA coding for the two overlapping nonstructural proteins of influenza virus. *Cell.* 1980;21(2):475–485.
154. Beauchemin C, Samuel J, Tuszynski J. A simple cellular automaton model for influenza A viral infections. *J Theor Biol.* 2005;232(2):223–234.
155. Josset L, Frobert E, Rosa-Calatrava M. Influenza A replication and host nuclear compartments: many changes and many questions. *J Clin Virol.* 2008;43(4):381–390.
156. Fortes P, Lamond AI, Ortín J. Influenza virus NS1 protein alters the subnuclear localization of cellular splicing components. *J Gen Virol.* 1995;76(Pt 4):1001–1007.
157. Jorba N, Juarez S, Torreira E, et al. Analysis of the interaction of influenza virus polymerase complex with human cell factors. *Proteomics.* 2008;8(10):2077–2088.
158. Mayer D, Molawi K, Martínez-Sobrido L, et al. Identification of cellular interaction partners of the influenza virus ribonucleoprotein complex and polymerase complex using proteomic-based approaches. *J Proteome Res.* 2007;6(2):672–682.

159. Plotch SJ, Krug RM. In vitro splicing of influenza viral NS1 mRNA and NS1-beta-globin chimeras: possible mechanisms for the control of viral mRNA splicing. *Proc Natl Acad Sci USA.* 1986;83(15):5444–5448.
160. Chua MA, Schmid S, Perez JT, Langlois RA, Tenoever BR. Influenza A virus utilizes suboptimal splicing to coordinate the timing of infection. *Cell Rep.* 2013;3(1):23–29.
161. Solnick D. Alternative splicing caused by RNA secondary structure. *Cell.* 1985;43 (3):667–676.
162. Buratti E, Baralle FE. Influence of RNA secondary structure on the pre-mRNA splicing process. *Mol Cell Biol.* 2004;24(24):10505–10514.
163. Krug RM, Yuan W, Noah DL, Latham AG. Intracellular warfare between human influenza viruses and human cells: the roles of the viral NS1 protein. *Virology.* 2003;309(2):181–189.
164. Hale BG, Randall RE, Ortín J, Jackson D. The multifunctional NS1 protein of influenza A viruses. *J Gen Virol.* 2008;89(Pt 10):2359–2376.
165. Wolff T, O'Neill RE, Palese P. NS1-Binding protein (NS1-BP): a novel human protein that interacts with the influenza A virus nonstructural NS1 protein is relocalized in the nuclei of infected cells. *J Virol.* 1998;72(9):7170–7180.
166. Shih SR, Nemeroff ME, Krug RM. The choice of alternative 5′ splice sites in influenza virus M1 mRNA is regulated by the viral polymerase complex. *Proc Natl Acad Sci USA.* 1995;92(14):6324–6328.
167. Shih SR, Krug RM. Novel exploitation of a nuclear function by influenza virus: the cellular SF2/ASF splicing factor controls the amount of the essential viral M2 ion channel protein in infected cells. *EMBO J.* 1996;15(19):5415–5427.
168. Qiu Y, Nemeroff M, Krug RM. The influenza virus NS1 protein binds to a specific region in human U6 snRNA and inhibits U6-U2 and U6-U4 snRNA interactions during splicing. *RNA.* 1995;1(3):304–316.
169. Wang W, Krug RM. U6atac snRNA, the highly divergent counterpart of U6 snRNA, is the specific target that mediates inhibition of AT-AC splicing by the influenza virus NS1 protein. *RNA.* 1998;4(1):55–64.
170. Perez JT, Varble A, Sachidanandam R, et al. Influenza A virus-generated small RNAs regulate the switch from transcription to replication. *Proc Natl Acad Sci USA.* 2010;107 (25):11525–11530.
171. Nemeroff ME, Barabino SM, Li Y, Keller W, Krug RM. Influenza virus NS1 protein interacts with the cellular 30 kDa subunit of CPSF and inhibits 3′end formation of cellular pre-mRNAs. *Mol Cell.* 1998;1(7):991–1000.
172. Satterly N, Tsai P-L, van Deursen J, et al. Influenza virus targets the mRNA export machinery and the nuclear pore complex. *Proc Natl Acad Sci USA.* 2007;104 (6):1853–1858.
173. Robertson JS, Schubert M, Lazzarini RA. Polyadenylation sites for influenza virus mRNA. *J Virol.* 1981;38(1):157–163.
174. Logg CR, Baranick BT, Lemp NA, Kasahara N. Adaptive evolution of a tagged chimeric gamma retrovirus: identification of novel cis-acting elements that modulate splicing. *J Mol Biol.* 2007;369(5):1214–1229.

CHAPTER NINE

Insights From Genomics Into Spatial and Temporal Variation in *Batrachochytrium dendrobatidis*

A.Q. Byrne*, J. Voyles†, G. Rios-Sotelo†, E.B. Rosenblum*,[1]
* Department of Environmental Science Policy and Management, University of California, Berkeley, CA, United States
† Department of Biology, University of Nevada, Reno, NV, United States
[1] Corresponding author. E-mail address: rosenblum@berkeley.edu

Contents

Abstract

Advances in genetics and genomics have provided new tools for the study of emerging infectious diseases. Researchers can now move quickly from simple hypotheses to complex explanations for pathogen origin, spread, and mechanisms of virulence. Here we focus on the application of genomics to understanding the biology of the fungal pathogen *Batrachochytrium dendrobatidis* (Bd), a novel and deadly pathogen of amphibians. We provide a brief history of the system, then focus on key insights into Bd variation garnered from genomics approaches, and finally, highlight new frontiers for future discoveries. Genomic tools have revealed unexpected complexity and variation in the Bd system suggesting that the history and biology of emerging pathogens may not be as simple as they initially seem.

Progress in Molecular Biology and Translational Science, Volume 142
ISSN 1877-1173
http://dx.doi.org/10.1016/bs.pmbts.2016.05.009

Batrachochytrium dendrobatidis (Bd) is a fungal pathogen of amphibians that causes the lethal disease chytridiomycosis. Since its initial discovery in 1998, Bd has been inextricably linked to precipitous amphibian declines occurring around the world[1–3] (Fig. 1). To date, no other pathogen is known to have had such a dramatic effect on such a broad range of host species and in so many different environments.[1] As a result, Bd-related declines have been dubbed "the most spectacular loss of biodiversity due to disease in recorded history".[4]

Bd is a member of an ancient lineage of fungi, the Chytidiomycota, which is in the order Rhizophydiales.[5] The Chytridiomycota are relatively diverse, and most chytrids are saprobes and are primarily aquatic or found within wet soils.[6] Bd was the first chytrid recognized to cause disease and mortality in vertebrate hosts.[7,8] A second related chytrid, *Batrachochytrium salamandrivorans* (Bsal), was recently isolated and found to cause lethal disease in salamanders,[9] but is not discussed in detail here.

Chytridiomycosis has garnered considerable interest from the scientific community since the discovery of Bd in 1998.[7,8] Early investigations used

Figure 1 A diseased frog (*Pristimantis cruentus*) found during a chytridiomycosis outbreak in Panama. Bd, the fungal pathogen that causes chytridiomycosis, has been linked to dramatic amphibian declines around the world.[7] Photo by Jamie Voyles.

ecological, physiological, and immunological methods to understand the effect of Bd on amphibians at the organismal, population, and community scales (reviewed in Refs. [10,11]). However, many questions about Bd origin, emergence, spread, and virulence remained unanswered for many years. In the last decade, molecular approaches have provided key insights into the history and virulence of this pathogen. From early phylogenetic studies with few genetic markers to recent whole genome analyses, genetic and genomic tools have revealed that chytridiomycosis is a far more complex disease system than initially appreciated.

Here, we highlight how molecular approaches have contributed to our understanding of Bd by providing a more mechanistic and nuanced understanding of disease ecology and the epidemiology of chytridiomycosis. We also outline some of the most important unanswered questions and highlight new frontiers for discovery. Throughout this review, we focus on how genomic tools can help reveal variation and complexity in the Bd system. A more in-depth understanding of Bd history, variation, and virulence will lead to a richer perspective on fungal pathogenesis, epidemiology and host–pathogen dynamics, and may also point to practical conservation solutions for emerging diseases of wildlife.

1. INSIGHTS FROM GENOMICS INTO Bd VARIATION

1.1 A Molecular Toolkit for Bd

The study of Bd genetics and genomics has evolved over the last decade as new technologies and genomic resources have become available. There are several key molecular approaches that have been applied to Bd research. Quantitative PCR is commonly used as a diagnostic assay to detect Bd from amphibian skin swabs.[12,13] The advent of a qPCR assay allowed researchers to document Bd presence and infection intensity from noninvasive skin swab samples collected in laboratory experiments, field studies[12,13] and from museum specimens.[14,15] To date, the use of swabs for population genetics has been limited by their low quality and/or quantity of DNA. Some studies have sequenced genetic markers from swabs, but these have been limited to the ribosomal DNA internal transcribed spacer (ITS)[16,17] or a small number of polymorphic repetitive loci.[18]

Population genetics and genomics studies have more commonly relied on Bd cultures that must be laboriously isolated and maintained. Many studies have used traditional Sanger sequencing of nuclear markers to characterize Bd genetic diversity.[16,19] However, whole genome resequencing became feasible after the first Bd genomes were sequenced and as sequencing costs decreased.[20,21] Whole genome sequencing provides thousands of markers for population genomics and allows analysis of structural variation in the Bd genome.[20,21] Whole genome resources for Bd have also made gene expression studies possible,[22,23] which have contributed to our understanding of the pathogenesis of chytridiomycosis. Insights from these molecular approaches into Bd variation at the phylogenetic and geographic levels and at the structural and functional levels are discussed later in the chapter.

1.2 Phylogenetic and Geographic Variation

One of the earliest applications of molecular tools to Bd research focused on resolving an early debate about the emergence and spread of Bd. Specifically, researchers weighed the evidence in support of the "Novel Pathogen Hypothesis" versus the "Endemic Pathogen Hypothesis" (NPH vs EPH[24]). The NPH postulates that Bd is a novel pathogen that recently spread around the world, possibly via carrier hosts, and caused epizootic events in naive amphibian communities.[24,25] In contrast, the EPH suggests that Bd has been historically endemic in amphibian populations, but could have recently become a lethal pathogen due to an environmental perturbation (eg, climate change[26]). At the outset, the answer to the NPH versus EPH question seemed relatively simple because chytridiomycosis appeared to fit a typical emerging disease model.[24] Specifically, Bd was not ubiquitous in amphibian populations, and researchers identified "fronts" or "waves" of disease in locations where mass mortality events occurred following the arrival of Bd (eg, Central America,[27] South America,[28] California[2,29]). In addition, researchers identified amphibian species that appeared to be tolerant of Bd,[30,31] suggesting that they could act as reservoir hosts and facilitate pathogen transmission in wild communities and in the global commercial trade.[25,32]

The first study to evaluate the emergence and spread of Bd with molecular tools used a relatively small number of sequenced loci to assess genetic diversity among 35 isolates.[33] Morehouse et al.[33] recovered only five variable sites in 10 loci surveyed. Based on the finding of low genetic variation among isolates and lack of discrimination among isolates from different parts of the world, Morehouse et al.[33] suggested that Bd was a recently emerged clone. Despite limited resolution due to few genetic markers, this study was formative for a wide range of subsequent research on Bd emergence and spread, specifically because it seemed to support the NPH.[24]

Subsequent molecular studies have revealed a more complex picture: recent emergence and weak geographic structuring for some Bd clades contrasts with deeply divergent, more endemic patterns for other clades.[20,21] These more recent studies have refined our understanding of phylogenetic diversity in Bd by sampling a wider panel of Bd isolates (ie, collected from diverse geographic locations and host species) and a wider panel of markers. Some of these studies have used traditional Sanger sequencing,[16,19] whereas others have relied on whole-genome sequencing.[20,21] Genome-scale approaches have been particularly fruitful and were made possible by two full Bd genomes sequenced by genome centers (JEL423 isolated from *Phyllomedusa lemur* in Panama and sequenced by the Broad Institute; and JAM81 isolated from *Rana muscosa* in California, United States and sequenced by the Joint Genome Institute). The studies that leveraged these genomes used resequencing approaches to obtain genomic data for global panels of Bd isolates collected from a variety of host species.[20,21] These studies used tens of thousands of variable sites, termed single nucleotide polymorphisms (SNPs), to obtain more resolution on the structure of the Bd phylogenetic tree.[20,21]

Taken together, recent studies have revealed previously unappreciated complexity within the Bd phylogeny. Bd contains a geographically widespread lineage that was named the "Global Panzootic Lineage" (or "BdGPL") that appeared to be associated with areas where chytridiomycosis epizootic events occurred.[20] Although BdGPL appears to be a globally distributed Bd clade, some weak geographic substructure has been reported. Schloegel et al.[34] sequenced 45 strains at 36 loci and named two subclades within the GPL. These authors suggested that there is a North American clade (GPLI) and a less genetically diverse globally dispersed clade (GPLII), although subsequent studies have indicated that GPLI is not restricted to North America.[21] Furthermore, Schloegel et al.[34] named a clade of highly divergent isolates from Brazil (Bd-Brazil). The occurrence of a fairly basal Bd-Brazil clade was corroborated by later sequencing studies,[21,35] and this clade may have been present in frog populations in Brazil as far back as the late 1800s.[36] Bd also contains at least two other geographically restricted clades: one from South Africa that may have been transported to Mallorca, Spain (Bd-CAPE) and one found only in Switzerland (Bd-CH).[21] In addition, molecular studies have revealed a number of isolates that appear to be hybrids,[20,21,34,35] raising important questions about the potential for interaction among strains.

Phylogenetic studies of Bd at a global scale have been complemented by molecular studies that focus on questions at finer spatial scales. For example, a study by Velo-Anton et al.[37] investigated the question of regional genetic diversity in Bd as it spread from North to South in the Americas. Velo-Anton et al.[37] found that Bd isolates showed a pattern of reduced heterozygosity and increased allele fixation along the established North to South invasion path, providing valuable information on the genetic signature of Bd invasion at a continental scale. Other regional population genetics studies have found signatures of long-term endemism of basal Bd lineages. Studies from Japan, China, and Korea have reported several divergent Bd haplotypes, suggesting a longer-term association of Bd with Asian amphibians.[16,17,37,38] However, studies from introduced bullfrogs in China and Korea also reported the presence of Bd-Brazil and BdGPL lineages.[17,37,38] Nevertheless, these results should be interpreted with caution given the difficulty interpreting ITS data (single Bd strains often contain many ITS haplotypes[34]) and because different molecular markers (ITS vs MLST) have given conflicting results.

At finer spatial scales, molecular approaches have elucidated patterns of regional spread and local disease dynamics. Evidence that a single genotype was introduced and spread throughout the Iberian Peninsula points to NPH dynamics in that system.[39] A population genetics study using microsatellite loci conducted in the alpine lakes of the Sierra Nevada Mountains showed that two Bd introductions may have occurred there,[19] although the reliance on relatively few markers make it challenging to draw firm conclusions. Findings such as these have shown that Bd may be introduced and spread rapidly in some regions, but Bd can also show signs of local endemism in other regions.

1.3 Structural and Functional Genomic Variation

Genome resequencing studies have been used to understand not only phylogenetic relationships among Bd isolates but also the evolutionary dynamics of the Bd genome itself. Overall, these studies reveal that Bd has a surprisingly dynamic genome and point to possible molecular mechanisms of virulence in this pathogen. One of the most dramatic patterns observed is that of chromosomal copy number variation (CNV)[21]; specifically, the number of copies of chromosomal segments varies both within and across isolates (from 1 to 5 copies[21]). Therefore, rather than being a strictly diploid species, Bd exhibits a high degree of variation in copy number, and this CNV may be one mechanism facilitating rapid genomic change.[20] For example, a recent study found variation in copy number even over extremely short time scales: a single isolate maintained in the lab exhibited changes in copy number at more than half of the largest chromosomal segments over only 30 generations.[40] Another large-scale pattern commonly observed in the Bd genome is loss of heterozygosity (LOH). James et al.[41] provided early evidence for LOH that can occur through mitotic recombination. Since then, other genetic and genomic studies have shown that LOH events can occur quickly and affect large regions of the genome.[20,21,37]

The signature of mitotic—or asexual—recombination is pervasive in the Bd genome, but there is also evidence that Bd may occasionally exhibit meiotic—or sexual—recombination. Originally, Morehouse et al.[33] suggested that Bd reproduces primarily asexually. However, even early molecular work recognized that the diversity among Bd isolates could not be fully explained by clonal reproduction.[19] It was hypothesized that sexual reproduction must occur at least occasionally in Bd.[41] Although sexual reproduction has never been directly observed, there is increasing evidence that hybridization may play an important role in dynamics of genome evolution in Bd. Several studies have found evidence that some Bd isolates resulted from recent hybridization.[20,21,35] In fact, some authors suggest that hybridization may have been important to the initial emergence of Bd,[20] although other studies have not explicitly supported this hypothesis.[42]

Genomic patterns such as CNV, LOH, and hybridization have not yet unequivocally been linked to variation in Bd virulence. However, there is mounting evidence that these processes may ultimately have important functional effects. For example, both Refsnider et al.[40] and Piovia-Scott et al.[43] showed that more virulent isolates exhibited elevated chromosomal copy numbers relative to less virulent isolates. These findings are consistent with research that has found functional effects of CNV in other fungal pathogens.[44,45]

Genomic data have also provided hypotheses about specific genomic regions that may play a role in Bd virulence. One method for identifying candidate genes involved in Bd pathogenicity has been to compare the genome of Bd with those of other pathogenic fungi. For example, Abramyan and Stajich[46] identified genes for cell surface proteins in Bd that are orthologs to virulence factors in a rice blast fungus (*Magnaporthe oryzae*). They also identified genes for molecules within a carbohydrate-binding module family that have been linked to pathogenicity in other fungal pathogens.[46]

Another method for identifying functionally relevant genes has been to look for gene family expansions in the Bd genome compared with the genomes of other non-Bd chytrids. For example, Joneson et al.[47] compared the Bd genome to that of the closely related, nonpathogenic species

Homolaphylctis polyrhiza (Hp, a saprobic fungus that is found in leaf litter and does not parasitize amphibians). Joneson et al.[47] found significant expansion in three protease families, which occurred after Bd and Hp diverged from their common ancestor. These protease families include metalloproteases, serine-type proteases, and aspartyl proteases, which have been implicated in pathogenesis in other fungal pathogens (ie, for adherence to, invasion of, and degradation of host cells).[48,49]

Gene expression studies have provided yet another approach for identifying genes that may be important for Bd pathogenesis.[22,23] Genome scale gene expression studies have revealed high degrees of condition-specific variation in Bd gene expression.[22,23] For example, Bd life stages exhibit striking differences in gene expression, providing candidate pathways that may be involved in growth, infection, and pathogenicity across life stages.[22] Bd has two main life stages[8,11,22] (Fig. 2). The dispersal stage is the infectious *zoospore*, which moves with a posterior flagellum. The zoospore encysts,

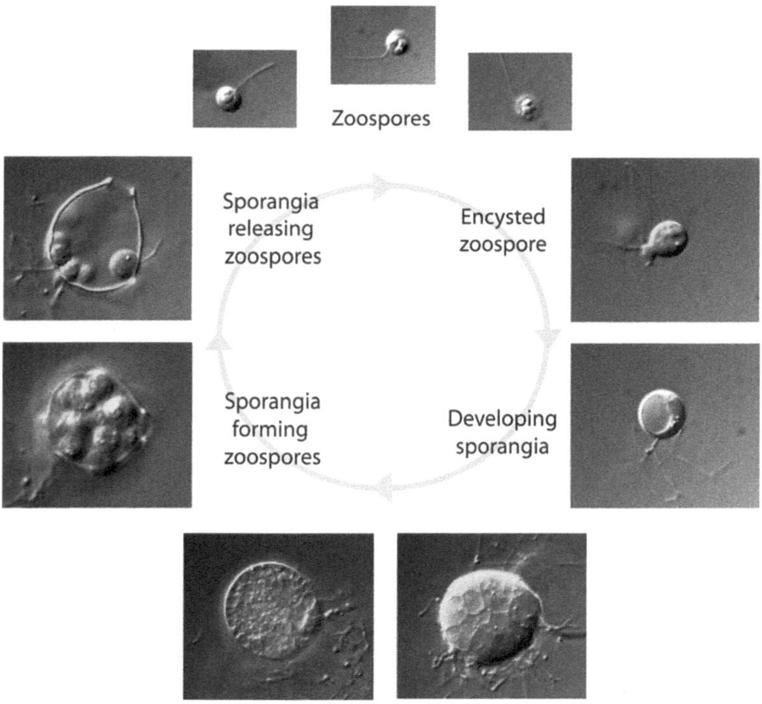

Figure 2 Life cycle of *Batrachochytrium dendrobatidis*. The dispersal stage is the infectious *zoospore*, and the reproductive stage is the *sporangium*. Bd infection of amphibians occurs in the epidermis and disrupts key physiological functions of the skin such as osmoregulation and electrolyte balance.[11]

absorbs the flagellum, and develops rhizoids. The maturing thallus then develops into the reproductive *zoosporangium* (ie, container for zoospores) in which the cytoplasm cleaves and forms flagellated zoospores.[8,11,22] When zoospores are mature, a discharge tube forms and zoospores are released.[8,11,22] The two key life stages (*zoospore* and *sporangia*) have highly differentiated expression profiles with more than half the genes in the genome exhibiting differential expression.[22] Although zoospores are transcriptionally less complex, several transcripts were more abundant in zoospores (including those transcribed from signaling and chitin-binding genes) and may influence Bd colonization.[22] Sporangia, on the other hand, were metabolically and transcriptionally complex, with many genes showing signatures of increased expression during this life stage.[22] Most notably, there was an increase in expression of multiple fungalysin metallopeptidases and serine-type peptidase genes in the sporangia stage, gene families that were previously hypothesized to be candidate virulence factors.[22]

Many Bd genes also show expression changes under different growth conditions. To date, studies on environmentally induced gene expression have focused primarily on Bd genes induced when Bd is colonizing amphibian host tissue.[23] Such studies can identify possible virulence factors that are upregulated during host-cell colonization, a key step in pathogenesis.[11,23] For example, one genome wide study of Bd gene expression on laboratory growth media and host amphibian skin showed that when Bd is exposed to amphibian skin, a number of putative Bd pathogenicity factors were upregulated.[23] In particular, metallo-, serine-, and aspartyl-proteases showed increased expression when Bd was cultured on amphibian skin, supporting findings from previous studies[23] and contributing to the search for putative virulence genes. Other Bd genes of interests that were upregulated on amphibian skin included crinkler and crinkler-like effectors, adhesin genes, and triglyceride lipases, again providing targets for future more focused studies on Bd pathogenesis.[22,23]

2. NEW FRONTIERS IN Bd GENOMICS

2.1 An Evolving Genomic Toolkit

Emerging molecular methodologies are revolutionizing the fields of disease ecology and evolutionary biology. Genomic studies are expanding in scale and power while decreasing in cost. By adopting emerging technologies, researchers can now address previously intractable questions in the Bd system. The application of new molecular tools to the Bd system will allow for a fine-scale understanding of genetic diversity across landscapes and through time and a more refined understanding of the functional role of Bd genes. Whole genome sequencing will continue to be an important approach, especially as sequencing costs continue to decrease. However, there are other promising applications of genomics to the Bd system.

Genome sequences from globally distributed Bd isolates provide a resource for identifying polymorphic regions of the Bd genome. Resequencing data can be used to selectively target diagnostic SNPs for the major and minor clade divisions in the Bd phylogeny or segregating SNPs within geographic regions.[20,21] Researchers have used genomic data from ascertainment panels to locate target regions for SNP genotyping in many different study systems.[50–52] However, SNP genotyping approaches can suffer from ascertainment bias.[53] New (microfluidic) multiplex PCR techniques can circumvent some of these issues and expand the scale of Bd studies. Hundreds of genomic regions (several hundred base pair regions rather than SNPs) can now be sequenced simultaneously and cost-effectively. Sample sizes can be increased dramatically because samples can be barcoded and pooled for sequencing, and hundreds or thousands of samples can be sequenced simultaneously. Therefore, researchers no longer need to use relatively costly whole genome sequencing to obtain population genomic data for Bd.

In the past, genomic approaches for Bd generally required pure Bd cultures. However, new multiplex PCR approaches allow for the amplification of many regions of the Bd genome with low quantities of starting DNA. These multiplex genotyping approaches are especially amenable to Bd DNA collected from skin swabs, eDNA samples, or from museum specimens (where DNA quality and quantity are suboptimal). A number of studies have reported successful genotyping of highly fragmented DNA with this approach using bird feathers;[54] historical fish scales;[55] and noninvasively collected mammal fur, feces, and urine.[56] Therefore, new technologies promise to unlock the genetic information stored in the vast libraries of swab samples that have been collected around the world over many decades.

Another powerful approach for the Bd system is RNA sequencing (RNA-seq), which has been used to characterize functional genomic responses in many model and nonmodel systems.[57,58] RNA-seq offers a number of benefits over the microarray techniques that have been used to study global Bd gene expression in the past. First, researchers no longer need to invest sequencing and bioinformatic resources toward microarray development (although having a transcriptome sequence is helpful for aligning reads from RNA-seq). Second, RNA-seq data can be used not only for gene expression, but also for population genomics inferences. Third, dual RNA-seq can be used to characterize multiple species responses in parallel.[59] In silico techniques can be used with RNA-seq data to separate host and pathogen transcripts and thus provide a powerful tool to investigate disease dynamics from both host and pathogen perspectives simultaneously.[59,60]

Finally, more manipulative molecular approaches may find application in the Bd system. For example, gene knockdown techniques such as posttranscriptional gene silencing via RNA interference[61] and/or genome editing experiments using techniques such as CRISPR/Cas9[62] can be used to test hypotheses about Bd gene function. Some Bd traits raise challenges for implementing these techniques. Relative to the model fungal organisms in which many of these techniques have been developed and applied (eg, filamentous fungi[63,64]), Bd has an unusual life cycle and complex genetic architecture.[11,20–22] Thus far, it has not been straightforward to develop a transformation system for Bd or to express Bd genes in other expression vectors. However, manipulative genetic approaches hold great promise for functional studies in Bd.

2.2 Applying Genomics to Understand Temporal Variation in Bd

Temporal dynamics represent an important frontier in Bd research. A focus on temporal dynamics over multiple time scales will provide key insights into epidemiological and evolutionary patterns in this host–pathogen system. Molecular tools can be applied to understand genetic variation in Bd over seasonal time scales, during and following epizootic events, and over evolutionary time scales. Molecular tools can also be applied to Bd in both natural and laboratory settings, providing new insights on temporal disease dynamics.

Bd-related amphibian declines have elicited concerted sampling efforts in many affected regions of the world.[2,27–29] As a result, many research groups have built a collection of Bd skin swabs and Bd isolates from natural systems before, during, and after epizootic events.[2,27–29] These swabs have primarily been used to document the prevalence and intensity of Bd in natural systems using qPCR.[12,13] However, the multiplex genotyping methods described previously, designed for use with low quality and quantity DNA samples, now provide an opportunity to obtain temporally stratified population genetic data from Bd skin swabs.

Genotyping swabs can provide insights into a number of unanswered temporal questions in the Bd system. For example, genotyping Bd swabs across seasons will provide vital information on possible annual fluctuations in the strain composition of the Bd community. Similarly, genotyping swabs that were collected during waves of colonization and spread will help reveal dynamics of genetic diversity over longer time scales (eg, whether Bd typically loses genetic diversity during colonization waves, as has been suggested in some systems).[37] Comparing genotypes of Bd in epizootic and enzootic environments will also be important to understand the hallmarks of genetic diversity as Bd becomes established in new regions and to determine whether there is strain turnover over time.

Another exciting application of multiplex genotyping techniques is to garner molecular data from preserved museum specimens. Molecular studies of museums specimens are increasingly common (reviewed in Ref. [65]), and recently published protocols offer the possibility of obtaining reliable DNA sequences from specimens preserved in formalin (a common preservation methods for amphibians).[66] Multiple studies have documented the presence of Bd on preserved museum specimens and have provided important information on the historical distribution of this pathogen.[14,15,36] However, these studies focused on historical presence/absence of Bd and did not genotype historical Bd samples. Genotyping assays designed with short target lengths are particularly amenable to the often-fragmented DNA present in historical samples. Population genetics analyses on swab data will provide a powerful investigation into the NPH versus EPH question at local and global scales, will shed light on the possible origin of the BdGPL, and will provide an understanding of changes in Bd genetic diversity in natural systems over longer time scales.

Although swabs represent a tremendous untapped resource, obtaining Bd isolates will continue to be essential for understanding evolution at the molecular level. At least for the near future, whole genome sequencing will still be relatively restricted to pure isolates that have high DNA quantity and quality. When researchers isolated and cryoarchived Bd from natural populations, more mechanistic temporal questions about selection can be addressed. For example, comparing the genomes of historical and contemporary Bd isolates may allow researchers to identify signatures of selection in the genome.

Temporal questions about genome evolution and fundamental evolutionary hypotheses about fungal disease dynamics can also be addressed using controlled laboratory experiments. Experimental evolution studies have provided important insights into microbial evolution and species interactions in many systems.[67,68] Bd is well suited for experimental evolution approaches, being easily cultivable and amenable to cryogenic freezing.[69] Prior studies have also shown that Bd can undergo rapid evolution in the lab, observable at both the genotypic and phenotypic levels.[67,70,71] Factors such as passage timing,[67] passage number,[71] and temperature[70] can potentially be agents of selection in the lab. Future experimental evolution studies can solidify putative links between genomic and phenotypic changes with well-designed replication across strains and treatment groups. Passaging Bd under different environmental conditions or different passage regimes (and periodically cryo-archiving isolates for future infection experiments) can reveal how quickly Bd virulence can evolve. Furthermore, exposing Bd to a wide range of selective environments, including re-isolating from host species,[72] will provide a more mechanistic understanding of how Bd can evolve in response to particular hosts or environments. Pairing experimental evolution studies with whole genome sequencing will allow researchers to identify regions of the genome that are correlated with observed phenotypic shifts and will likely reveal more specific molecular targets for selection and virulence in the Bd genome.

2.3 Applying Genomics to Understand Spatial Variation in Bd

Many of the molecular approaches described previously to study temporal variation in Bd can also be applied to future studies of spatial variation in Bd. Although a number of studies have used genetic and genomic approaches to understand Bd genetic diversity at the global scale,[20,21,34] Bd dynamics across finer spatial scales remain understudied. To date, studies of fine-scale genetic diversity in Bd have relied on relatively few samples and/or relatively few genetic markers.[16,17]

Key questions about spatial variation in Bd can now be addressed with whole genome sequencing from cultures and with multiplex genotyping from swab samples and museum specimens. Specifically, spatial variation in Bd can be assessed at multiple nested scales. For example, researchers can study genetic variation of Bd within a single host individual, within a population of a single host species, within a community of multiple host species, and within geographic regions.

Understanding the spatial population structure of Bd at nested scales will help reveal how different Bd strains are distributed across environments and among species and how different biotic or abiotic factors correlate with Bd genetic variation. Large sample sizes across environmental conditions and host species could provide insight into the selective pressures acting on Bd in natural systems. Studies of spatial genetic variation could also help resolve dynamics of Bd dispersal over both short and long distances. Comparing Bd genotypes in vulnerable species to Bd genotypes found in the environment (from eDNA samples) or potential reservoir species could also help reveal pathways of infection and transmission across species. For example, if Bd genotypes collected from single host species are typically more similar to each other than Bd genotypes collected from other hosts in the same community, it may suggest conspecific contact as a primary means of Bd dispersal. Conversely, if Bd genotypes from all host species in a community tend to be similar to each other and to Bd genotypes collected from the environment via eDNA sampling (as done in Schmidt et al.,[73] Chestnut et al.[74]), it may suggest the primacy of environmental reservoirs for Bd transmission. Ultimately, population genomics assays with large sample sizes will provide answers to outstanding questions about Bd dispersal, transmission, and evolution across spatial scales.

2.4 Applying Genomics to Understand Functional Variation in Bd

Building on the work of previous functional genomics Bd studies that characterized gene expression across life stages[22] and growth substrates,[23] future studies can capture additional functional variation and better link genotype to phenotype in Bd. Through the wider adoption of genomic methods in experimental and natural studies of Bd, functional genomics can help resolve outstanding questions about virulence in Bd.

To date, no published studies have characterized Bd gene expression from live hosts, across different host species, or in natural systems. Expanding the current scope of transcriptomic investigations in these ways will provide insights on the specificity of Bd gene expression in different environments. Studies of Bd in more natural conditions could also benefit from a dual RNA-seq approach,[59] whereby gene expression in the host and the pathogen are measured simultaneously. Integrating host and pathogen perspectives in this way could reveal how different host species respond to infection and how Bd gene expression varies under different host conditions.

Studies of Bd in the laboratory will also remain critical for understanding molecular and functional mechanisms of Bd virulence. A particularly intriguing area for gene expression approaches in the Bd system is the interaction of Bd with the amphibian immune system. Previous research suggests that Bd may be able to suppress the host immune system.[75] Some susceptible host species seem to lack a robust immunogenetic response to Bd,[76] although immunogenetic responses have been shown to vary widely among susceptible species.[77,78] To better understand the mechanisms Bd uses to evade or suppress the amphibian immune system, genomics approaches could be integrated with lymphocyte challenge assays, which are now a common tool for phenotypically profiling the ability of different Bd strains to kill amphibian immune cells.[75,79] Implementing a dual RNA-Seq approach[59] using lymphocytes cultured with different Bd strains could provide valuable information on the interaction of host and pathogen at the cellular level and could potentially identify genes induced during the host–pathogen interaction.

In addition to characterizing how Bd interacts with host immune defenses, understanding the functional response of Bd to antifungal treatments is an interesting avenue for future research. Many anti-Bd treatments have been proposed or experimentally tested, including antifungal chemicals, bacteria that produce antifungal compounds, and heat.[80–82] Characterizing Bd's transcriptional response to these treatments could provide insights into how different treatments affect the pathogen at a mechanistic level and could be important for designing new antifungal treatment strategies.

Finally, a critical limitation of gene expression work on Bd to date has been the difficulty of unambiguously linking molecular mechanisms to specific virulence traits. A number of studies have proposed specific gene families, specific patterns of gene expression, or specific genomic processes that could contribute to Bd virulence.[40,43,46,47] However, no studies have robustly linked genotypic and phenotypic variation in Bd. Two approaches offer great promise for the development of a more mechanistic understanding of Bd virulence. First, genome wide association studies could be useful, but only if replicated sets of isolates can be found with and without particular virulence traits. Second, genetic manipulation will ultimately be necessary to conclusively demonstrate the role of particular genes or gene families in Bd virulence. As described previously, there are a number of challenges for applying tools such as RNA interference[61] or CRISPR[62] to the Bd system, but ultimately it will be important to apply manipulative techniques and use insights gained from molecular studies of model fungi to Bd.

3. CONCLUSIONS

Molecular approaches have already provided key insights into the history and virulence of Bd. What was once thought to be a simple story of an emerging pathogen has proven to be a more complex tale. Genomics studies have played a pivotal role in describing phylogenetic and functional variation in Bd and will continue to do so as new molecular approaches are adopted.

The future of genomics research on Bd will now be shaped by how researchers apply emerging molecular techniques to samples collected across space and time (Table 1). It is now possible to couple a wide array of molecular approaches (eg, multiplex PCR genotyping, whole genome sequencing, gene expression analysis, genome manipulation) with a diversity of sampling approaches (eg, swab samples, pure cultures, museum specimens, eDNA samples). For example, multiplex PCR genotyping can be applied to low DNA quantity samples (eg, swabs collected from live animals, museum specimens, eDNA samples) to unlock a wealth of data about Bd diversity across both spatial and temporal gradients. In addition, whole genome DNA and RNA sequencing of cultured Bd isolates collected from the wild and experimentally manipulated in the lab will refine our understanding of evolutionary change in Bd at the molecular level. Furthermore, functional genomics experiments and genomic manipulations can shed light on mechanisms of virulence in this pathogen.

Molecular approaches can advance fundamental knowledge about host–pathogen interactions and can also inform conservation practice. For example, understanding phylogenetic, geographic, and host-associated patterns of Bd variation could help researchers predict virulence of newly invaded Bd isolates. Similarly, identifying regions of the Bd genome under selection and determining the molecular basis of Bd virulence could open new avenues for vaccine development or other chytridiomycosis treatments. Ultimately, developing a mechanistic understanding of spatial and temporal dynamics of Bd in natural systems will be necessary to predict future disease outcomes and mitigate further loss of amphibian biodiversity.

Our focus here has been on application of molecular tools to the study of Bd, but genomics data must be interpreted in light of other complementary datasets to provide a robust understanding of host–pathogen dynamics in chytridiomycosis. Genomics, ecological, evolutionary, epidemiological, physiological, immunological, and mathematical approaches that focus on

Table 1 Common Sampling Methods for Obtaining Genetic Material From Bd and Accompanying Challenges for Genomic Analysis.

Sampling Method	Associated Challenges	Key Genomic Approaches	Potential Novel Contributions
Field-collected swabs	Low quantity/quality DNA	Multiplex PCR genotyping	Genetic variation in Bd across spatial and temporal scales in natural systems
eDNA	Very low quantity/quality DNA	Multiplex PCR genotyping	Bd transmission between hosts and the environment
Museum specimens	Low quantity/quality DNA (highly fragmented in formalin-fixed specimens)	Multiplex PCR genotyping	Bd genetic diversity over historical timescales
Pure cultures isolated from field-collected amphibians	Laborious to isolate Bd from live animals, often requires animal sacrifice	Whole genome sequencing, RNA sequencing	Bd genome evolution and genomic regions under selection in natural systems
Pure cultures manipulated in laboratory experiments	Unintended effects of laboratory maintenance	Whole genome sequencing, RNA sequencing, genomic manipulation	Bd virulence genes and experimental evolution of Bd genomes

Recent advances in molecular techniques have the potential to address these challenges and reveal previously hidden variation in Bd across scales through space and time.

both host and pathogen perspectives must be integrated. When grounded in a natural history perspective, genomics tools can contribute powerfully toward advancing fundamental knowledge and guiding conservation of imperiled amphibians.

REFERENCES

1. Fisher MC, Garner TWJ, Walker SF. Global emergence of *Batrachochytrium dendrobatidis* and amphibian chytridiomycosis in space, time, and host. *Annu Rev Microbiol.* 2009;63:291–310.
2. Briggs CJ, Knapp RA, Vredenburg VT. Enzootic and epizootic dynamics of the chytrid fungal pathogen of amphibians. *Proc Natl Acad Sci USA.* 2010;107(21):9695–9700.
3. Crawford AJ, Lips KR, Bermingham E. Epidemic disease decimates amphibian abundance, species diversity, and evolutionary history in the highlands of central Panama. *Proc Natl Acad Sci USA.* 2010;107(31):13777–13782.
4. Skerratt LF, Berger L, Speare R, et al. Spread of chytridiomycosis has caused the rapid global decline and extinction of frogs. *Ecohealth.* 2007;4(2):125–134.
5. Letcher. Peter M, Powell Martha J, Churchill Perry F, Chambers JG. Ultrastructural and molecular phylogenetic delineation of a new order, the Rhizophydiales (Chytridiomycota). *Mycol Res.* 2006;1:898–915.
6. James TY, Porter D, Leander CA, Vilgalys R, Longcore JE. Molecular phylogenetics of the Chytridiomycota supports the utility of ultrastructural data in chytrid systematics. *CanJ Bot.* 2000;78(3):336–350.
7. Berger L, Speare R, Daszak P, et al. Chytridiomycosis causes amphibian mortality associated with population declines in the rain forests of Australia and Central America. *Proc Natl Acad Sci USA.* 1998;95(15):9031–9036. Available from: http://www.pubmedcentral.nih.gov/articlerender.fcgi?artid=21197&tool=pmcentrez&rendertype=abstract.
8. Longcore JE, Pessier AP, Nichols DK. *Batrachochytrium dendrobatidis* gen. et sp. nov., a chytrid pathogenic to amphibians. *Mycologia.* 1999;91(2):219–227.
9. Martel A, Spitzen-van der Sluijs A, Blooi M, et al. *Batrachochytrium salamandrivorans* sp. nov. causes lethal chytridiomycosis in amphibians. *Proc Natl Acad Sci USA.* 2013;110 (38):15325–15329.
10. Kilpatrick AM, Briggs CJ, Daszak P. The ecology and impact of chytridiomycosis: an emerging disease of amphibians. *Trends Ecol Evol.* 2009;25(2):1–10.
11. Voyles J, Rosenblum EB, Berger L. Interactions between *Batrachochytrium dendrobatidis* and its amphibian hosts: a review of pathogenesis and immunity. *Microbes Infect.* 2011;13(1):25–32.
12. Boyle DG, Boyle DB, Olsen V, Morgan JA, Hyatt AD. Rapid quantitative detection of chytridiomycosis (*Batrachochytrium dendrobatidis*) in amphibian samples using real-time Taqman PCR assay. *Dis Aquat Organ.* 2004;60(2):141–148.
13. Hyatt a D, Boyle DG, Olsen V, et al. Diagnostic assays and sampling protocols for the detection of *Batrachochytrium dendrobatidis. Dis Aquat Organ.* 2007;73(3):175–192.
14. Zhu W, Bai C, Wang S, et al. Retrospective survey of museum specimens reveals historically widespread presence of *Batrachochytrium dendrobatidis* in China. *Ecohealth.* 2014;11(2):241–250.
15. Cheng TL, Rovito SM, Wake DB, Vredenburg VT. Coincident mass extirpation of neotropical amphibians with the emergence of the infectious fungal pathogen *Batrachochytrium dendrobatidis. Proc Natl Acad Sci USA.* 2011;108(23):9502–9507.
16. Goka K, Yokoyama J, Une Y, et al. Amphibian chytridiomycosis in Japan: distribution, haplotypes and possible route of entry into Japan. *Mol Ecol.* 2009;18(23): 4757–4774.

17. Bai C, Liu X, Fisher MC, Garner TWJ, Li Y. Global and endemic Asian lineages of the emerging pathogenic fungus *Batrachochytrium dendrobatidis* widely infect amphibians in China. *Divers Distrib.* 2012;18(3):307–318.
18. Garland S, James TY, Blair D, Berger L, Skerratt LF. Polymorphic repetitive loci of the amphibian pathogen *Batrachochytrium dendrobatidis. Dis Aquat Organ.* 2011;97(1):1–9.
19. Morgan J, a T, Vredenburg VT, Rachowicz LJ, et al. Population genetics of the frog-killing fungus *Batrachochytrium dendrobatidis. Proc Natl Acad Sci USA.* 2007;104(34): 13845–13850.
20. Farrer RA, Weinert LA, Bielby J, et al. Multiple emergences of genetically diverse amphibian-infecting chytrids include a globalized hypervirulent recombinant lineage. *Proc Natl Acad Sci USA.* 2011;108(46):18732–18736.
21. Rosenblum EB, James TY, Zamudio KR, et al. Complex history of the amphibian-killing chytrid fungus revealed with genome resequencing data. *Proc Natl Acad Sci USA.* 2013;110:9385–9390.
22. Rosenblum EB, Stajich JE, Maddox N, Eisen MB. Global gene expression profiles for life stages of the deadly amphibian pathogen *Batrachochytrium dendrobatidis. Proc Natl Acad Sci USA.* 2008;105(44):17034–17039.
23. Rosenblum EB, Poorten TJ, Joneson S, Settles M. Substrate-specific gene expression in *Batrachochytrium dendrobatidis*, the chytrid pathogen of amphibians. *PLoS One.* 2012;7 (11):e49924.
24. Rachowicz LJ, Hero J-M, Alford RA, et al. The novel and endemic pathogen hypotheses: competing explanations for the origin of emerging infectious diseases of wildlife. *Conserv Biol.* 2005;19(5):1441–1448.
25. Schloegel LM, Ferreira CM, James TY, et al. The North American bullfrog as a reservoir for the spread of *Batrachochytrium dendrobatidis* in Brazil. *Anim Conserv.* 2010;13:53–61.
26. Pounds JA, Bustamante MR, Coloma LA, et al. Widespread amphibian extinctions from epidemic disease driven by global warming. *Nature.* 2006;439(7073):161–167.
27. Lips KR, Brem F, Brenes R, et al. Emerging infectious disease and the loss of biodiversity in a Neotropical amphibian community. *Proc Natl Acad Sci USA.* 2006;103 (9):3165–3170.
28. Lips KR, Diffendorfer J, Mendelson III JR, Sears MW. Riding the wave: reconciling the roles of disease and climate change in amphibian declines. *Plos Biology.* 2008;6 (3):441–454.
29. Vredenburg VT, Knapp RA, Tunstall TS, Briggs CJ. Dynamics of an emerging disease drive large-scale amphibian population extinctions. *Proc Natl Acad Sci USA.* 2010;107 (21):9689–9694.
30. Daszak P, Strieby A, Cunningham AA, Longcore JE, Brown CC, Porter D. Experimental evidence that the bullfrog (*Rana catesbeiana*) is a potential carrier of chytridiomycosis, an emerging fungal disease of amphibians. *Herpetol J.* 2004;14(4):201–207.
31. Garner TW, Perkins MW, Govindarajulu P, Seglie D, Walker S, Cunningham AA, Fisher MC. The emerging amphibian pathogen *Batrachochytrium dendrobatidis* globally infects introduced populations of the North American bullfrog, *Rana catesbeiana. Biol Lett.* 2006;2:455–459.
32. Mazzoni R, Cunningham AA, Daszak P, Apolo A, Perdomo E, Speranza G. Emerging pathogen of wild amphibians in frogs (*Rana catesbeiana*) farmed for international trade. *Emerg Infect Dis.* 2003;9(8):995–998.
33. Morehouse EA, James TY, Ganley ARD, Vilgalys R, Berger L, Murphy PJ, Longcore JE. Multilocus sequence typing suggests the chytrid pathogen of amphibians is a recently emerged clone. *Mol Ecol.* 2003;12:395–403.
34. Schloegel LM, Toledo LF, Longcore JE, et al. Novel, panzootic and hybrid genotypes of amphibian chytridiomycosis associated with the bullfrog trade. *Mol Ecol.* 2012;21 (21):5162–5177.

35. Jenkinson TS, Betancourt Román CM, Lambertini C, et al. Amphibian-killing chytrid in Brazil comprises both locally endemic and globally expanding populations. *Mol Ecol.* 2016. Epub ahead of print.
36. Rodriguez D, Becker CG, Pupin NC, Haddad CFB, Zamudio KR. Long-term endemism of two highly divergent lineages of the amphibian-killing fungus in the Atlantic Forest of Brazil. *Mol Ecol.* 2014;23(4):774–787.
37. Velo-Antón G, Rodríguez D, Savage AE, Parra-Olea G, Lips KR, Zamudio KR. Amphibian-killing fungus loses genetic diversity as it spreads across the New World. *Biol Conserv.* 2012;146(1):213–218.
38. Bataille A, Fong JJ, Cha M, et al. Genetic evidence for a high diversity and wide distribution of endemic strains of the pathogenic chytrid fungus *Batrachochytrium dendrobatidis* in wild Asian amphibians. *Mol Ecol.* 2013;22(16):4196–4209.
39. Walker SF, Bosch J, Gomez V, et al. Factors driving pathogenicity vs. prevalence of amphibian panzootic chytridiomycosis in Iberia. *Ecol Lett.* 2010;13(3):372–382.
40. Refsnider JM, Poorten TJ, Langhammer PF, Burrowes PA, Rosenblum EB. Genomic correlates of virulence attenuation in the deadly amphibian chytrid fungus, *Batrachochytrium dendrobatidis*. *G3 (Bethesda).* 2015;5(11):2291–2298.
41. James TY, Litvintseva AP, Vilgalys R, et al. Rapid global expansion of the fungal disease Chytridiomycosis into declining and healthy amphibian populations. *PLoS Pathog.* 2009;5(5):e1000458.
42. Farrer RA, Henk DA, Garner TWJ, Balloux F, Woodhams DC, Fisher MC. Chromosomal copy number variation, selection and uneven rates of recombination reveal cryptic genome diversity linked to pathogenicity. In: Heitman J, ed. *PLoS Genet.* 2013;9(8):e1003703.
43. Piovia-Scott J, Pope K, Joy Worth S, et al. Correlates of virulence in a frog-killing fungal pathogen: evidence from a California amphibian decline. *ISME J.* 2014;9(7):1–9.
44. Hu G, Wang J, Choi J, Jung WH, Liu I, Litvintseva AP, Bicanic T, Aurora R, Mitchell TG, Perfect JR, Kronstad JW. Variation in chromosome copy number influences the virulence of *Cryptococcus neoformans* and occurs in isolates from AIDS patients. *BMC genomics.* 2011;12(1):526.
45. Sillo F, Garbelotto M, Friedman M, Gonthier P. Comparative genomics of sibling fungal pathogenic taxa identifies adaptive evolution without divergence in pathogenicity genes or genomic structure. *Genome Biol Evol.* 2015;7(12):evv209.
46. Abramyan J, Stajich JE. Species-specific chitin-binding module 18 expansion in the amphibian pathogen *Batrachochytrium dendrobatidis*. *MBio.* 2012;3(3):1–9.
47. Joneson S, Stajich JE, Shiu S-H, Rosenblum EB. Genomic transition to pathogenicity in chytrid fungi. *PLoS Pathog.* 2011;7(11):e1002338.
48. Burmester A, Shelest E, Glöckner G, et al. Comparative and functional genomics provide insights into the pathogenicity of dermatophytic fungi. *Genome Biol.* 2011;12(1):R7.
49. Vu K, Tham R, Uhrig JP, et al. Invasion of the central nervous system by *Cryptococcus neoformans* requires a secreted fungal metalloprotease. *MBio.* 2014;5(3):1–13.
50. Gunderson KL, Steemers FJ, Lee G, Mendoza LG, Chee MS. A genome-wide scalable SNP genotyping assay using microarray technology. *Nat Genet.* 2005;37(5): 549–554.
51. Moran JL, Bolton AD, Tran PV, et al. Utilization of a whole genome SNP panel for efficient genetic mapping in the mouse. *Genome Res.* 2006;16(3):436–440.
52. Groenen MA, Megens HJ, Zare Y, Warren WC, Hillier LW, Crooijmans RP, Vereijken A, Okimoto R, Muir WM, Cheng HH. The development and characterization of a 60K SNP chip for chicken. *BMC Genomics.* 2011;12(1):1.
53. Albrechtsen A, Nielsen FC, Nielsen R. Ascertainment biases in SNP chips affect measures of population divergence. *Mol Biol Evol.* 2010;27(11):2534–2547.

54. Ruegg KC, Anderson EC, Paxton KL, et al. Mapping migration in a songbird using high-resolution genetic markers. *Mol Ecol.* 2014;23(23):5726–5739.
55. Smith MJ, Pascal CE, Grauvogel ZA, Habicht C, Seeb JE, Seeb LW. Multiplex preamplification PCR and microsatellite validation enables accurate single nucleotide polymorphism genotyping of historical fish scales. *Mol Ecol Resour.* 2011;11(s1): 268–277.
56. Kraus RH, Vonholdt B, Cocchiararo B, Harms V, Bayerl H, Kühn R, Förster DW, Fickel J, Roos C, Nowak C. A single||nucleotide polymorphism||based approach for rapid and cost||effective genetic wolf monitoring in Europe based on noninvasively collected samples. *Mol Ecol Resour.* 2015;15(2):295–305.
57. Mortazavi A, Williams BA, McCue K, Schaeffer L, Wold B. Mapping and quantifying mammalian transcriptomes by RNA-Seq. *Nat Method.* 2008;5(7):621–628.
58. Djebali S, Davis CA, Merkel A, Dobin A, Lassmann T, Mortazavi A, Xue C. Landscape of transcription in human cells. *Nature.* 2012;489(7414):101–108.
59. Westermann AJ, Gorski SA, Vogel J. Dual RNA-seq of pathogen and host. *Na Rev Microbiol.* 2012;10(9):618–630.
60. Asai S, Rallapalli G, Piquerez SJ, Caillaud MC, Furzer OJ, Ishaque N, Jones JD. Expression profiling during *Arabidopsis*/downy mildew interaction reveals a highly-expressed effector that attenuates responses to salicylic acid. *PLoS Pathog.* 2014;10(10): e1004443.
61. Mouyna I, Henry C, Doering TL. Latgé J-P. Gene silencing with RNA interference in the human pathogenic fungus *Aspergillus fumigatus. FEMS Microbiol Lett.* 2004;237 (2):317–324.
62. Doudna JA, Charpentier E. The new frontier of genome engineering with CRISPR-Cas9. *Science.* 2014;346(6213):1258096.
63. Jiang D, Zhu W, Wang Y, Sun C, Zhang KQ, Yang J. Molecular tools for functional genomics in filamentous fungi: recent advances and new strategies. *Biotechnol Adv.* 2013;31(8):1562–1574.
64. Liu R, Chen L, Jiang Y, Zhou Z, Zou G. Efficient genome editing in filamentous fungus Trichoderma reesei using the CRISPR/Cas9 system. *Cell Discov.* 2015;1:15007.
65. Holmes MW, Hammond TT, Wogan GOU, et al. Natural history collections as windows on evolutionary processes. *Mol Ecol.* 2016;25(4):864–881.
66. Hykin SM, Bi K, McGuire JA. Fixing formalin: a method to recover genomic-scale DNA sequence data from formalin-fixed museum specimens using high-throughput sequencing. *PLoS One.* 2015;10(10):1–16.
67. Voyles J, Johnson LR, Briggs CJ, et al. Experimental evolution alters the rate and temporal pattern of population growth in *Batrachochytrium dendrobatidis*, a lethal fungal pathogen of amphibians. *Ecol Evol.* 2014;4(18):3633–3641.
68. Wichman HA, Brown CJ. Experimental evolution of viruses: microviridae as a model system. *Philos Trans R Soc Lond B Biol Sci.* 2010;365(1552):2495–2501.
69. Boyle DG, Hyatt AD, Daszak P, et al. Cryo-archiving of *Batrachochytrium dendrobatidis* and other chytridiomycetes. *Dis Aquat Organ.* 2003;56(1):59–64.
70. Voyles J, Johnson LR, Briggs CJ, et al. Temperature alters reproductive life history patterns in *Batrachochytrium dendrobatidis*, a lethal pathogen associated with the global loss of amphibians. *Ecol Evol.* 2012;2(9):2241–2249.
71. Langhammer PF, Lips KR, Burrowes PA, Tunstall T, Palmer CM, Collins JP. A fungal pathogen of amphibians, *Batrachochytrium dendrobatidis*, attenuates in pathogenicity with in vitro passages. *PLoS One.* 2013;8(10):e77630.
72. Brem FMR, Parris MJ, Padgett-Flohr GE. Re-isolating *Batrachochytrium dendrobatidis* from an amphibian host increases pathogenicity in a subsequent exposure. *PLoS One.* 2013;8(5):e61260.

73. Schmidt BR, Kéry M, Ursenbacher S, Hyman OJ, Collins JP. Site occupancy models in the analysis of environmental DNA presence/absence surveys: a case study of an emerging amphibian pathogen. *Method Ecol Evol.* 2013;4(7):646–653.
74. Chestnut T, Anderson C, Popa R, Blaustein AR, Voytek M, Olson DH, Kirshtein J. Heterogeneous occupancy and density estimates of the pathogenic fungus *Batrachochytrium dendrobatidis* in waters of North America. *PloS one.* 2014;9(9):e106790.
75. Fites JS, Ramsey JP, Holden WM, et al. The invasive chytrid fungus of amphibians paralyses lymphocyte responses. *Science.* 2013;342:366–369.
76. Rosenblum EB, Poorten TJ, Settles M, Murdoch GK. Only skin deep: shared genetic response to the deadly chytrid fungus in susceptible frog species. *Mol Ecol.* 2012;21(3):3110–3120.
77. Ellison AR, Savage AE, DiRenzo GV, Langhammer P, Lips KR, Zamudio KR. Fighting a losing battle: vigorous immune response countered by pathogen suppression of host defenses in the chytridiomycosis-susceptible frog *Atelopus zeteki*. *G3 (Bethesda).* 2014;4(July):1275–1289.
78. Ellison AR, Tunstall T, Direnzo GV, et al. More than skin deep: Functional genomic basis for resistance to amphibian chytridiomycosis. *Genome Biol Evol.* 2014;7(1):286–298.
79. Rollins-Smith LA, Fites JS, Reinert LK, Shiakolas AR, Umile TP, Minbiole KPC. Immunomodulatory metabolites released by the frog-killing fungus *Batrachochytrium dendrobatidis*. *Infect Immun.* 2015;83(12):4565–4570.
80. Berger L, Speare R, Pessier A, Voyles J, Skerratt LF. Treatment of chytridiomycosis requires urgent clinical trials. *Dis Aquat Organ.* 2010;92(2–3):165–174.
81. Martel A, Van Rooij P, Vercauteren G, et al. Developing a safe antifungal treatment protocol to eliminate *Batrachochytrium dendrobatidis* from amphibians. *Med Mycol.* 2011;49(2):143–149.
82. Woodhams DC, Geiger CC, Reinert LK, Rollins-Smith LA, Lam B, Harris RN, Voyles J. Treatment of amphibians infected with chytrid fungus: learning from failed trials with itraconazole, antimicrobial peptides, bacteria, and heat therapy. *Dis Aquat Organ.* 2016;98(1):11–25.

CHAPTER TEN

Translational Activities to Enable NTD Vaccines

S.A. Gray*, R.N. Coler†,‡, D. Carter*,†,‡, A.A. Siddiqui§,¶,1

* PAI Life Sciences, Seattle, WA, United States
† Infectious Disease Research Institute, Seattle, WA, United States
‡ Department of Global Health, University of Washington, Seattle, WA, United States
§ Department of Internal Medicine, Texas Tech University School of Medicine, Lubbock, TX, United States
¶ Center of Tropical Medicine and Infectious Diseases, Texas Tech University School of Medicine, Lubbock, TX, United States
1 Corresponding author. E-mail address: afzal.siddiqui@ttuhsc.edu

Contents

Abstract

There is an urgent need to develop new vaccines for tuberculosis, HIV/AIDS, and malaria, as well as for chronic and debilitating infections known as neglected tropical diseases (NTDs). The term "NTD" emerged at the beginning of the new millennium to describe a set of diseases that are characterized as (1) poverty related, (2) endemic to the tropics and subtropics, (3) lacking public health attention and inadequate industrial investment, (4) having poor research funding and a weak research and development (R&D) pipeline, (5) usually associated with high morbidity but low mortality, and (6) often having no safe

Progress in Molecular Biology and Translational Science, Volume 142
ISSN 1877-1173
http://dx.doi.org/10.1016/bs.pmbts.2016.05.004

and long-lasting treatment available. Many additional challenges to the current control and elimination programs for NTDs exist. These include inconsistent performance of diagnostic tests, regional differences in access to treatment and in treatment outcome, lack of integrated surveillance and vector/intermediate host control, and impact of ecological climatic changes particularly in regions where new cases are increasing in previously nonendemic areas. Moreover, the development of NTD vaccines, including those for schistosomiasis, leishmaniasis, leprosy, hookworm, and Chagas disease are being led by nonprofit product development partnerships (PDPs) working in partnership with academic and industrial partners, contract research organizations, and in some instances vaccine manufacturers in developing countries. In this review, we emphasize global efforts to fuel the development of NTD vaccines, the translational activities needed to effectively move promising vaccine candidates to Phase-I clinical trials and some of the hurdles to ensuring their availability to people in the poorest countries of Africa, Asia, Latin America, and the Caribbean.

1. INTRODUCTION

Tropical infectious diseases continue to be a major cause of morbidity and mortality worldwide, especially in developing countries. These diseases include the more widely studied diseases such as malaria and tuberculosis; cumulatively these diseases infect nearly one-third of the world's population. Indeed, research efforts to develop vaccines for these diseases are well underway and several candidate vaccines are far along in the pipeline. As of 2016, here are 18 NTDs recognized by World Health Organization (WHO) consisting of 11 parasitic, 4 bacterial, and 3 viral diseases. These are defined as "chronically endemic and epidemic-prone tropical diseases, which have a very significant negative impact on the lives of poor populations and remain critically neglected in the global health agenda" (http://www.who.int/neglected_diseases/diseases/en/). Among these are the causative agents of schistosomiasis, sleeping sickness, Chagas disease, river blindness, disease caused by hookworm infection, leishmaniasis, and lymphatic filariasis (Table 1). Although less common than the big three (HIV, malaria, and TB), these neglected NTDs still cause significant morbidity in 149 countries.

To date, there are no vaccines approved for prevention of the majority of the NTDs, with the exception of the rabies vaccine. A second vaccine, the Bacillus Calmette Guerin (BCG) vaccine for *Mycobacterium tuberculosis*, has been shown to provide at least partial protection to leprosy. Treatments for the 11 parasitic diseases are usually antiparasitic drugs, where they exist, and many of these treatments suffer from problems including drug resistance,

Table 1 Impact of neglected tropical diseases.

Disease category	Disease	Number infected (est.)	Deaths per year (est.)	Current treatment
Parasitic NTDs	Soil-transmitted Helminths [(1) Roundworm; (2) Whipworm; (3) Hookworms]	Up to 2–3 billion combined	Up to 135,000 combined	Albendazole, Mebendazole, Praziquantel
	Schistosomiasis/ Bilharzia	400–600 million	280,000	Oxamniquine, Praziquantel
	Leishmaniasis (Kala-Azar)	12 million	20,000–50,000	Pentavalent antimonials, Amphoteracin B, Miltefosine
	Lymphatic filariasis	120 million	<1000	Ivermectin and Albendazole, Diethylcarbamazine and Albendazole; Moxidectin
	River blindness/ Onchocerciasis	25 million	Rare/unlikely	Ivermectin (except with Loa loa coinfection); Doxycycline
	Chagas disease	6–7 million	14,000	Benznidazole, Nifurtimox
	Guinea worm/ Dracunculiasis	126 in 2014	None	No drug treatment; Physical removal of worms through the skin
	Echinococcosis (*Echinococcus spp* Tapeworm)	1 million at any one time	Rare/unlikely	Surgery and long-term Albendazole and/or Praziquantel
	Cysticercosis (*T. solium* Tapeworm)	Up to 50 million	Rare/unlikely	Albendazole, Praziquantel, Niclosamide, anticonvulsants, steroids, surgery
Viral NTDs	Dengue	390 million/ year	10,000–25,000	None for virus; control of pain and swelling
	Chikungunya	1.7 million	Rare	None for virus; control of pain and swelling
	Rabies	15 million/ year	26,000–55,000	Vaccine available for human and animals

(Continued)

Table 1 Impact of neglected tropical diseases.—cont'd.

Disease category	Disease	Number infected (est.)	Deaths per year (est.)	Current treatment
Bacterail NTDs	Yaws	Unknown	Rare/unlikely	Antibiotics (Benzathine Penicillin)
	Buruli Ulcer	5,000–6,000/year	Rare/unlikely	Antibiotics (rifampicin, clarithromycin, moxifloxacin) and surgery
	Leprosy (Hansen's Disease)	2 million	Rare/indirectly	Antibiotics
	Trachoma	40 million	Rare/unlikely	Antibiotics

toxicity, and limited efficacy with regard to various parasitic stages. An alternative to drug treatment would be a vaccine capable of protecting against certain growth stages of the single causative agent of the NTD, and another effective strategy could be the development of multivalent vaccine(s) against multiple parasites. Indeed, there are promising candidate vaccines to nearly every NTD shown in Table 1, although most have not progressed beyond early clinical testing. This review will focus on the translational activities needed to effectively move a promising vaccine candidate to Phase-I clinical trials: traditionally this has been a significant bottleneck in the translational life cycle of a candidate vaccine.

People most impacted by NTDs are the poor, living in countries with underdeveloped infrastructure in rural settings. These people have very low median incomes and often have only limited access to health care. Given the large expense required to develop a drug or vaccine, and the need for large drug companies to make a profit, this presents a unique socioeconomic conundrum that has further limited development of vaccines for many of the NTDs. This review will pay particular attention to unique processes and hurdles encountered when developing a vaccine for NTDs versus vaccines developed for more common diseases, such as influenza, that infect people of all socioeconomic groups.

1.1 Parasitic NTDs

1.1.1 Soil-Transmitted Helminths

Perhaps the greatest burden of disease caused by parasites is due to a group of soil-transmitted helminths (STH) which includes roundworms

(Ascariasis), whipworms (Trichuriasis), and hookworms (*Necatoramericanus* and *Ancylostoma duodenale*).[1–3] Estimates suggest that these three parasites infect ~400–600 million people each, causing an estimated ~0.3 million deaths annually. Perhaps more significant is the impact of these diseases with regard to disability-adjusted life years (DALYs) lost, estimated at ~3.2 million DALYs, as well as the propagation of poverty due to limitations incurred during disease states. A bivalent hookworm vaccine based on two proteins, *Na*-GST-1 and *Na*-APR-1, is currently in Phase-I trials.[4–6] Recent efforts are aimed at inclusion of additional antigens from both *Ascaris* and *Trichuris* in hopes of generating a pan-antihelmintic vaccine.[7]

1.1.2 Schistosomiasis

Despite more than a decade of a mass drug administration (MDA) campaign with praziquantel, schistosomiasis is estimated to infect more than 200 million people in 74 countries and to cause an estimated 280,000 deaths each year. There are only three vaccine candidates currently in Phase-I to Phase-III trials (reviewed in Ref. [8]). These include *Schistosoma haematobium* 28-kD glutathione S-transferase (rSh28GST), *Schistosoma mansoni* 14-kDa fatty acid-binding protein (Sm14) and *S. mansoni* tetraspanin, a 9-kDa surface antigen (Sm-TSP-2). However, these three vaccines target only a single *Schistosoma* species. Current trials are focused on combination therapy with praziquantel (PZQ) and the vaccine, but no single vaccine currently in trials is likely to eliminate established male and female worms as well as eggs.[8] A newer vaccine candidate, SchistoShield based on *S. mansoni* calcium-activated surface membrane protein (Sm-p80), is soon to enter clinical trials and targets three major species of schistosomes that cause intestinal/hepatic schistosomiasis (*S. mansoni, Schistosoma japonicum*) as well as urinary schistosomiasis (*S. haematobium*). The uniqueness of SchistoShield vaccine is that it significantly reduces the egg-induced pathology.[9,10] This is significant since observation of eggs in feces is the primary method for diagnosis, and a recent finding that for every egg-positive individual, there is at least one egg-negative infected individual.[11]

1.1.3 Leishmaniasis

An estimated 350 million people are threatened by leishmaniasis disease with ~51,000 deaths per year.[2,12] Leishmaniasis causes annual total DALYs up to 3.32 million, and is the second most important parasitic disease next to schistosomiasis. Concentrated in poverty-stricken countries within Southeast Asia, East Africa, and Latin America, it is also endemic in several Mediterranean countries. Four forms of disease caused by as many as 20 species of *Leishmania* parasite exist.[13] Cutaneous (CL) and mucocutaneous leishmaniasis are characterized by skin lesions and mucous membrane involvement whereas visceral leishmaniasis (VL), a more serious form, causes lesions in various organs. VL can also evolve into a cutaneous form known as post-kala-azar dermal leishmaniasis (PKDL). Treatment is difficult due to multiple vectors and reservoirs, and successful vaccines would need to treat all forms of the disease.[14] Although the prevalence of VL is relatively low compared to CL, the annual mortality rate of more than 200,000 and the newly emerging epidemiology of coinfection with HIV or other infectious agents are of concern. Complete disease elimination and development of new tools to combat VL are difficult given the uniquely clustered disease distribution, complex life cycle of the parasites, and diverse routes of transmission by vectors. Development of vaccines for reservoir dogs against leishmaniasis could also be one of the strategies to control the spread of the parasite and reduce the incidence in humans.[14] Currently, a canine VL vaccine, LEISHMUNE is commercially available and many more are in Phase-III field trials. Several human vaccine candidates targeted toward inducing the production of interferon-γ (IFNγ), interleukin-2 (IL-2), and tumor necrosis factor (TNF) by polyfunctional CD4 + T cells are currently being explored and are in different stages of early development. The multicomponent vaccines LEISH-F2/LEISH-111f + MPL-SE and LEISH-F3 + GLA-SE were the first-defined vaccine candidates to progress to human clinical trials in healthy volunteers for CL and VL.[14,15]

1.1.4 River Blindness

The parasite *Onchocerca volvulus* is transmitted by the bite from a blackfly and causes a rarely fatal, but severely debilitating disease of the skin and eye.[16,17] Nearly 37 million people are currently infected, despite this diseases being treatable with a single dose of ivermectin widely available through a donation program with Merck & Co., Inc. (Mectizan).[18] Additional MDA campaigns (reviewed in Ref. [18]) have led to a reduction in ~8.6 million, fewer DALYs at a cost savings of $27/DALY. Despite these MDAs, there continues to be a rise in drug resistance and loiasis, a parasitic eye disease caused by Loa loa, in which ivermectin treatment will result in severe adverse effects.[19] The result has led to formation of The Onchocerciasis Vaccine for Africa (TOVA) initiative aimed at accelerating the development of a vaccine for river blindness.[18] This vaccine, focused on three validated antigens (Ov-103, Ov-RAL-2, and Ov-CPI-2 M) is just beginning with Phase-I trials with Phase-II trials expected to begin in 2020.[17,18]

1.1.5 Chagas Disease

A disease mostly found in Mexico and Latin America, Chagas disease is caused by the parasite *Trypanosoma cruzi* are mainly transmitted by contact with feces/urine of infected blood-sucking triatomine bugs, also called the "kissing" bug.[20,21] These bugs usually bite an exposed area of skin such as the face, and the bug defecates close to the bite. The parasites enter the body when the person instinctively smears the bug feces into the bite, the eyes, the mouth, or into any skin break. Left untreated, the disease can cause chronic infection of the heart muscle and is recognized as a leading cause of heart disease in Latin America.[22] Early disease is readily treatable with drugs; however, the longer a person is infected the more difficult the disease is to treat. Many people are asymptomatic, but about 20–30% will develop chronic infection and heart disease and death.[22] Drug treatment has had limited success for several reasons: (1) treatment may not eradicate parasites from the host; (2) there are significant side effects and the drugs are contraindicated in pregnancy; and (3) the cost of therapy for a chronic patient is prohibitive, ranging from US$50 to 8000 per year in the United States and form US$3000 to 14,580 in Mexico.[22] As an NTD and disease of the poor, a 2012 estimate suggested that ~22% of chronic patients never seek treatment for the above reasons.[22]

1.1.6 Other NTDs

There are at least 11 additional NTDs that cause significant morbidity and mortality. Human echinococcosis (hydatidosis or hydatid disease) is caused by the larval stages of cestodes (tapeworms) of the genus *Echinococcus*.[23] *E. granulosus* causes cystic echinococcosis, the form most frequently encountered. *E. multilocularis* causes alveolar echinococcosis. *E. vogeli* and *E. oligarthrus* causes polycystic echinococcosis.[24] Cystic echinococcus results in cyst formation in the liver, spleen, and other organs. If left undiagnosed and untreated, alveolar echinococcus can be fatal as parasites disseminate throughout the body via blood and lymph. Treatment is a combination of antiparasitic drugs and surgical resection of the cysts. Human cysticercosis is acquired by ingesting *Taenia solium* eggs shed in the feces of a human *T. solium* tapeworm carrier, and thus can occur in populations that neither eat pork nor share environments with pigs. Eggs following ingestion develop into cysts throughout the body. Disease symptoms vary depending on where the cysts form and can vary from asymptomatic to neurological and life threatening (neurocysticercosis). Similar to Echinococcus, treatment is through antiparasitic drugs and/or surgical resection of cysts. Dracunculiasis is caused by commonly known guinea worm, *Dracunculus medinensis*. This disease has almost been eradicated, left in only one to two countries (only 126 cases remain as of 2014). This was achieved by eliminating the intermediate host of the parasite, a copepod, from the drinking water supply.

Three of the NTDs are viruses: (1) rabies is a virus transmitted most commonly by dog bites and, if left untreated, can infect the nervous system resulting in more than 99% fatality. A vaccine is available that has been found effective both pre- and postexposure. According to the CDC, between 16,000 and 39,000 people are vaccinated each year as a precaution after animal bites (http://www.cdc.gov/vaccines/hcp/vis/vis-statements/rabies.html). However, rabies continues to infect a small population of people every year, primarily due to nonvaccination of dogs in part due to the rapid turnover of animal populations. (2) Dengue virus is transmitted by mosquitoes and causes disease similar to the flu. There are four serotypes, and survival following infection with a single serotype will result in lifelong protection from that serotype. However, dengue-immune people who encounter additional serotypes are at risk for developing severe dengue and a condition termed breakbone fever.[25] Significant vaccine progress has been made with dengue, with at least seven candidate vaccines currently in Phase-I trials and one vaccine in Phase-II trials.[26] (3) Chikungunya virus is a recently emerging virus spread by mosquitoes and causing chronic, severe, and debilitating arthralgia accompanied by fever, rash, and occasionally death.[27] Despite its discovery in 1952, chikungunya began a rapid spread throughout Africa, southern Asia, and the Pacific islands, and into the North and South America. No vaccine or rapid, field-based diagnostic is available.[28] As of 2015, at least three distinct vaccines (one VLP and two protein based) are being produced in plants with demonstrated promise in macaques and initiation of preclinical trials.[29]

Four of the 18 NTDs are diseases are caused by bacteria including leprosy, buruli ulcer, trachoma, and yaws. (1) Leprosy, or Hansen's disease, is caused by the bacterium *Mycobacterium leprae*. There is currently no vaccine for leprosy, although the BCG (Bacillus Calmetta Guèrin) vaccine confers at least partial protection and is used for this purpose in India and Brazil. Today, leprosy can be treated and often cured using steroids and antibiotics. (2) Buruli ulcer is caused by a similar species of *Mycobacterium* (*M. ulcerans*) and causes a chronic debilitating skin and soft tissue infection that can lead to permanent disfigurement and disability. Like leprosy, the BCG vaccine offers partial protection and the infection is readily treatable with antibiotics. (3) Trachoma is an infection of the eyes by the bacterium *Chlamydia trachomatis* and can cause severe scarring on the inside of the eyelids leading to corneal damage and blindness. Trachoma is easily prevented by proper hygiene and is readily treatable by antibiotics, yet it remains the leading cause of preventable blindness. (4) Yaws is a chronic infection that affects mainly the skin, bone, and cartilage following infection with the bacterium *Treponema pallidum*, which is more commonly associated with the venereal disease syphilis. There are no vaccines for each of the bacterial NTDs, and little progress or efforts are aimed at a vaccine due in part to low incidences of disease, limited mortality, and unavailable and ineffective antibiotic treatments.

2. CURRENT LANDSCAPE OF VACCINES FOR NTDs

2.1 Why the Lack of Vaccines to NTDs?

It has been estimated that approximately one-third of the world's population lives on less than US$2 per day and 1.3 billon people live below the World Bank poverty figure of US$1.25/day.[26] Further estimates suggest that virtually all of these people are currently infected with at least one of the 17 NTDs, with the majority of disease primarily due to the soil-transmitted helminths (STH) and schistosomiasis.[26] Despite available chemotherapies, diseases with very high burdens, including STH and schistosomiasis, have been prioritized for development of both preventative and postexposure vaccines.[26] Additional NTDs to which a vaccine is urgently needed include dengue, chikungunya, leishmaniasis, Chagas disease, and onchocerciasis.[26]

There is no single reason that vaccine development for NTDs has been slow, but they include: (1) how to finance cGMP vaccine development; (2) how to identify a target population for vaccine testing; (3) how to define vaccine success and achieve sterilizing immunity, and (4) how to define the target vaccine product profile. These issues are described in more detail and summarized in Table 2.

2.1.1 Financing

Perhaps the single biggest hurdle for an NTD vaccine is how to finance the costs of developing the vaccine from the research laboratory to cGMP production. Arguably, the costs and activities required for developing an NTD vaccine are identical to that of a mainstream vaccine. Typical research laboratories do not have the resources-financial or laboratory equipment-to carry out all of the steps needed to develop a vaccine. Costs to develop a typical

Table 2 Key issues limiting development of NTD vaccines.

Issue	Description
Financing	• Costs to develop an NTD vaccine are equal to the costs for mainstream vaccines with a much lower likelihood of recouping those investments following approval of the vaccine • Developmental cost burdens are mostly absorbed by first world countries • Vaccine cost needs to be less than US$1.00 per dose • Burden of vaccination cost must not be put on target population • Need to identify PDPs and PPPs to develop the vaccine from basic research to cGMP production
Identification of a target population	• Incidences of many NTDs are low, so how to identify a target population • NTD vaccines will need to be tested on people from endemic disease regions • Religious beliefs may limit vaccination of target populations • Biases or fear of western medicine may limit vaccination within target populations • Nearly two-third of people infected with one NTD are coinfected with at least one more NTD • Treatments for some NTDs are contraindicated for a second NTD (see Onchocerciasis and Loa loa coinfections) • Treatment of HIV infected, or immune compromised individuals may not be possible
Achieving sterilizing immunity	• Difficult to achieve due to many life-stages, multiple subspecies of parasites, and numerous reservoirs in the environment • Vaccine must be both prophylactic as well as therapeutic to address both uninfected and infected populations • Full eradication is probably unachievable without near 100% compliance • Identification of correlates of protection may be difficult
Defining a target vaccine product profile	• Induce high levels of antigen-specific Abs with fewest possible doses to maximize attrition rates • Simplistic vaccine delivery method (pill form is preferred over injectable form) • Unique cold-chain requirements for NTD vaccines; must be stable for longer periods of time at higher temperatures and higher humidity • Access to endemic regions or people are more challenging due to remote and more primitive locations • Many drug and vaccine treatments are contraindicated for very young, elderly, immune-compromised, or pregnant populations

vaccine can easily exceed $100 million to $1 billion, and pharmaceutical companies are required to recoup these costs by bringing the vaccine to market. In the case of NTD vaccines, this is much more of a challenge given the low economic nature of the target population where a successful vaccine would need to cost less than $1 per dose. However, with NTD vaccines the target population is unlikely to be able to pay for the vaccination itself. Therefore, the developmental costs must be absorbed by first world countries without the guarantee of a return on investment (ROI). It has been suggested that the lack of interest by the pharmaceutical industry to develop a vaccine to NTDs is primarily because there is no guarantee for ROI. In the pharmaceutical industry, it is well established that there are significantly more efforts aimed at developing new antiparasitic drugs rather than vaccines.[30–32] This bias toward drug development may be to ensure an ROI given that drugs may treat infected people, but there will always be newly acquired infections to ensure that drugs will be needed. On the other hand, an effective vaccine would minimize, or eliminate altogether, any need for drugs targeting the NTD. The vaccines in development outlined in Table 3 are among the first of their kind targeting diseases that affect the bottom billion people. Interestingly, with the exception of the dengue TV vaccine developed by Sanofi-Pasteur, all of these vaccines have been developed without any support from the pharmaceutical industry.[26] Efforts to develop vaccines to NTDs have primarily been fueled by nonprofit organizations called PDPs. Examples of PDPs include institutions such as the Infectious Disease Research Institute (IDRI), the Sabin Vaccine Institute PDP, Foundation for Innovative New Diagnostics (FIND), and Programs for Appropriate Treatments in Health (PATH). These PDPs are typically funded by a combination of federal and private grants and contracts from organizations such as the NIH and the Bill and Melinda Gates Foundation (BMGH).

A 2014 study analyzed 650 clinical studies (drugs and vaccines) with all NTDs which were dominated by three diseases: leishmaniasis, arbovirus, and dengue underscoring the lack of focus on the high-priority NTDs.[33] Of these studies, researchers found that ~59% were funded by research centers and public organizations, 24% were funded by pharmaceutical companies, and ~9% funded by nongovernmental organizations and foundations.[33] In recent years, many pharmaceutical companies have begun to include public health objectives in their ethical responsibilities which has led to the development of Public–Private Partnerships (PPPs).[33]

To spur increased interest in NTDs, the US Food and Drug Administration initiated a program in 2007 which awarded Priority

Table 3 Vaccine progress for high-priority NTDs.

Disease	Antigen	Progress	Expression system
STH	Na-GST-1	Preclinical	*Pichia*
	Na-APR-1 (M74)	Preclinical	Tobacco
	Na-Asp-2	Phase I	*Pichia*
Schistosomiasis	*Sm*-Tsp-2	Phase I	*Pichia*
	*Sm*14	Phase I	*Pichia*
	*Sh*28GST	Phase II	*Pichia*
	Sm-p80	Preclinical	*E. coli*
	*Sj*97	Preclinical	*E. coli*
	*Sm*29	Preclinical	*E. coli*
Leishmaniasis	NH36		
	LEISH-F1/F2/F3	Phase I/II	*E. coli*
	Lutzomyia sand fly antigens	Preclinical	*E. coli* and *Pichia*
River blindness	Ov-103	Preclinical	*Pichia*
	Ov-RAL-2	Preclinical	*Pichia*
	Ov-CPI-2M	Preclinical	*Pichia*
Chagas disease	Tc24	Preclinical	*Pichia*
	TSA-1		DNA vaccine
Dengue	7 total (Tetra-Vax-DV; TDEN; DENVax; TDEN-PIV; DENV-1 PIV; TVDV	Phase I	Live attenuated virus or inactivated virus
	TV (Sanofi-Pasteur)	Phase III	Live attenuated virus

Review Vouchers (PRV) to companies willing to focus efforts on drugs and vaccines for NTDs. These PRVs, which can be bought or sold as commodities, guaranteed the holder to a 6-month priority review shaving between 4 and 12 months off the standard review timeline.[34] Economic models predict this could translate to a cost savings of between $50 and 500 million, based on the indication. Unfortunately, this program was widely underutilized due to a host of restrictions including (1) requirements that the NTD studied being one of the "big three" (malaria, TB, HIV), and (2) requiring focusing on drugs and not vaccines.[34,35] Lack of participation in the PRV program led to a follow-on US Senate Bill entitled the *Creating Hope Act of 2011*, which aimed to relax many of the restrictions in the initial program and to remove barriers to engaging with the program.[34,36]

Mathematical modeling studies have shown that the costs required to develop a vaccine for Chagas disease could easily be recouped just by not having to treat with antiparasitic drugs.[37] The models showed that even with development costs reaching US$400 million, and taking up to 30 years for development before reaching the target population, a vaccine would only need to reach 17% of a target population (of 6 million infected people) for the vaccine to achieve a positive ROI. If development costs are reduced to $100 million, and the vaccine took only 15 years to reach the market, then a positive ROI could be reached by vaccination as few as 3% of the 6 million infected people.[37] A similar study showed that a leishmaniasis vaccine could generate a positive ROI even if the duration of protection were only 5 years and the vaccine had as little as 50% efficacy.[13] Given that most of the parasitic NTDs are treatable with the same drugs, it is likely that a vaccine for any NTD would generate a positive ROI regardless of development costs.

2.1.2 Identification of a Target Population

With any vaccine, a significant hurdle is to identify a target population for clinical vaccine trials. This is especially difficult with NTD vaccines that are developed in first world countries but must be tested in endemic regions on susceptible people. For each NTD, it will be necessary to determine whether the vaccine is designed for prophylactic treatment, therapeutic treatment, or both. If the vaccine is intended for prophylaxis, then the trials must be performed with uninfected people. This is difficult given that in endemic regions, many of the people may have been infected previously, may be infected but do not exhibit signs and symptoms, or may be coinfected with additional NTDs. Many of the vaccine trials are performed in parallel with drug treatment. This presents a problem in the case of pregnant women, in which drug treatments are contraindicated.

2.1.3 Achieving Sterilizing Immunity

In the case of parasitic NTDs, most of the causative agents have very complex life cycles involving multiple growth stages (cysts, worms, larvae), multiple vectors (flies, mosquitoes, beetles), and numerous natural reservoirs (animals, snails). As an example, a review of six schistosomiasis vaccines highlights some of the issues that make achieving sterilizing immunity difficult.[8] For example, of the six vaccines in development only two target all three *Schistosoma* species, whereas the remaining four vaccines target a single species. Some NTDs have multiple subspecies that cause the same disease, or cause different variations of disease. In the case of Schistosomiasis, there are multiple species of the parasite, of which three species cause distinct disease in different parts of the world. The ideal vaccine would treat the three most prevalent species, however of the six vaccines in preclinical development or in Phase-I trials only two have demonstrated efficacy against the three key subspecies. Two of these vaccines demonstrated activity on both eggs and worms, and only one exhibits potent cross-species protection in both naïve and infected populations.[8]

2.1.4 Defining a Target Vaccine Product Profile

In the case of schistosomiasis, the optimal vaccine would: (1) eliminate adult worms, (2) eliminate eggs in tissues and stool, (3) prophylactically prevent new infections, and (4) be effective against all species of *Schistosoma*.[8] In the case of the six vaccines in development for schistosomiasis, only two vaccines address both the egg and adult worm burden for all three species.[8] Similar cases exist for leishmaniasis, in which a vaccine must be effective for up to 20 species of leishmania capable of causing disease and must treat both VL and CL forms of the disease.[12,15] Mathematical models suggest that the duration of protection would need to be at least 5 years to generate a positive ROI, and the longer the duration of protection the fewer percentage of population would need to be vaccinated.[2,13,37] Additional considerations when identifying a target population include age, sex, pregnancy status, type of infection (ie, VL versus CL in leishmaniasis), immunosuppression (HIV, cancer, etc.), and potential coinfections. Also to be considered are the more intangible considerations such as religious beliefs, cultural biases, the remote nature of villages and tribes, and fears of western medicine which also may limit administration of a vaccine to a large percent of the population.

Even if a successful vaccine were developed, significant hurdles still remain in getting the vaccine from the manufacturer to the patient. The cold chain is a significant problem whenever a protein or drug is shipped from a first world to a third world country. Rural and developing parts of the world often do not have access to refrigeration for shipping or storage of the vaccine. This necessitates developing thermostable versions of the vaccine product, often by means such as lyophilization. For NTD vaccines, the majority will be shipped to Zone IV countries in which extreme temperatures and relative humidity will be encountered throughout the supply chain. There are very specific regulations and recommendations for testing vaccines under these conditions (reviewed in Refs. [38,39]). Route of administration, number of vaccinations and boosts required to achieve full protection, and duration of protection are all of significant concerns with NTDs. Simpler routes of administration, such as pill form versus parenteral administration, may be preferable. If the number of doses can be reduced to two or even one, the vaccine would be cheaper and attrition rates would be much lower. Attrition rates have significantly hampered treatment for tuberculosis, given that drug regimens require treatments at specific intervals for 6 months to 2 years.[40]

3. DEVELOPMENTAL PROCESS FOR NTD VACCINES

Although the main activities for developing an NTD vaccine are the same as for a mainstream vaccine, it is the developmental cost that is perhaps the most significant barrier. This means the GMP production process must be cheap, robust, and achieved by the most simple means possible. The manufacturing process must be readily scalable and amenable to bulk production, and the purification strategy should be robust and easily transferable to GMP manufacturing. The final vaccine product must be stable, lyophilizable, and easily administered.

3.1 Type of Vaccines

Vaccines vary on the basis of the organism/disease being targeted. Viral vaccines such as rabies and dengue may be whole inactivated virus, virus-like particles (VLP), attenuated virus, or chimeric viruses. The same is true of bacterial vaccines that are usually whole attenuated pathogens, pathogen-derived extracts, or recombinant proteins. By contrast, nearly every vaccine in development for parasitic NTDs is recombinant parasitic protein-based or a combination of DNA prime/recombinant protein boost. The major activities associated with developing a low-cost, protein-based vaccine to an NTD, the major steps of which are summarized in Fig. 1.

3.2 Choice of Expression System

Most proteins produced in the research laboratory rely on *E. coli*-based expression systems using reasonably crude methods not always amenable to scale up. The first decision that must be made is to identify an expression system that: (1) will yield large amounts of protein (>100 mg/L); (2) is cheap to express and purify; (3) is functional with regard to eliciting a protective immune response; (4) is transferable to cGMP manufacture; (5) is scalable to much higher volumes (20–1000 L) (reviewed in Ref. [41]). Often the first step in development of vaccine antigens is a lateral comparison of multiple expression systems such as:

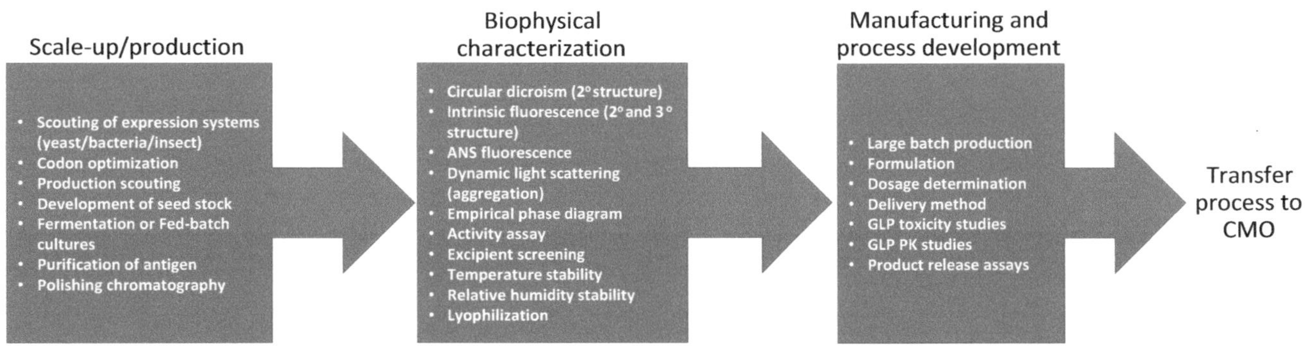

Figure 1 Key activities to enable vaccine production at contract manufacturing organizations.

3.2.1 *Bacterial Expression*

The most common bacterial expression system is the Gram-negative bacterium *E. coli*. Two drawbacks to this system include lower yields relative to other systems, and the contamination of the final protein with endotoxin. Endotoxin, also referred to as lipopolysaccharide, is a cell membrane component of all Gram-negative bacteria that often binds tightly to recombinant proteins during purification and must be removed prior to use for vaccines. Endotoxin removal is often a very costly process and depending on the difficulty of the removal process may also lead to significant protein losses. Alternative bacterial expression systems include Gram-positive bacteria including Brevibacillus or Lactococcus. Gram-positive bacteria do not contain endotoxin; however, the expression systems are far less widely used and only a handful of expression vectors have been developed for these organisms. One characteristic of bacterial expression is that the expressed target protein does not contain posttranslational modifications such as O- or N-linked glycosylation or myristoylation.

3.2.2 *Yeast Expression*

One of the most cost-effective methods for recombinant protein production are yeast expression systems, typically *Pichiapastoris* or *Saccharomyces cerevisiae*. Both are attractive for protein expression for multiple reasons including: (1) the ability to achieve very high culture optical densities (ie, >100 ODs) under specific fermentation conditions; (2) the fact that protein expression levels in yeast fermentation cultures typically are much higher than bacterial systems, ranging up to 100 mg/L; (3) yeast expression systems will perform posttranslational modifications including O- or N-linked glycosylation. Ability to achieve very high ODs coupled with very high expression levels have resulted in yeast being one of the most commonly preferred methods for manufacture of vaccine candidate proteins. Examples of vaccines produced in yeast are summarized in Table 2 including several in Phase-I–III vaccines in testing for schistosomiasis, onchocerciasis, soil-transmitted helminths, Chagas Disease, and leishmaniasis.

3.2.3 *Cell Culture Expression*

The most commonly used system for heterologous protein expression is mammalian cell cultures derived from cancer cell lines. The main advantage of a mammalian cell culture is that proteins will have posttranslational modifications very similar to human proteins. Expression levels are variable, and typically lower that with yeast-based systems, but some examples have been reported of protein yields of 1 g/L. Another type of cell culture employs insect cell lines that have been adapted to in vitro growth. This method is less widely used and typically is chosen for difficult-to-express proteins. Drawbacks with cell-culture expression include higher cost and difficulties with scalability relative to bacterial or yeast expression systems.

3.2.4 Plant Cell Expression

Another protein-expression system utilizes the tobacco plant in which the gene for the protein of interest is introduced using a plant viral vector that inserts the gene stably into the plant genome. Plant expression has been used for numerous therapeutic antibodies and recently is the chosen system for expressing one component of the human hookworm vaccine currently in Phase-I trials[6] as well as for the chikungunya vaccines currently in preclinical development and one enter Phase-II trials.[29,42]

3.3 Scale-Up of Expression

Optimal expression levels are typically achieved through a multistep process including: (1) codon-optimization; (2) addition of solubility-enhancing tags; (3) identification of optimal induction conditions; and (4) development of fermentation protocols.

3.3.1 Primary Expression Screen

The primary goals of the expression screen are to identify the clonal populations that (1) express the highest level of soluble protein per liter and (2) maintain the desired level of expression following multiple passages. Following multiple propagations, most subclones will exhibit variations in levels of recombinant protein leading to an eventual decrease in expression over time. Identification of a stable, high-producing subclone typically includes comparing 100–1000 colonies for protein expression levels followed by propagating the identified high expressor to confirm stable and consistent expression levels.

3.3.2 Development of Seed Stock

The seed stock is a collection of 100–1000 glycerol stocks all derived from a single stable and highly expressing clone identified through the expression screen. The seed stock ensures consistent high-level protein production and minimizes variances that may occur between fermentation batches. Each new protein purification run is derived from a fresh seed stock, and therefore run #1000 should result in protein yields comparable to run #1.

3.3.3 Scale-Up and Fermentation

Most small-scale bacterial or yeast expression is carried out under aerobic conditions. Under these conditions, growth of the strain continues until key nutrients in the media have been exhausted and/or waste products, such as acids and alcohols, have increased to inhibitory levels.[43,44] To achieve very high cell densities, one strategy is the fed-batch fermentation method (reviewed in Ref. [45]) in which anaerobic cultures are constantly supplied with fresh nutrients while maintaining constant temperature, oxygen levels, and pH. In order to transfer the protein production process to cGMP manufacture, it is important to perform small-scale fed-batch fermentation runs to optimize culture conditions and adjust parameters based on the data from this run. Single-use mini bioreactors, such as the ambr250 (TAP Biosystems) have been shown to be effective as a means to transition from small-scale aerobic expression to fermentation.[46] Successful small-scale fermentation activities can proactively identify the conditions needed for downstream large batch fermentation efforts to result in a reproducible and robust fermentation process. The numerous challenges encountered when transitioning from aerobic expression to small-scale fermentation and then to industrial-scale fermentation are beyond the scope of this review but are reviewed elsewhere.[46–48]

3.4 Large-Scale Purification

Following expression, a robust process for purification of the protein antigen must be developed. Especially with NTDs, it is of paramount importance that the purification process be simple and cheap to minimize downstream costs of manufacture. Some of the factors to consider in purification are discussed below. It is typical, but not always necessary, to remove affinity tags such as GST or hexahistidine tags. Often, these tags can be removed after purification using enzymatic cleavage or the protein can be expressed without tags in native form. However, purification of proteins without tags often requires development of complex, multistep purification strategies using a variety of anion and/or cation exchange steps. Although this may lead to very high levels of purity, these additional steps are not always amenable to bulk or cheap production of proteins. Following development of a robust purification scheme, a standard operating procedure can be written and the process transferred to a Certified Manufacturing Organization for cGMP manufacture of the vaccine antigens.

3.5 Characterization of Antigens

3.5.1 Biophysical Characterization

The following steps are typically required for characterization of the final vaccine antigen. Each of these steps is required to assess the stability of the antigens when stored for prolonged time or when shipped to areas of extreme temperatures and humidity. Some of the characterization methods include (1) freeze/thaw stability, (2) short-term and long-term stability, and (3) accelerated long-term stability. In these assays, the antigen is exposed to extreme conditions and then characterized for appearance and degradation by SDS-PAGE and mass spectrometry. Primary, secondary, and tertiary structures are assessed using circular dicroism, intrinsic fluorescence, and ANS fluorescence.[6] Aggregation is common as antigens age, denature, and degrade and these can be assessed using assays such as Dynamic light scattering or analytical ultracentrifugation (AU).[6] Finally, the biophysical data can be used to generate an Empirical Phase Diagram such as the one generated for the Hookworm vaccine antigen Na-APR-1 (M74).[6] An EPD is a visual representation of the stability of a protein and is compiled using data obtained from each of the aforementioned biophysical assays.

3.5.2 Lyophilization and Excipient Screening

Lyophilization, or freeze-drying, is a method of preservation by drying the protein under vacuum. Lyophilized proteins are much more amenable to long-term storage and are more resistant to temperature extremes and humidity, both of which are concerns for antigens that will be shipped to countries with limited cold chain transport. The lyophilized proteins can also be stored under inert gases to prevent oxidation. Excipient screening involves testing numerous compounds, typically sugars such as mannitol or sorbitol, which will act to stabilize proteins during storage, freeze-thaw, and during lyophilization.

4. VACCINE EMBODIMENT—GOING FROM AN ANTIGEN TO A DEPLOYABLE VACCINE

During the development and manufacturing of the antigen, the deployable form of the vaccine—that is, combinations with adjuvants and what delivery system will be used—must also be considered.

4.1 Adjuvants

In research settings, to elicit potent immune responses, complete Freund's or other research grade adjuvants are often used; however, due to reactogenicity, these will not be suitable for ultimate human use and therefore are often compared to human grade adjuvants prior to moving further into clinical testing.[49] Aluminum salts such as aluminum hydroxide (eg, Alhydrogel) and aluminum phosphate (Adjuphos) are often used due to extensive experience with these adjuvants.[50] However, especially with "clean" modern proteins that are antigen candidates, these adjuvants in humans generally only produce weak responses of short duration that are generally T_H2 biased and not sufficient for protection against certain pathogens.[51] Therefore, the addition of TLR ligands to the adjuvant formulation is now becoming more common, as with GSK's malaria and cervical cancer vaccines, which contain the TLR4 ligand MPL,[52–54] and many others currently in development like IDRI's GLA and SLA-based formulations.[10,55,56] The addition of TLR agonists to the adjuvant formulation enables interferon production, which accelerates hypermutation resulting in more functional antibody being produced.[57] These signals are amplified when the ligand is coformulated with inflammasome activators such as squalene emulsions or alum.[58]

While the addition of modern adjuvants may be critical to a successful vaccine, additional factors are important to consider. First, adjuvants add costs and configurational complexity; therefore, adjuvants should only be included when necessary. Second, potency and toxicological testing of the vaccine product will have to take into account the adjuvant used and testing may have to be performed with and without the adjuvants to provide a potency and safety baseline.[59] Additionally, the potency test has to take into account what the adjuvant is doing; for this reason many times an in vivo test is performed and the T_H1 bias of the adjuvanted vaccine is used as a measure for successful adjuvantation of the vaccine.[59]

4.2 Dosing and Administration

The dosing configuration is also an important consideration when using an adjuvanted vaccine. Logistical problems for single vial configurations are compounded by two vial configurations and further exacerbated if the antigen and the adjuvant have to be stored in separate vials at different temperatures. A preferred configuration therefore would be a heat stable single vial formulation. Since this is often not achievable and since the cost of the vial itself can drive up the price of the vaccine, prefilled syringes are often considered. Here, the syringe—which is required for dosing in any case—is prefilled with the adjuvant and the vial with antigen is provided separately. The contents of the syringe are injected into the vial, mixed with the antigen formulation, and withdrawn just prior to injection. Compatibility of each component will be needed to be tested depending on the deployment scenario. In the two-component scenario, a short-term kinetic study is performed demonstrating that the two components do not have detrimental effects on each other for a maximum of a few days since they should not be mixed until right before the injection is envisioned. In a single vial configuration, the two components are assayed in real time to determine a proper shelf life for the product as well as in accelerated conditions (higher temperatures and relative humidity) to provide data on what would happen to the vaccine during short-term temperature excursions. When demonstrating compatibility, measurements often include biochemical and biophysical elements thought to be critical for potency such as particle size, antigen conformation, antigen intactness, and TLR agonist strength.

In summary, global health vaccine developers should prefer simple vaccine configurations and only include an adjuvant if needed, but not shy from adding one if appropriate since the vaccines could otherwise fail in the field due to lack of durable and appropriate immune responses. When selecting an adjuvant these researchers should keep in mind antigen compatibility, heat stability, scalability and cost as well as the final dosing configuration.

5. CONCLUDING REMARKS

The coming decade offers tremendous potential for the development of new NTD vaccines that advance into clinical trials, as well as vaccines for some of the major childhood killers, including malaria. Previous Ebola and current Zika and Chikungunya virus outbreaks illustrate the fact that increased and sustained investment is required in R&D to drive discovery and development of vaccines for neglected diseases (and neglected populations). Although there are significant scientific challenges ahead for the product and clinical development of vaccines for NTDs, the more formidable challenges may be socioeconomic and/or political. Specifically, there is no obvious roadmap for licensure and global access to ensure that vaccines are made accessible to those most in need. Achieving these milestones would represent important steps toward achieving the Millennium Development

Goals; the enormous disease and economic burdens from NTDs provide a compelling reason for vaccine development to treat or prevent NTDs.

ACKNOWLEDGMENTS

There is no conflict of interest with funding agencies that have supported the research of the authors. This work was supported by grants from Bill and Melinda Gates Foundation grant (OPP1097535) to AAS and from the NIAID/NIH SBIR (R43/R44 AI103983) to DC and AAS.

REFERENCES

1. Bottazzi ME. The human hookworm vaccine: recent updates and prospects for success. *J Helminthol.* 2015;89(5):540–544.
2. Hotez PJ, Bottazzi ME, Strych U. New vaccines for the world's poorest people. *Annu Rev Med.* 2015;67:405–417.
3. Hotez PJ, et al. The human hookworm vaccine. *Vaccine.* 2013;31(Suppl 2):B227–B232.
4. Curti E, et al. Optimization and revision of the production process of the *Necator americanus* glutathione S-transferase 1 (Na-GST-1), the lead hookworm vaccine recombinant protein candidate. *Hum Vaccin Immunother.* 2014;10(7):1914–1925.
5. Goud GN, et al. Expression of the *Necator americanus* hookworm larval antigen Na-ASP-2 in *Pichia pastoris* and purification of the recombinant protein for use in human clinical trials. *Vaccine.* 2005;23(39):4754–4764.
6. Seid CA, et al. Expression, purification, and characterization of the *Necator americanus* aspartic protease-1 (Na-APR-1 (M74)) antigen, a component of the bivalent human hookworm vaccine. *Hum Vaccin Immunother.* 2015;11(6):1474–1488.
7. Zhan B, et al. Advancing a multivalent 'Pan-anthelmintic' vaccine against soil-transmitted nematode infections. *Expert Rev Vaccines.* 2014;13(3):321–331.
8. Molehin AJ, et al. Development of a schistosomiasis vaccine. *Expert Rev Vaccines.* 2016;15 (5):619–627.
9. Karmakar S, et al. Cross-species protection: *Schistosoma mansoni* Sm-p80 vaccine confers protection against *Schistosoma haematobium* in hamsters and baboons. *Vaccine.* 2014;32 (11):1296–1303.
10. Karmakar S, et al. Use of an Sm-p80-based therapeutic vaccine to kill established adult schistosome parasites in chronically infected baboons. *J Infect Dis.* 2014;209(12): 1929–1940.
11. Hotez PJ, et al. The global burden of disease study 2010: interpretation and implications for the neglected tropical diseases. *PLoS Negl Trop Dis.* 2014;8(7):e2865.
12. Jain K, Jain NK. Vaccines for visceral leishmaniasis: a review. *J Immunol Methods.* 2015;422:1–12.
13. Bacon KM, et al. The potential economic value of a cutaneous leishmaniasis vaccine in seven endemic countries in the Americas. *Vaccine.* 2013;31(3):480–486.
14. Duthie MS, et al. The development and clinical evaluation of second-generation leishmaniasis vaccines. *Vaccine.* 2012;30(2):134–141.
15. Coler RN, et al. From mouse to man: safety, immunogenicity and efficacy of a candidate leishmaniasis vaccine LEISH-F3+GLA-SE. *Clin Transl Immunol.* 2015;4(4):e35.
16. Cano J, et al. The global distribution and transmission limits of lymphatic filariasis: past and present. *Parasit Vectors.* 2014;7:466.
17. Hess JA, et al. Vaccines to combat river blindness: expression, selection and formulation of vaccines against infection with *Onchocerca volvulus* in a mouse model. *Int J Parasitol.* 2014;44(9):637–646.

18. Hotez PJ, et al. The Onchocerciasis Vaccine for Africa—TOVA—Initiative. *PLoS Negl Trop Dis.* 2015;9(1):e0003422.
19. Kelly-Hope LA, et al. Innovative tools for assessing risks for severe adverse events in areas of overlapping Loa loa and other filarial distributions: the application of micro -stratification mapping. *Parasit Vectors.* 2014;7:307.
20. Hotez PJ, et al. An unfolding tragedy of Chagas disease in North America. *PLoS Negl Trop Dis.* 2013;7(10):e2300.
21. Ribeiro I, et al. New, improved treatments for Chagas disease: from the R&D pipeline to the patients. *PLoS Negl Trop Dis.* 2009;3(7):e484.
22. Dumonteil E, et al. Accelerating the development of a therapeutic vaccine for human Chagas disease: rationale and prospects. *Expert Rev Vaccines.* 2012;11(9):1043–1055.
23. Lightowlers MW. Cysticercosis and echinococcosis. *Curr Top Microbiol Immunol.* 2013;365:315–335.
24. Barnes TS, et al. Challenges for diagnosis and control of cystic hydatid disease. *Acta Trop.* 2012;123(1):1–7.
25. Bhatt S, et al. The global distribution and burden of dengue. *Nature.* 2013;496(7446): 504–507.
26. Beaumier CM, et al. New vaccines for neglected parasitic diseases and dengue. *Transl Res.* 2013;162(3):144–155.
27. Rougeron V, et al. Chikungunya, a paradigm of neglected tropical disease that emerged to be a new health global risk. *J Clin Virol.* 2015;64:144–152.
28. Mardekian SK, Roberts AL. Diagnostic options and challenges for dengue and chikungunya viruses. *Biomed Res Int.* 2015;2015:834371.
29. Salazar-Gonzalez JA, Angulo C, Rosales-Mendoza S. Chikungunya virus vaccines: current strategies and prospects for developing plant-made vaccines. *Vaccine.* 2015;33 (31):3650–3658.
30. Andrews KT, Fisher G, Skinner-Adams TS. Drug repurposing and human parasitic protozoan diseases. *Int J Parasitol Drugs Drug Resist.* 2014;4(2):95–111.
31. Jones CL, et al. Drug discovery for the developing world: progress at the Novartis Institute for Tropical Diseases. *Nat Rev Drug Discov.* 2015;14(6):442–444.
32. Nwaka S, Hudson A. Innovative lead discovery strategies for tropical diseases. *Nat Rev Drug Discov.* 2006;5(11):941–955.
33. di Procolo P, Jommi C. Current pipelines for neglected diseases. *PLoS Negl Trop Dis.* 2014;8(9):e3092.
34. Stefanakis R, et al. Analysis of neglected tropical disease drug and vaccine development pipelines to predict issuance of FDA priority review vouchers over the next decade. *PLoS Negl Trop Dis.* 2012;6(10):e1803.
35. FDA, U.F.a.D.A. FDA draft guidance for industry: tropical disease priority review vouchers, FDA. Washington DC, Government Printing Office; 2008
36. Robertson AS, et al. The impact of the US priority review voucher on private-sector investment in global health research and development. *PLoS Negl Trop Dis.* 2012;6(8): e1750.
37. Lee BY, et al. Modeling the economic value of a Chagas' disease therapeutic vaccine. *Hum Vaccin Immunother.* 2012;8(9):1293–1301.
38. Lucas TI, Bishara RH, Seevers RH. A stability program for the distribution of drug product. *Pharm Technol.* 2004;July:68–73.
39. Seevers RH. Regulatory strategy for long-term stability conditions to support submission in zone IV countries. *Pharm Outsourcing.* 2011;12(2):1–7.
40. Kaufmann SH, et al. Progress in tuberculosis vaccine development and host-directed therapies—a state of the art review. *Lancet Respir Med.* 2014;2(4):301–320.
41. Sugiki T, Fujiwara T, Kojima C. Latest approaches for efficient protein production in drug discovery. *Expert Opin Drug Discov.* 2014;9(10):1189–1204.

42. Schwameis M, et al. Chikungunya vaccines in development. *Hum Vaccin Immunother.* 2016;12(3):716–731.
43. Hewitt CJ, et al. Studies related to the scale-up of high-cell-density *E. coli* fed-batch fermentations using multiparameter flow cytometry: effect of a changing microenvironment with respect to glucose and dissolved oxygen concentration. *Biotechnol Bioeng.* 2000;70(4):381–390.
44. Lee J, et al. Control of fed-batch fermentations. *Biotechnol Adv.* 1999;17(1):29–48.
45. Cos O, et al. Operational strategies, monitoring and control of heterologous protein production in the methylotrophic yeast *Pichiapastoris* under different promoters: a review. *Microb Cell Fact.* 2006;5:17.
46. M. Ngibuini, How single-use, mini bioreactors could revolutionize bioprocess scale-up. Available from: http://www.bioprocessintl.com/upstream-processing/bioreactors/automated-mini-bioreactor-technology-microbial-mammalian-cell-culture-flexible-strategy-optimize-early-process-development-biologics-vaccines-2/
47. Formenti LR, et al. Challenges in industrial fermentation technology research. *Biotechnol J.* 2014;9(6):727–738.
48. Jones SD, Kanarek A, Levine H. Reliable scale-up for biopharmaceutical production. BioProcess Technology Consultants, 2012; Available from: http://www.chem.info/article/2012/10/reliable-scale-biopharmaceutical-production
49. Roeffen W, et al. Transmission-blocking activity of antibodies to Plasmodium falciparum GLURP.10C chimeric protein formulated in different adjuvants. *Malar J.* 2015;14 (1):443.
50. Gupta PK, et al. Production and preclinical evaluation of Plasmodium falciparum MSP-119 and MSP-311 chimeric protein, PfMSP-Fu24. *Clin Vaccine Immunol.* 2014;21(6): 886–897.
51. Orr MT, et al. Adjuvant formulation structure and composition are critical for the development of an effective vaccine against tuberculosis. *J Control Release.* 2013;172 (1):190–200.
52. Didierlaurent AM, et al. AS04, an aluminum salt- and TLR4 agonist-based adjuvant system, induces a transient localized innate immune response leading to enhanced adaptive immunity. *J Immunol.* 2009;183(10):6186–6197.
53. Garcon N, Van Mechelen M. Recent clinical experience with vaccines using MPL- and QS-21-containing adjuvant systems. *Expert Rev Vaccines.* 2011;10(4):471–486.
54. Morrison C. Landmark green light for Mosquirix malaria vaccine. *Nat Biotechnol.* 2015;33(10):1015–1016.
55. McKay PF, et al. Glucopyranosyl lipid A adjuvant significantly enhances HIV specific T and B cell responses elicited by a DNA-MVA-protein vaccine regimen. *PLoS One.* 2014;9(1):e84707.
56. Patra KP, et al. Alga-produced malaria transmission-blocking vaccine candidate Pfs25 formulated with a human use-compatible potent adjuvant induces high-affinity antibodies that block Plasmodium falciparum infection of mosquitoes. *Infect Immun.* 2015;83 (5):1799–1808.
57. Wiley SR, et al. Targeting TLRs expands the antibody repertoire in response to a malaria vaccine. *Sci Transl Med.* 2011;3(93):93ra69.
58. Desbien AL, et al. Squalene emulsion potentiates the adjuvant activity of the TLR4 agonist, GLA, via inflammatory caspases, IL-18, and IFN-gamma. *Eur J Immunol.* 2015;45 (2):407–417.
59. World Health Organization. *WHO Guidelines on the nonclinical evaluation adjuvants and adjuvanted vaccines.* Geneva, Switzerland: World Health Organization; 2013.

INDEX

E

K

L

M

N

O

P

V

W

Y

Z

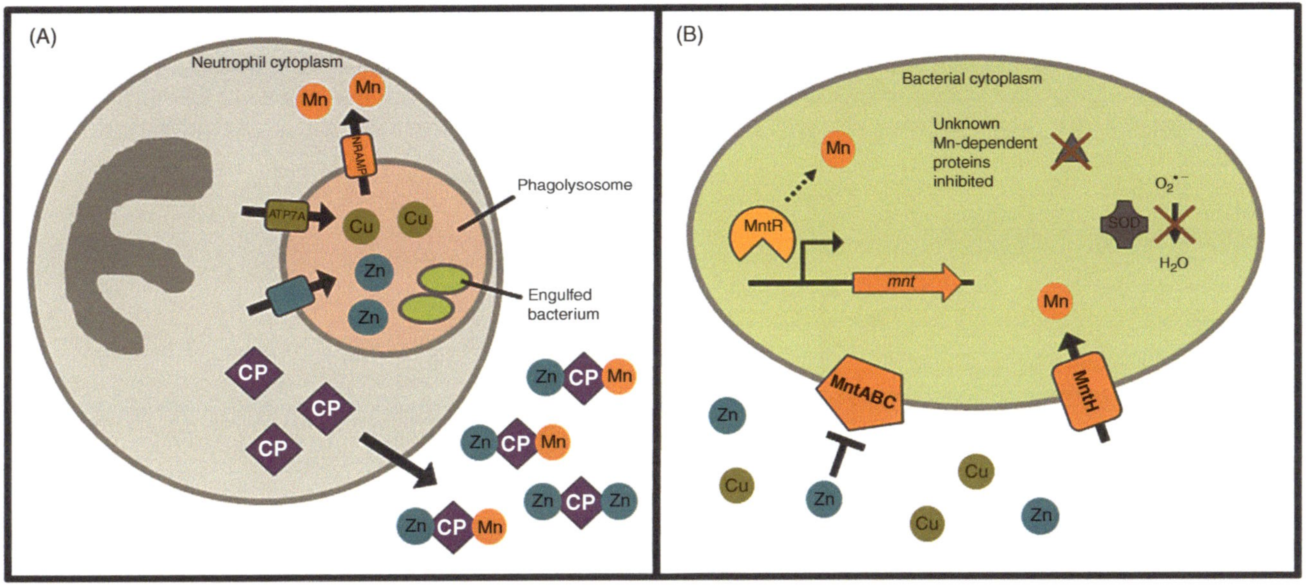

J.L. Kelliher and T.E. Kehl-Fie, Figure 1 The host and pathogens compete for manganese during infection. (A) Diagram of the mechanisms utilized by the host to limit the ability of invaders to obtain manganese. In response to microbial invaders, host cells, primarily neutrophils, release calprotectin (CP), which limits the ability of extracellular pathogens to obtain manganese (and zinc). In the phagolysosomal membrane, NRAMP transporters remove manganese (Mn) from the phagolysosome. Additionally, zinc (Zn) and copper (Cu) are imported into this compartment, inhibiting the activity of bacterial manganese importers. (B) Diagram of how bacteria respond to manganese limitation and the processes that are disrupted by host- imposed manganese limitation. In response to manganese starvation, the MntR regulon is derepressed, and the expression of dedicated manganese importers such as MntH and MntABC increases. Despite the expression of high-affinity manganese importers by invading pathogens, the host remains capable of imposing manganese starvation, which inactivates manganese-dependent superoxide dismutase (SOD) and unknown essential enzymes.

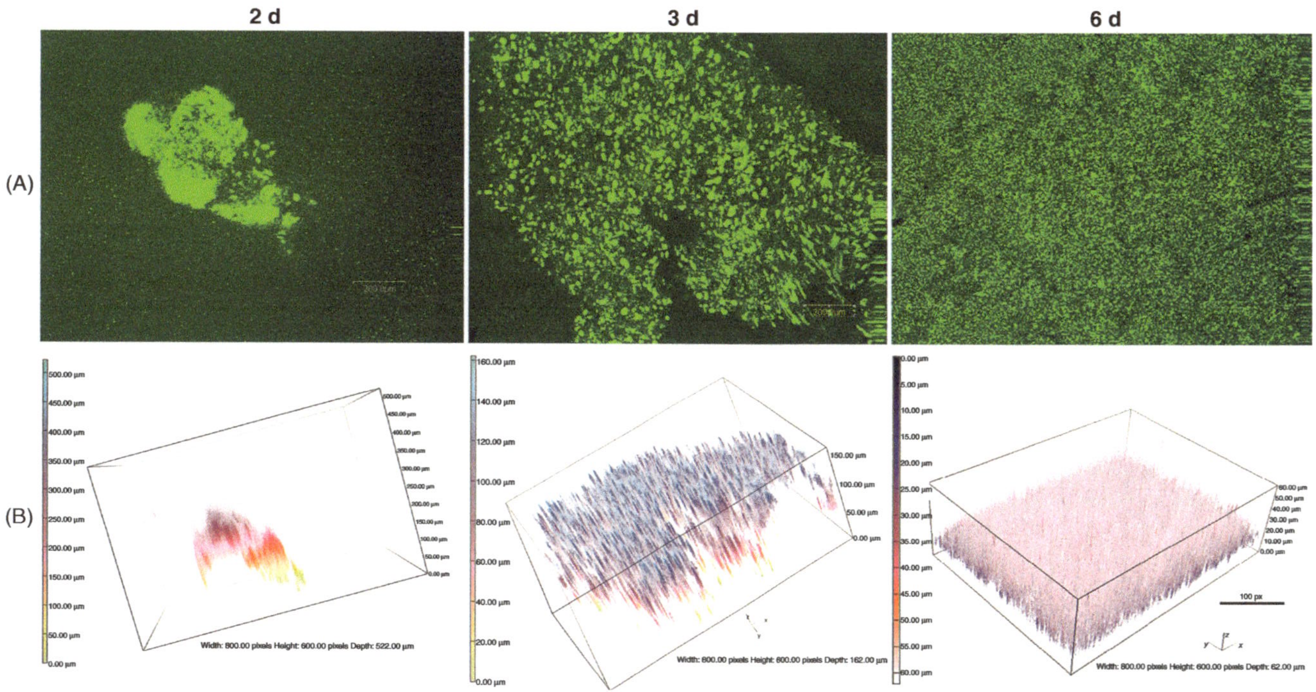

J.A. Colmer-Hamood *et al.*, Figure 1 PAO1 BLS vary structurally over time. PAO1/pMRP9-1 [plasmid carrying the gene for gray (green in the web version) fluorescent protein] was inoculated into ASM+ to an initial OD_{600} of 0.02–0.03. One-milliliter aliquots were distributed into wells of 24-well microtiter plate and the plate was incubated at 37°C in static conditions under 20% EO_2 for up to 6 days without changing the medium. (A) CLSM micrographs of BLS at 2, 3, and 6 d postinoculation; magnification, 10X; bars, 200.00 nm. (B) The 3-D architecture of the BLS shown in (A). Boxes, 800.00 px W × 600.00 px H; bars, 100 px. *Haley et al.; licensee BioMed Central Ltd. 2012.*[177] *Figure used without changes as permitted under the terms of the Creative Commons Attribution License (http://creativecommons.org/licenses/by/2.0).*

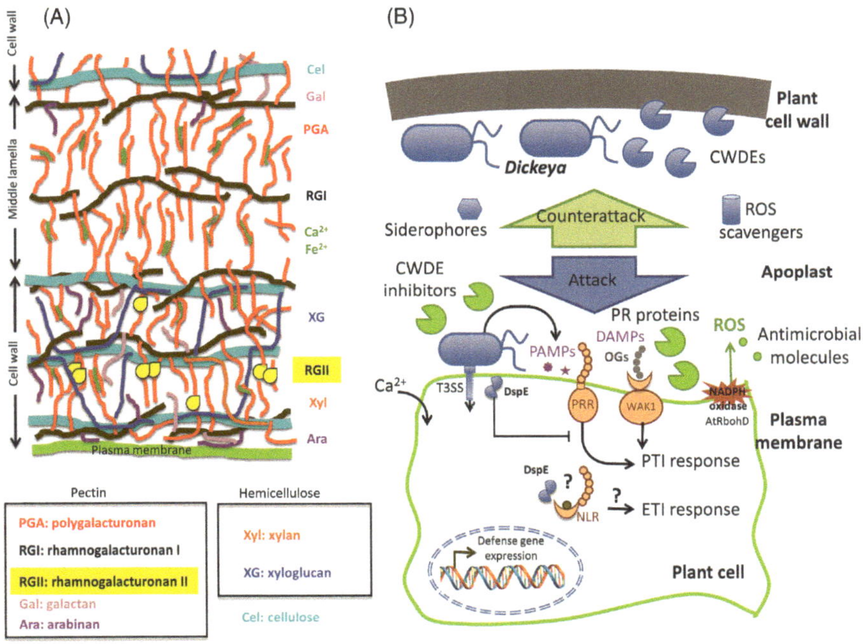

S. Reverchon *et al.*, Figure 1 Structure of the plant cell wall (A) and the sketch of Plant–*Dickeya* interaction (B). *Dickeya* propagate in the extracellular spaces of plant tissues. The molecules released from bacteria (PAMPs) into the apoplast and oligogalacturonides (DAMPs) resulting from pectin cleavage by bacterial pectinases are recognized, respectively, by plant receptors (PRRs) and by the wall-associated kinase (WAK1) leading to elicitation of the PAMP/DAMP-triggered immunity (PTI). Plant produces ROS, antimicrobial molecules, and pathogenesis-related (PR) proteins, some of which are CWDE inhibitors. To cope with ROS, bacteria produce the indigoidine ROS scavenger. *Dickeya* deliver the DspE effector in the host cell by the T3SS secretion pilus. DspE is supposed to suppress PTI and facilitates bacterial propagation *in planta*. Inside the apoplast, *D. dadantii* produces siderophores, which allow the bacterium to cope with the restricted iron bioavailability in the plant. These siderophores also manipulate plant immunity.

N. Hugouvieux-Cotte-Pattat, Figure 1 Different niches and lifestyles of *Dickeya*. *Dickeya* species are considered as facultative saprophytes that cycle between saprophytic and virulent lifestyles. They are found in a variety of environmental niches (plants, soil, water, and insects). Plant colonization corresponds to the initial establishment of the bacteria at the surface of leaves, stems, or roots. *D. dadantii* can also infect insects and it is able to kill the pea aphid *Acyrtosiphon pisum*. The role of insects in bacterial transmission is poorly documented. Bacteria reside as epiphytes on plant surfaces where they remain latent until conditions become favorable for the disease development. The phytopathogenic cycle is usually divided into two main phases. During the first few hours, the bacteria begin to slowly multiply without provoking any symptoms. This "asymptomatic" phase is followed by a transition stage corresponding to a strong metabolic change with an induction of pectate lyase production. The "symptomatic" phase corresponds to the expansion of soft rot due to the pectate lyase action. After experimental infection of chicory by *D. dadantii* 3937, the total plant maceration is observed in less than 48 h (photos: 4 h and 24 h postinfection).

FSC
www.fsc.org
MIX
Papier | Fördert gute Waldnutzung
FSC® C083411

Kolibri 360 GmbH
Ferdinand-Jühlke-Straße 7
99095 Erfurt, Deutschland
produktsicherheit@kolibri360.de